Lecture Notes in Mathematics

1294

Editors:
J.-M. Morel, Cachan
F. Takens, Groningen
B. Teissier, Paris

Martine Queffélec

Substitution Dynamical Systems – Spectral Analysis

Second Edition

Martine Queffélec
Université Lille 1
Laboratoire Paul Painlevé
59655 Villeneuve d'Ascq CX
France
martine@math.univ-lille1.fr

ISBN: 978-3-642-11211-9 e-ISBN: 978-3-642-11212-6
DOI: 10.1007/978-3-642-11212-6
Springer Heidelberg Dordrecht London NewYork

Lecture Notes in Mathematics ISSN print edition: 0075-8434
ISSN electronic edition: 1617-9692

Library of Congress Control Number: 2010920970

Mathematics Subject Classification (2000): 37, 42, 11, 28, 47, 43, 46, 82

© Springer-Verlag Berlin Heidelberg 2010
This work is subject to copyright. All rights are reserved, whether the whole or part of the material is concerned, specifically the rights of translation, reprinting, reuse of illustrations, recitation, broadcasting, reproduction on microfilm or in any other way, and storage in data banks. Duplication of this publication or parts thereof is permitted only under the provisions of the German Copyright Law of September 9, 1965, in its current version, and permission for use must always be obtained from Springer. Violations are liable to prosecution under the German Copyright Law.

The use of general descriptive names, registered names, trademarks, etc. in this publication does not imply, even in the absence of a specific statement, that such names are exempt from the relevant protective laws and regulations and therefore free for general use.

Cover design: SPi Publisher Services

Printed on acid-free paper

springer.com

Preface for the Second Edition

This revised edition initially intended to correct the misprints of the first one. But why does it happen now, while the subject extensively expanded in the past twenty years, and after the publication of two major books (among other ones) devoted to dynamical systems [88] and automatic sequences [14]? Let us try to explain why we got convinced to do this new version. On the one hand, the initial account of the LNM 1294 offered a basis on which much has been built and, for this reason, it is often referred to as a first step. On the other hand, the two previously quoted books consist in impressive and complete compilations on the subject [14, 88]; this was not the spirit of our LNM, almost self-contained and "converging" to the proof of a specific result in spectral theory of dynamical systems. From this point of view, those three books might appear as complementary ones.

This having been said, reproducing the corrected LNM identically would have been unsatisfactory : a lot of contributions have concurred to clarify certain aspects of the subject and to fix notations and definitions; also a great part of the raised questions have now been solved. Mentioning these improvements seemed to us quite necessary. Therefore, we chose to add some material to the first introductory chapters, which of course does not (cannot) reflect the whole progress in the field but some interesting directions. Moreover, two applications of substitutions - more generally of combinatorics of words - to discrete Schrödinger operators and to continued fraction expansions clearly deserved to take place in this new version : two additional appendices summarize the main results in those fields.

The initial bibliography has been inflated to provide a much more up-to-date list of references. This renewed bibliography is still far from being exhaustive and we should refer the interested reader to the two previously cited accounts.

In recent contributions, the terminology has changed, emphasizing on the morphism property. However, we chose to keep to the initial terminology, bearing in mind the fact that this is definitely a second edition.

Lille
December 2009

Martine Queffélec

Preface for the First Edition

Our purpose is a complete and unified description of the spectrum of dynamical systems arising from substitution of constant length (under mild hypotheses). The very attractive feature of this analysis is the link between several domains : combinatorics, ergodic theory and harmonic analysis of measures.

The rather long story of these systems begins perhaps in 1906, with the construction by A. Thue [234] of a sequence with certain non-repetition properties (rediscovered in 1921 by M. Morse [190]):

$$0\,1\,1\,0\,1\,0\,0\,1\,1\,0\,0\,1\,0\,1\,1\,0\cdots$$

This sequence (called from now on the Thue-Morse sequence) can be obtained by an obvious iteration of the substitution $0 \to 01$, $1 \to 10$, or else, as an infinite block product : $01 \times 01 \times 01 \times \cdots$, where $B \times 01$, for any $0-1$ block B, means : repeat B and then $\tilde{B}$, the block deduced from B by exchanging 0 and 1. Also, if $S_2(n)$ denotes the sum of digits of n in the 2-adic expansion, $u = (u_n)$ with

$$u_n = e^{i\pi S_2(n)}$$

is the ± 1 Thue-Morse sequence.

The Thue-Morse sequence admits a strictly ergodic (= minimal and uniquely ergodic) orbit closure and a simple singular spectrum, as observed by M. Keane [143].

The various definitions of the Thue-Morse sequence lead to various constructions of sequences, and thus, of dynamical systems:

– substitution sequences [55, 63, 68, 104, 188] then [71, 119, 132, 135, 173, 174, 189, 208], ...

– a class of $0-1$ sequences introduced by M. Keane, called generalized Morse sequences [143], admitting in turn extensions [175, 176] then [102, 154–156, 162], ...

– q-multiplicative sequences, with $q = (q_n)$, q_n integer ≥ 2 [59] then [166, 202], ...

In this account, we restrict our attention to the first category of sequences, but, in case of bijective substitutions (chapter 9), we deal with particular G-Morse sequences and q-multiplicative sequences.

Ergodic and topological properties of substitution dynamical systems have been extensively studied; criteria for strict ergodicity [68,188], zero entropy [68,209], rational pure point spectrum [68,173,174], conditions for presence of mixed spectrum [68] and various mixing properties [71] are main investigations and results in these last years. But, except in some examples ([135, 143], ...), no descriptive spectral analysis of the continuous part of the spectrum has been carried out.

Indeed, not so many dynamical systems lead themselves to a comprehensive computation of spectral invariants. I mean, mainly, maximal spectral type and spectral (global) multiplicity (see [214] for a rather complete historical survey). Of course, transformations with purely discrete spectrum are quite well-known [240], and in this case, the spectrum is simple. In the opposite direction, a countable Lebesgue spectrum occurs in ergodic automorphisms of compact abelian groups as in K-automorphisms (see [61]). A very important class of dynamical systems, with respect to the spectral analysis, consists of gaussian dynamical systems. Guirsanov proved a conjecture of Kolmogorov [110]: the maximal spectral type of a gaussian dynamical system is equivalent to e^{σ}, where σ denotes the spectral measure of the process; and its spectral multiplicity has been shown by Vershik to be either one - with singular spectrum - or infinite ([237,238], see also [89]). Then arose the question of whether finite multiplicity ≥ 2 (or ≥ 1 for Lebesgue spectrum) was possible, and the last results in multiplicity theory have been mostly constructions of suitable examples. I just quote the last three important ones : Robinson E.A. Jr in [214] exhibits, for every $m \geq 1$, a measure-preserving transformation with singular spectrum and spectral multiplicity m. On the other hand, Mathew and Nadkarni in [177, 178] construct, for every $N \geq 2$, a measure-preserving transformation with a Lebesgue spectrum of multiplicity $N\phi(N)$ (ϕ Euler totient function). In these examples, the transformations are group extensions. Recently, M. Lemanczyk obtained every even Lebesgue multiplicity [160].

Turning back to substitution dynamical systems, we prove the following : for a substitution of length q over the alphabet A (or q-automaton [55]), the spectrum is generated by $k \leq \text{Card } A$ probability measures which are strongly mixing with respect to the q-adic transformation on $\mathbf{T}$; in most examples, these measures are specific generalizations of Riesz products, which is not so surprising because of the self-similarity property inherent in this study. (Note that such Riesz products play a prominent part in distinguishing normal numbers to different bases [136]; see also [50, 198], ...).

Earlier Ledrappier and Y. Meyer already realized classical Riesz products as the maximal spectral type of some dynamical system.

The generating measures of the spectrum of some q-automaton are computable from a matrix of correlation measures, indeed a matrix Riesz product, whose rank gives rise to the spectral multiplicity. For example, the continuous part of the Rudin-Shapiro dynamical system is Lebesgue with multiplicity 2, while, by using the mutual singularity of generalized Riesz products (analyzed in chapter 1), we get various singular spectra with multiplicity 1 or 2, as obtained by Kwiatkowski and Sikorski ([156], see also [101, 102]). For substitutions of nonconstant length, no spectral description seems accessible at present but we state a recent characterization of eigenvalues established by B. Host [119] and list some problems.

We have aimed to a self-contained text, accessible to non-specialists who are not familiar with the topic and its notations. For this reason, we have developed with all details the properties of the main tools such that Riesz products, correlation measures, matrices of measures, nonnegative matrices and even basic notions of spectral theory of unitary operators and dynamical systems, with examples and applications.

More precisely, the text gets gradually more specialized, beginning in chapter 1 with generalities on the algebra $M(\mathbf{T})$ and its Gelfand spectrum Δ. We introduce generalized Riesz products and give a criterion for mutual singularity.

Chapter 2 is devoted to spectral analysis of unitary operators, where all fundamental definitions, notations and properties of spectral objects can be found. We prove the representation theorem and two versions of the spectral decomposition theorem.

We restrict ourselves, in chapter 3, to the unitary operator associated with some measure-preserving transformation and we deduce, from the foregoing chapter, spectral characterizations of ergodicity and of various mixing properties (strong, mild, weak). As an application of D-ergodicity (ergodicity with respect to a group of translations [47]), we discuss spectral properties of some skew products over the irrational rotation [100, 103, 140, 212].

In chapter 4, we investigate shift invariant subsets of the shift space (subshifts), such like the orbit closure of some sequence. Strict ergodicity can be read from the given sequence, if taking values in a finite alphabet. The correlation measure of some sequence - when unique - belongs to the spectral family; hence, from earlier results, we derive spectral properties of the sequence. We give a classical application to uniform distribution modulo 2π (Van der Corput's lemma) and we discuss results around sets of recurrence [25, 35, 93, 219].

From now on we are concerned with substitution sequences. All previously quoted results regarding substitution dynamical systems are proved in chapters 5–6, sometimes with a different point of view and unified notations (strict ergodicity, zero entropy, eigenvalues and mixing properties). We are needing the Perron-Frobenius theorem and, for sake of completeness, we give too a proof of it.

Till the end of the account, the substitution is supposed to have a constant length. We define, in chapter 7, the matrix of correlation measures Σ and we show how to deduce the maximal spectral type from it. Then we prove elementary results about matrices of measures which will be used later.

In chapter 8, we realize Σ as a matrix Riesz product and this fact provides a quite simple way to compute it explicitly. Applying the techniques immediately, we treat the first examples : Morse sequence, Rudin-Shapiro sequence, and a class of sequences arising from commutative substitutions (particular G-Morse sequences), admitting generalized Riesz product as generating measures.

An important class of substitutions is studied in chapter 9 without complete success. It would be interesting in this case to get a more precise estimate of the spectral multiplicity, which is proved to be at least 2 for substitutions over a nonabelian group.

Finally, the main results on spectral invariants in the general case are obtained in chapters 10–11 by using all the foregoing. We have to consider a bigger matrix of correlation measures, involving occurrences of pairs of given letters instead of simple ones, which enjoys the fundamental strong mixing property and provides the maximal spectral type of the initial substitution.

The spectral multiplicity can be read from the matrix Σ, as investigated with the Rudin-Shapiro sequence and some bijective substitution. We obtain in both cases a Lebesgue multiplicity equal to 2, while N-generalized Rudin-Shapiro sequences admit a Lebesgue multiplicity $N\phi(N)$ [203, 211].

In an appendix, we suggest an extension to automatic sequences over a compact nondiscrete alphabet. We give conditions ensuring strict ergodicity of the orbit closure.

As explained before, we preferred to develop topics involving spectral properties of measures and for this reason, the reader will not find in this study a complete survey of substitutions. A lot of relevant contributions have been ignored or perhaps forgotten : we apologize the mathematicians concerned.

Paris *Martine Queffélec*
July 1987

Contents

Chapter 1
The Banach Algebra $M(\mathbf{T})$

This first chapter is devoted to the study of the Banach algebra $M(\mathbf{T})$. This study will be brief because we need only little about $M(\mathbf{T})$, and there exist excellent books on the subject, in which all the proofs will be found [123, 141, 218, 232]. We introduce the technics of generalized characters to precise the spectral properties of measures such as generalized Riesz products, which will nicely appear later as maximal spectral type of certain dynamical systems.

1.1 Basic Definitions

We consider $\mathbf{U}$ the multiplicative compact group of complex numbers of modulus one, and $\mathbf{T} = \mathbf{R}\backslash 2\pi\mathbf{Z}$ that we identify with $\mathbf{U}$ by the map $\lambda \to e^{i\lambda}$; $\mathbf{T}$ is equipped with the Haar measure m, identified this way with the normalized Lebesgue measure $\frac{1}{2\pi}dx$ on $[-\pi,\pi]$.

1. The elements of the character group $\Gamma = \hat{\mathbf{T}}$, isomorphic to $\mathbf{Z}$, will be considered sometimes as integers, with addition, sometimes as multiplicative functions on $\mathbf{T}$, and, in this case, we denote by γ_n instead of n the element $t \mapsto e^{int}$.
2. $M(\mathbf{T})$ is the algebra of the regular Borel complex measures on $\mathbf{T}$, equipped with the convolution product of measures, defined by

$$\mu * \nu(E) = \int_{\mathbf{T}} \mu(E-t)\, d\nu(t)$$

   for $\mu, \nu \in M(\mathbf{T})$ and E any measurable subset of $\mathbf{T}$.
   $M(\mathbf{T})$ is a Banach algebra for the norm

$$||\mu|| = \int d|\mu|,$$

   $|\mu|$ being the total variation of μ.

M. Queffélec, *Substitution Dynamical Systems – Spectral Analysis: Second Edition*, Lecture Notes in Mathematics 1294, DOI 10.1007/978-3-642-11212-6_1,

© Springer-Verlag Berlin Heidelberg 2010

The Fourier coefficients of $\mu \in M(\mathbf{T})$ are, by definition,

$$\hat{\mu}(n) = \int_{\mathbf{T}} e^{int} d\mu(t) = \int \gamma_n d\mu, \quad n \in \mathbf{Z}$$

and satisfy : $||\hat{\mu}||_\infty := \sup_{n\in\mathbf{Z}} |\hat{\mu}(n)| \leq ||\mu||$. The *Fourier spectrum* of μ is the set of integers $n \in \mathbf{Z}$ for which $\hat{\mu}(n) \neq 0$.

3. The measure μ is positive if $\mu(E) \geq 0$ for every measurable set E, and, in this case, the sequence $(\hat{\mu}(n))$ is *positive definite*, namely

$$\sum_{1\leq i,j\leq n} z_i \overline{z_j} \hat{\mu}(i-j) \geq 0$$

for any finite complex sequence $(z_i)_{1\leq i\leq n}$.
Conversely, the *Bochner theorem* asserts that a positive definite sequence $(a_n)_{n\in\mathbf{Z}}$ is the Fourier transform of a positive measure on $\mathbf{T}$.
Positive measures of total mass one are *probability measures*.

4. We recall that μ is a *discrete* measure if $\mu = \sum a_j \delta_{t_j}$, ($\delta_t$ being the unit mass at $t \in \mathbf{T}$) and that μ is a *continuous* measure if $\mu\{t\} = 0$ for all $t \in \mathbf{T}$. $M_d(\mathbf{T})$ is the sub-algebra of discrete measures in $M(\mathbf{T})$ and $M_c(\mathbf{T})$, the convolution-ideal of all continuous measures on $\mathbf{T}$. Every $\mu \in M(\mathbf{T})$ can be uniquely decomposed into a sum

$$\mu = \mu_d + \mu_c$$

where $\mu_d \in M_d(\mathbf{T})$ and $\mu_c \in M_c(\mathbf{T})$ respectively are the discrete part and the continuous part of μ.
There is a necessary and sufficient condition for a measure μ to be continuous, which involves the Fourier transform of μ :

Lemma 1.1 (Wiener). *Let $\mu \in M(\mathbf{T})$. Then :*

$$\mu \in M_c(\mathbf{T}) \Longleftrightarrow \lim_{N\to\infty} \frac{1}{2N+1} \sum_{n=-N}^{N} |\hat{\mu}(n)|^2 = 0$$

and in this case, we have $\lim_{N\to\infty} \frac{1}{2N+1} \sum_{n=-N+K}^{N+K} |\hat{\mu}(n)|^2 = 0$ *uniformly in K.*

5. Let $\mu, \nu \in M(\mathbf{T})$; we say that μ is *absolutely continuous with respect to* ν and we write $\mu \ll \nu$ if $|\mu|(E) = 0$ as soon as $|\nu|(E) = 0$, for any measurable set E. Then, by the Radon-Nikodym property, $\mu = f \cdot \nu$ where $f \in L^1(\nu)$ is referred to as the density of μ with respect to ν, usually denoted by $d\mu/d\nu$. Let us define

$$L(\nu) = \{\mu \in M(\mathbf{T});\ \mu \ll \nu\} \tag{1.1}$$

So we are allowed to identify $L(\nu)$ with $L^1(\nu)$. The measures μ, ν are said to be *equivalent*, and we write $\mu \sim \nu$, if $\mu \ll \nu$ and $\nu \ll \mu$.
In the opposite direction, we say that μ and ν are *mutually singular*, and we write $\mu \perp \nu$, if there exists a measurable set E such that

$$|\mu(E)| = ||\mu|| \quad \text{and} \quad |\nu|(E) = 0.$$

The measure μ is said to be *singular* if $\mu \perp m$, m the Haar measure on $\mathbf{T}$.

Every $\mu \in M(\mathbf{T})$ can be uniquely decomposed into a sum

$$\mu = \mu_a + \mu_s$$

where $\mu_a \ll m$ and μ_s is singular (respectively the absolutely continuous part and the singular part of μ).

Affinity between two measures. Let μ and ν be two positive measures in $M(\mathbf{T})$ and $\lambda \in M(\mathbf{T})$ be such that both $\mu \ll \lambda$ and $\nu \ll \lambda$. The affinity $\rho(\mu,\nu)$ of the measures μ and ν is the quantity

$$\rho(\mu,\nu) = \int_{\mathbf{T}} \Big(\frac{d\mu}{d\lambda}\Big)^{1/2}\Big(\frac{d\nu}{d\lambda}\Big)^{1/2} d\lambda, \tag{1.2}$$

obviously independent of the choice of λ. Note that

$$\rho(\mu,\nu) = 0 \text{ if and only if } \mu \perp \nu.$$

The measure μ is said to have *independent powers* if $\mu^n \perp \mu^m$ whenever $n \neq m$, $n,m \in \mathbf{N}$, where $\mu^n := \mu * \cdots * \mu$ n times. Such a measure is singular : if not, the absolutely continuous part ν of μ satisfies $0 \neq \nu^n \ll \mu^n$ for all $n \in \mathbf{N}$; thus, if $m \geq n$, $\nu^m \ll \mu^m$, $\nu^m \ll \nu^n \ll \mu^n$ and $\mu^m \not\perp \mu^n$.

It is less obvious to give conditions on the Fourier transform of μ, ensuring the absolute continuity of μ (with respect to m). Of course, it is necessary for μ to satisfy : $\lim_{|n|\to\infty} \hat{\mu}(n) = 0$.

This condition is not sufficient and we shall use the notation $M_0(\mathbf{T})$ for the ideal of all measures μ whose Fourier transform vanishes at infinity. Sometimes, those measures are called *Rajchman measures*, and a nice survey on them appears in [172].

6. $M(\mathbf{T})$ is identified with the dual space $C(\mathbf{T})^*$ of the continuous functions on $\mathbf{T}$. Let (μ_n) and μ in $M(\mathbf{T})$. From Fejer's theorem, μ_n converges to μ in the weak-star topology of $M(\mathbf{T})$, $\sigma(M(\mathbf{T}),C(\mathbf{T}))$, if and only if

$$\hat{\mu}_n(\gamma) \to \hat{\mu}(\gamma) \quad \text{for every} \quad \gamma \in \Gamma.$$

We shall write : $w^* - \lim_{n\to\infty} \mu_n = \mu$. Recall that the unit ball of $M(\mathbf{T})$ is a weak-star compact set.

The following proposition will be used in chapter 4 (see [59]).

Proposition 1.1. *Let (μ_n) and (ν_n) be two sequences of positive measures on $\mathbf{T}$ such that $\mu_n \to \mu$ and $\nu_n \to \nu$ in the weak-star topology of $M(\mathbf{T})$. Then*

$$\limsup_n \rho(\mu_n,\nu_n) \leq \rho(\mu,\nu),$$

where ρ denotes the affinity defined in (1.2)

Proof. The proof involves the Cauchy-Schwarz inequality applied with some suitable partition of unity. We assume, without loss of generality, that μ and ν are probability measures and we fix a probability measure λ dominating both μ and ν. We put $M_0 = \{\frac{d\mu}{d\lambda} = 0\}$, $N_0 = \{\frac{d\nu}{d\lambda} = 0\} \backslash M_0$ and we consider for $j \in \mathbf{Z}$ and some fixed $\varepsilon > 0$

$$U_j = \{x \in \mathbf{T} \backslash (M_0 \cup N_0),\ (1+\varepsilon)^j \frac{d\mu}{d\lambda}(x) < \frac{d\nu}{d\lambda}(x) \leq (1+\varepsilon)^{j+1} \frac{d\mu}{d\lambda}(x)\}. \quad (1.3)$$

Clearly, the sequence (U_j), supplemented by M_0 and N_0, provides an infinite partition of $\mathbf{T}$. In particular, $\sum_j \mu(U_j) < \infty$ and we fix J such that

$$\sum_{|j| \geq J} \mu(U_j) \leq \varepsilon^2. \quad (1.4)$$

From now on, we denote by $V_0, V_1, V_2, \ldots, V_{2J}, V_{2J+1}$ the finite Borel partition

$$M_0, N_0, U_{-J+1}, \ldots, U_{J-1}, \cup_{|j| \geq J} U_j$$

of $\mathbf{T}$. Note that, for every $2 \leq j \leq 2J$,

$$(1+\varepsilon)^{j-1-J} \mu(V_j) \leq \nu(V_j) \leq (1+\varepsilon)^{j-J} \mu(V_j), \quad (1.5)$$

by integrating the inequalities (1.3) on U_j with respect to λ. For each j, $0 \leq j \leq 2J+1$, let us choose by regularity an open set $\omega_j \supset V_j$ such that

$$\mu(\omega_j) \leq (1+\varepsilon)^{1/2} \mu(V_j),\ \nu(\omega_j) \leq (1+\varepsilon)^{1/2} \nu(V_j);$$

let $(f_j)_{0 \leq j \leq 2J+1}$ be a continuous partition of unity subordinate to the open covering $(\omega_j)_{0 \leq j \leq 2J+1}$. Clearly we have

$$\int_{\mathbf{T}} f_j \, d\mu \leq \mu(\omega_j) \leq (1+\varepsilon)^{1/2} \mu(V_j) \quad (1.6)$$

as well as

$$\int_{\mathbf{T}} f_j \, d\nu \leq \nu(\omega_j) \leq (1+\varepsilon)^{1/2} \nu(V_j). \quad (1.7)$$

We deduce that $\rho(\mu_n, \nu_n) :=$

$$\begin{aligned}
\int_{\mathbf{T}} \left(\frac{d\mu_n}{d\lambda}\right)^{1/2} \left(\frac{d\nu_n}{d\lambda}\right)^{1/2} d\lambda &= \sum_{j=0}^{2J+1} \int_{\mathbf{T}} \left(f_j \frac{d\mu_n}{d\lambda}\right)^{1/2} \left(f_j \frac{d\nu_n}{d\lambda}\right)^{1/2} d\lambda \\
&\leq \sum_{j=0}^{2J+1} \left(\int_{\mathbf{T}} f_j \, d\mu_n\right)^{1/2} \left(\int_{\mathbf{T}} f_j \, d\nu_n\right)^{1/2}
\end{aligned}$$

by the Cauchy-Schwarz inequality, and the weak* convergence hypothesis, combined with (1.6) and (1.7), leads to

$$\begin{aligned}\limsup_n \rho(\mu_n, \nu_n) &\leq \lim_n \sum_{j=0}^{2J+1} \Big(\int_{\mathbf{T}} f_j \, d\mu_n\Big)^{1/2} \Big(\int_{\mathbf{T}} f_j \, d\nu_n\Big)^{1/2} \\ &= \sum_{j=0}^{2J+1} \Big(\int_{\mathbf{T}} f_j \, d\mu\Big)^{1/2} \Big(\int_{\mathbf{T}} f_j \, d\nu\Big)^{1/2} \\ &\leq (1+\varepsilon)^{1/2} \sum_{j=0}^{2J+1} \sqrt{\mu(V_j)\nu(V_j)}.\end{aligned}$$

We are left to compare this latter term with the affinity $\rho(\mu, \nu)$. Actually, since $V_0 = M_0$ and $V_1 = N_0$, and by the choice of J in (1.4), we have

$$\sum_{j=0}^{2J+1} \sqrt{\mu(V_j)\nu(V_j)} \leq \sum_{j=2}^{2J} \sqrt{\mu(V_j)\nu(V_j)} + \varepsilon.$$

Now, from (1.5) and (1.3),

$$\begin{aligned}\sum_{j=2}^{2J} \sqrt{\mu(V_j)\nu(V_j)} &\leq \sum_{j=2}^{2J} (1+\varepsilon)^{(j-J)/2} \mu(V_j) \\ &\leq (1+\varepsilon)^{1/2} \sum_{j=2}^{2J} \int_{V_j} \Big(\frac{d\mu}{d\lambda}\Big)^{1/2} \Big(\frac{d\nu}{d\lambda}\Big)^{1/2} d\lambda \\ &\leq (1+\varepsilon)^{1/2} \rho(\mu, \nu).\end{aligned}$$

Finally,

$$\limsup_n \rho(\mu_n, \nu_n) \leq (1+\varepsilon)(\rho(\mu, \nu) + \varepsilon)$$

and the result follows by letting ε tend to zero. □

A very important example of measure is given as a weak-star limit point of some sequence of absolutely continuous measures.

Fundamental Example (Classical Riesz products). Consider a sequence (a_j) of real numbers of modulus ≤ 1; the polynomial $1 + a_j \cos(3^j t)$ is non-negative for every $t \in \mathbf{R}$, hence

$$P_N(t) = \prod_{j=0}^{N-1} (1 + a_j \cos(3^j t)) \geq 0, \quad \text{for every } N \geq 1$$

and $||P_N||_{L^1} = \widehat{P_N}(0)$. The sequence (3^j) is *dissociate* i.e. every integer n has at most one representation of the form

$$n = \sum \varepsilon_j 3^j, \quad \text{with } \varepsilon_j = -1, 0, 1.$$

This is due to the following fact : if $\eta_j \in \{0, \pm 1, \pm 2\}$,

$$|\pm 3^n + \eta_{n-1}3^{n-1} + \cdots + \eta_1 3 + \eta_0| \geq 3^n - 2(3^{n-1} + \cdots + 3 + 1) = 1.$$

It follows that

$$||P_N||_{L^1} = \widehat{P_N}(0) = 1,$$

$$\widehat{P_N}(n) = 0 \ \text{ if } n \notin \Big\{ \sum_{j<N} \varepsilon_j 3^j,\ \varepsilon_j = -1, 0, 1 \Big\}$$

while

$$\widehat{P_N}(\sum_{j<N} \varepsilon_j 3^j) = \prod_{j<N} \left(\frac{a_j}{2}\right)^{|\varepsilon_j|},\ \ \varepsilon_j = -1, 0, 1.$$

Moreover,

$$P_{N+1}(t) = P_N(t)(1 + \frac{a_N}{2} e^{i3^N t} + \frac{\overline{a_N}}{2} e^{-i3^N t})$$

and P_{N+1} contains P_N as a partial sum.

The sequence of measures $(P_N \cdot m)$ admits a unique weak-star cluster point in $M(\mathbf{T})$, therefore a weak-star limit ρ, which is a probability measure with Fourier coefficients

$$\hat{\rho}(n) = \begin{cases} \prod \left(\frac{a_j}{2}\right)^{|\varepsilon_j|} & \text{if } n = \sum \varepsilon_j 3^j, \quad \varepsilon_j = -1, 0, 1 \\ 0 & \text{otherwise} \end{cases} \tag{1.8}$$

Such measures, called *Riesz products*, have been constructed by F. Riesz, looking for a probability measure with more or less prescribed Fourier transform. They obey a very simple purity law : ρ is a singular measure if and only if $\sum_{n \in \mathbf{Z}} |a_n|^2 = \infty$; otherwise, $\rho \in L^2(\mathbf{T})$ [248].

Now consider (a_n) and (b_n) two sequences of real numbers in $[-1, 1]$; we denote by ρ_a (respectively ρ_b) the classical Riesz product

$$\rho_a = \prod_{j=1}^{\infty} (1 + a_j \cos 3^j t)$$

(resp. $\rho_b = \prod_{j=1}^{\infty}(1 + b_j \cos 3^j t)$). The mutual singularity, or equivalence problem, of two such measures has been investigated by Brown & Moran, and Peyrière [48,201]; here is a special case of their result.

Proposition 1.2. *With same notations,*

1. *If* $\sum_j |a_j - b_j|^2 = \infty$ *then* $\rho_a \perp \rho_b$;
2. *If* $\sum_j |a_j - b_j|^2 < \infty$ *and* $\limsup |a_n| < 1$ *then* $\rho_a \sim \rho_b$.

1.2 The Gelfand Spectrum Δ of $M(\mathbf{T})$

We outline in this section the properties of the Gelfand spectrum Δ of the Banach algebra $M(\mathbf{T})$, following the description of Šreider. Of course Δ contains Γ, and this new tool leads to new decompositions of the algebra $M(\mathbf{T})$ as we shall see.

1.2.1 Generalized Characters

1. Let f be an element of the dual space of the Banach space $M(\mathbf{T})$; f acts on $\mu \in M(\mathbf{T})$ in the following way :

$$(f,\mu) =: f(\mu) = \int_{\mathbf{T}} f(t)\, d\mu(t), \quad \text{for } f \in C(\mathbf{T}).$$

When restricted to $L(\mu) \simeq L^1(\mu)$ (see (1.1)), f can be identified with a function f_μ of $L^\infty(\mu)$. Therefore, considering $M(\mathbf{T})$ as the projective limit of the spaces $L^1(\mu), \mu \in M(\mathbf{T})$, f can be viewed as a family $(f_\mu)_{\mu \in M(\mathbf{T})}$ of elements of $L^\infty(\mu)$, enjoying the following properties :

$$(a)\ f_\nu = f_\mu \quad \nu-\text{a.e.} \quad \text{if } \nu \ll \mu$$

$$(b)\ \sup_{\mu \in M(\mathbf{T})} ||f_\mu||_{L^\infty(\mu)} < \infty$$

and by (a), $\quad (f,\nu) = \int f_\mu\, d\nu \quad \text{if } \nu \ll \mu.$

2. Now, let us denote by Δ the Gelfand spectrum of the Banach algebra $M(\mathbf{T})$ i.e. the set of continuous homomorphisms $M(\mathbf{T}) \to \mathbf{C}$. Thus, a *character* $\chi \in \Delta$ is an element of the dual space $M(\mathbf{T})^*$, of norm 1, such that

$$\chi(\mu * \nu) = \chi(\mu)\chi(\nu), \quad \mu, \nu \in M(\mathbf{T}).$$

Following 1. above, χ can be identified with a family $(\chi_\mu)_{\mu \in M(\mathbf{T})}$ of elements of $L^\infty(\mu)$ enjoying the following properties :

$$\begin{aligned} &(i)\ \ \chi_\nu = \chi_\mu \quad \nu-\text{a.e.} \quad \text{if } \nu \ll \mu \\ &(ii)\ \ \sup_{\mu \in M(\mathbf{T})} ||\chi_\mu||_{L^\infty(\mu)} = 1 \\ &(iii)\ \chi_{\mu*\nu}(x+y) = \chi_\mu(x)\chi_\nu(y) \quad \mu \otimes \nu - \text{a.e.} \end{aligned} \tag{1.9}$$

the last property resulting from the multiplicativity property of χ. By (i),

$$\chi(\mu) = \int \chi_\mu\, d\mu = \hat{\mu}(\chi).$$

3. Conversely, any family (χ_μ) satisfying (i), (ii), (iii) defines an element χ of Δ by formula (1.9).

1.2.2 Basic Operations

1. $\Gamma = \hat{\mathbf{T}}$ can be imbedded into Δ by the formula

$$\gamma(\mu) = \int \gamma \, d\mu = \hat{\mu}(\gamma), \quad \gamma \in \Gamma.$$

2. An element χ of Δ *operates* on $M(\mathbf{T})$ in this way : If $\mu \in M(\mathbf{T})$, $\chi \cdot \mu$, or simply $\chi\mu$, is the measure whose Fourier transform is

$$(\chi\mu)\hat{}(\gamma) = (\gamma\mu)\hat{}(\chi) = \int \gamma\chi_\mu \, d\mu, \quad \gamma \in \Gamma.$$

3. Endowed with the product

$$(\chi, \phi) \to (\chi\phi)_\mu = \chi_\mu \phi_\mu,$$

Δ is a commutative *semi-group*, and

$$(\chi\mu)\hat{}(\phi) = (\phi\mu)\hat{}(\chi) = \hat{\mu}(\chi\phi), \quad \mu \in M(\mathbf{T}).$$

4. We define $\overline{\chi}$, the *conjugate* of χ by the rule

$$(\overline{\chi})_\mu = \overline{\chi_{\overline{\mu}}}\ ;$$

we still have $\overline{\chi} \in \Delta$ and $|\chi|^2 = \chi\overline{\chi}$.
5. A character χ of Δ is said to be *positive* if the function χ_μ is ≥ 0 for every positive measure μ in $M(\mathbf{T})$.

Of course, $|\chi|^2$ is positive for any $\chi \in \Delta$.

Remark 1.1. Most often, we shall be concerned with a finite or countable set of measures (μ_n), and therefore, we shall restrict ourselves to a simpler space $L^1(\omega)$, where ω, for example, is the measure $\sum_{n\geq 0} 2^{-n}\frac{\mu_n}{||\mu_n||}$, $\mu_0 = \delta_0$, so that $\mu_n \ll \omega$ for all $n \in \mathbf{N}$. Then, a character ϕ reduces to a norm-1 function in $L^\infty(\omega)$ satisfying the functional equation :

$$\phi(x+y) = \phi(x)\phi(y) \quad \omega \otimes \omega - \text{a.e.}.$$

1.2.3 Topologies on $\Delta(\mu)$

Let μ be a fixed measure in $M(\mathbf{T})$. Taking into account the previous remark, we focus on

$$\Delta(\mu) = \{\chi_\mu, \chi \in \Delta\},$$

the set of all characters restricted to $L^\infty(\mu)$ (i.e. restricted to μ). We shall consider two topologies on $\Delta(\mu)$: the topology of $L^1(\mu)$, named further the *strong topology*, and the *weak-star topology* $\sigma(L^\infty(\mu), L^1(\mu))$, for which $\Delta(\mu)$ is compact.

$\overline{\Gamma}(\mu)$ denotes the weak-star closure of Γ in the unit ball of $L^\infty(\mu)$, in other words, $\phi \in \overline{\Gamma}(\mu)$ if and only if there exists a sequence $(\gamma_j) \subset \Gamma$ such that

$$\hat{\mu}(\gamma_j\gamma) \to \hat{\mu}(\phi\gamma) \quad \text{for each } \gamma \in \Gamma.$$

Note that, extracting a subsequence if necessary, $\gamma_j\overline{\gamma_{j+1}}$ converges to $|\phi|^2$.

Remark 1.2. If we equip Δ with the Gelfand topology, and if $\overline{\Gamma}$ is the closure of Γ in Δ for this topology, $\overline{\Gamma}(\mu)$ is nothing else than the set of restrictions $\{\chi_\mu, \chi \in \overline{\Gamma}\}$.

1.2.4 Constants in $\Delta(\mu)$

From now on, we assume that μ is a probability measure. We are looking for conditions on such a measure μ, involving the action of generalized characters on μ, which ensure the singularity of this measure.

The character χ is said to be *constant* on μ if there exists a constant C, $|C| \leq 1$, such that $\chi_\mu = C \quad \mu$-a.e.. This property is easily seen to be equivalent to the following :

$$\hat{\mu}(\chi\gamma) = \hat{\mu}(\chi)\hat{\mu}(\gamma), \quad \text{for every } \gamma \in \Gamma. \tag{1.10}$$

A probability measure μ which admits a constant character $\chi_\mu = C \quad \mu$-a.e., with $C^n \neq 1$ for every $n \in \mathbf{N}^*$, has independent powers (section 1.1 § 5). This claim turns out to be a direct consequence of the following simple remark :

Lemma 1.2. *Let μ and ν be probability measures and suppose that there exists $\chi \in \Delta$ such that $\chi_\mu = a \quad \mu$-a.e. and $\chi_\nu = b \quad \nu$-a.e., with a, b different constants. Then, μ and ν are mutually singular.*

Proof. In the opposite case, there would exist a measure $\tau \neq 0$ with $\tau \ll \mu$ and $\tau \ll \nu$ so that $\chi_\tau = \chi_\mu = \chi_\nu \quad \tau$-a.e. by property (1.9) (i) of χ; this is impossible if $a \neq b$. □

Since $\chi_{\mu^n} = C^n \quad \mu^n$-a.e. (property (1.9) (iii)), we get the claim by applying this remark to μ^n and μ^m, $n \neq m$.

Example 1.1. The *auto-similar* Riesz product $\rho = \prod_{j=0}^\infty (1 + r\cos 3^j t)$ where $r \in \mathbf{R}$, $0 < |r| \leq 1$, has independent powers. In fact, we observe that, for $n \geq n(m)$,

$$\hat{\rho}(3^n + m) = \hat{\rho}(m)\hat{\rho}(1) = \frac{r}{2}\hat{\rho}(m)$$

for every $m \in \mathbf{Z}$, which we write

$$\hat{\rho}(\gamma_{3^n}\gamma) = \frac{r}{2}\hat{\rho}(\gamma), \qquad \text{for every } \gamma \in \Gamma.$$

Now, if ϕ is any weak-star cluster point of the sequence (γ_{3^n}) in $\overline{\Gamma}(\rho)$, we have

$$\hat{\rho}(\phi\gamma) = \frac{r}{2}\hat{\rho}(\gamma)$$

for every $\gamma \in \Gamma$, which means, by (1.10), that $\phi = \frac{r}{2}$ ρ-a.e.. Therefore, ρ has independent powers.

Example 1.2. More generally, let us consider a probability measure μ with the following property : for some given integer $q \geq 2$,

$$\lim_{n\to\infty} \hat{\mu}(aq^n + b) = \hat{\mu}(a)\hat{\mu}(b), \quad \text{for every } a, b \in \mathbf{Z}. \tag{1.11}$$

In terms of generalized characters, (1.11) means, a being fixed, that the constant $\hat{\mu}(a)$ is the weak-star limit point of the sequence (γ_{aq^n}) in $\overline{\Gamma}(\mu)$. Consequently, for such a probability measure, all the constants $\hat{\mu}(a)$, $a \in \mathbf{Z}$, are in $\overline{\Gamma}(\mu)$.
This property (1.11), shared by the auto-similar Riesz products, will be interpreted in chapter 3 as a strong mixing property with respect to the transformation : $x \mapsto qx \mod 2\pi$ on $\mathbf{T}$.

1.3 Generalized Riesz Products

There are many possible ways to generalize Riesz products but we shall deal with a special type of one's, which will arise as maximal spectral type of dynamical systems; so we always refer to the following definition-proposition.

Definition 1.1. Let q be an integer ≥ 2 and $z_0, \ldots, z_{q-1}$ be q complex numbers of modulus 1. We consider the polynomials

$$P(t) = \frac{1}{q}\left|\sum_{k=0}^{q-1} z_k e^{ikt}\right|^2$$

and, for every $N \geq 1$,

$$P_N(t) = \prod_{n=0}^{N-1} P(q^n t).$$

Then, the sequence (P_N) converges weakly* (i.e. in the weak-star topology) to a probability measure ρ_z that we call a *generalized Riesz product.*

Proof. We may assume that $z_0 = 1$. The Fourier spectrum of P belongs to $\{-q+1, \ldots, q-1\}$, with coefficients

$$\widehat{P}(k) = \frac{1}{q}\sum_{j=0}^{q-k-1} z_{j+k}\overline{z_j}, \qquad 0 \le k \le q-1$$

and

$$\widehat{P}(-k) = \overline{\widehat{P}(k)}.$$

As for classical Riesz products, we have $P_N \ge 0$, $||P_N||_{L^1} = \hat{P}_N(0) = 1$, but we have no longer P_N as a partial sum of P_{N+1}. Nevertheless, we can conclude.

The Fourier spectrum of P_N belongs to $\{-q^N+1,\dots,q^N-1\}$, and $\hat{P}_N(-k) = \overline{\widehat{P}_N(k)}$. Now, if $0 \le k < q^{N+1}$, we write $k = aq^N + b$, $0 \le a \le q-1$, $0 \le b < q^N$, so that

$$\widehat{P}_{N+1}(k) = \widehat{P}_N(b)\widehat{P}(a) + \widehat{P}_N(b-q^N)\widehat{P}(a+1), \tag{1.12}$$

corresponding to the two possible decompositions of k into k_1+k_2, with k_1 in the Fourier spectrum of P_N and k_2 in the Fourier spectrum of P. When $0 \le k < q^N$ already, it follows that,

$$\widehat{P}_{N+1}(k) = \widehat{P}_N(k) + \widehat{P}_N(k-q^N)\widehat{P}(1). \tag{1.13}$$

In addition, a direct computation will show that

$$|\widehat{P}_N(k)| \le \frac{q^N-k}{q^N} \quad \text{for} \quad 0 < k < q^N. \tag{1.14}$$

Indeed, let us write $P_N = \frac{1}{q^N}|Q_N|^2$ with $Q_N(t) = Q(t)Q(qt)\cdots Q(q^{N-1}t)$ and $Q(t) = \sum_{k=0}^{q-1} z_k e^{ikt}$. It is easily seen that, for all $N \ge 1$,

$$Q_N(t) = \sum_{k=0}^{q^N-1} \alpha(k)e^{ikt}$$

where the sequence α is *strongly q-multiplicative* i.e. satisfies

$$\alpha(aq^n+b) = \alpha(a)\alpha(b), \quad \text{for } a \ge 0,\ 0 \le b < q^n.$$

Note that α is unimodular, which is clear inductively. Now

$$\widehat{P}_N(k) = \frac{1}{q^N}\sum_{n=0}^{q^N-k-1} \alpha(n+k)\overline{\alpha(k)}, \quad -q^N < k < q^N,$$

whence the inequality (1.14). This inequality, combined with (1.13), implies the weak-star convergence of the sequence (P_N) to a probability measure ρ_z. □

Proposition 1.3. *Let $(z_k)_{0\le k\le q-1}$ and $(z'_k)_{0\le k\le q-1}$ be two vectors of complex numbers of modulus 1. If there exists an integer a, $0 < a \le q-1$, for which $\hat{\rho}_z(a) \ne \hat{\rho}_{z'}(a)$, then the generalized Riesz products are mutually singular.*

Proof. This is, once more, an immediate consequence of lemma 1.2. We consider D_N the subgroup of $\mathbf{T}$, generated by $\frac{2\pi}{q^N}$, and ω_N the Haar measure of D_N. We set

$$\rho_N = \rho * \omega_N,$$

omitting the "z" for a moment. Note that $\widehat{\omega}_N(k) = 0$ if $q^N \not| k$ and $= 1$ otherwise. Hence, for each $N \in \mathbf{N}$,

$$\hat{\rho}_N(k) = \begin{cases} \hat{\rho}(k) & if \quad q^N \text{ divides } k \\ 0 & \text{otherwise} \end{cases}$$

and

$$\rho = P_N \rho_N.$$

This leads to the relation

$$\hat{\rho}(aq^N + b) = \widehat{P}_N(b)\hat{\rho}(aq^N) + \widehat{P}_N(b - q^N)\hat{\rho}(aq^N + q^N), \tag{1.15}$$

for $a, b \in \mathbf{N}$, $b < q^N$.

If we can prove that $\hat{\rho}(aq) = \hat{\rho}(a)$ for every $a \in \mathbf{Z}$, we obtain from (1.15) that

$$\lim_{N\to\infty} \hat{\rho}(aq^N + b) = \hat{\rho}(b)\hat{\rho}(a),$$

which means that the constants $\hat{\rho}(a), a \in \mathbf{Z}$ belong to $\overline{\Gamma}(\rho)$ (see example 1.2). The assumption on the measures ρ_z and $\rho_{z'}$, together with the lemma 1.2, will infer the mutual singularity of those two measures.

Thus, we shall establish by induction on N that

$$\widehat{P}_{N+1}(qa) = \widehat{P}_N(a), \tag{1.16}$$

and the proof will be complete.

By relation (1.12), (1.16) is immediate for $N = 1$. Now set $k = aq^{N-1} + b$, with $0 \leq a \leq q-1$ and $0 \leq b < q^{N-1}$, so that $0 \leq k < q^N$ and the same relation gives

$$\begin{aligned} \widehat{P}_{N+1}(qk) = \widehat{P}_{N+1}(aq^N + qb) &= \widehat{P}_N(qb)\widehat{P}(a) + \widehat{P}_N(qb - q^N)\widehat{P}(a+1) \\ &= \widehat{P}_{N-1}(b)\widehat{P}(a) + \widehat{P}_{N-1}(b - q^{N-1})\widehat{P}(a+1) \end{aligned}$$

by the induction hypothesis; finally, by using (1.12) again, we get

$$\widehat{P}_{N+1}(qk) = \widehat{P}_N(k)$$

which was the claim. □

The measure ρ_z is said to be *invariant* under the transformation $x \mapsto qx \mod 2\pi$ on $\mathbf{T}$.

The following assertion will be more convenient since the assumption relates directly to the sequence (z_k).

Corollary 1.1. *Let* $(z_k), (z'_k), \rho_z, \rho_{z'}$ *as in the previous proposition. If* $\left|\sum_{k=0}^{q-1} z_k e^{ikt}\right| \neq \left|\sum_{k=0}^{q-1} z'_k e^{ikt}\right|$ *for some* $t \in \mathbf{T}$*, the generalized Riesz products are mutually singular.*

Proof. If not, by making use of the previous proposition, we may suppose that $\hat{\rho}_z(a) = \hat{\rho}_{z'}(a)$ for every $a \in \{0,1,\ldots,q-1\}$. If we put $P_z(t) = \frac{1}{q}\left|\sum_{k=0}^{q-1} z_k e^{ikt}\right|^2$ and $P_{z'}(t) = \frac{1}{q}\left|\sum_{k=0}^{q-1} z'_k e^{ikt}\right|^2$, we have to prove that $P_z = P_{z'}$. By (1.15) we get

$$\hat{\rho}_z(a) = \hat{P}_z(a) + \hat{P}_z(a-q)\hat{\rho}_z(1), \tag{1.17}$$

and the analogous for z'. It follows that the polynomial $R = P_z - P_{z'}$ satisfies

$$\hat{R}(a) = -\hat{R}(a-q)\xi \tag{1.18}$$

where ξ is the common value $\hat{\rho}_z(1) = \hat{\rho}_{z'}(1)$.

A simple iteration of (1.18) leads to

$$\hat{R}(a) = \hat{R}(a)|\xi|^2, \quad a \in \{0,1,\ldots,q-1\}$$

whence the two cases :

If $|\xi| < 1$, R is identical to zero and we are done.

If $|\xi| = 1$, we go back to (1.17); since $\xi = \hat{\rho}_z(1) = \hat{P}_z(1) + \hat{P}_z(1-q)\hat{\rho}_z(1)$, or

$$\xi = \hat{P}_z(1) + \hat{P}_z(1-q)\xi$$

we get an expression of ξ in the form :

$$\xi = \frac{z_1 + z_2\overline{z_1} + \cdots + z_{q-1}\overline{z_{q-2}}}{q - \overline{z_{q-1}}},$$

and the same with (z'_k)

Now, $|\xi| = 1$ means $|z_1 + z_2\overline{z_1} + \cdots + z_{q-1}\overline{z_{q-2}}| = |q - \overline{z_{q-1}}| \geq q-1$. Necessarily, $z_{j+1}\overline{z_j} = e^{i\theta}$ for $j = 0,1,\ldots,q-2$ and $z_{q-1} = 1$. From that, $\xi = e^{i\theta}$, and $z_k = z'_k$ for $0 \leq k < q$; once again $P_z = P_{z'}$ and the corollary is proved. □

Remark 1.3. More generally, the polynomials P, P_N can be considered with $|z_j| \leq 1$ only; in this case, the sequence (P_N) converges weakly* to a positive measure with total mass $\frac{1}{q}\sum_{k=0}^{q-1}|z_k|^2$, and the previous properties are essentially preserved.

1.4 Idempotents in Δ and Decompositions of $M(\mathbf{T})$

We already met a decomposition of $M(\mathbf{T})$ (section 1.1 § 4) that we summarized by the following direct sum :

$$M(\mathbf{T}) = M_d(\mathbf{T}) \oplus M_c(\mathbf{T}), \tag{1.19}$$

$M_d(\mathbf{T})$ being a sub-algebra, and $M_c(\mathbf{T})$ an ideal of $M(\mathbf{T})$. Also

$$M_c(\mathbf{T}) = M_{cs}(\mathbf{T}) \oplus L^1(\mathbf{T}), \tag{1.20}$$

where $M_{cs}(\mathbf{T})$ denotes the continuous singular measures. We aim for a new decomposition of the algebra $M(\mathbf{T})$, which should contain (1.19) as a particular case. Specific characters will play an important role in it.

A character $h \in \Delta$ is said to be *idempotent* if $h^2 = h$. It is said *idempotent on the measure* μ if only $h_\mu^2 = h_\mu$ μ-a.e..

We can exhibit idempotents on the measure μ, by taking limits of sequences of characters in the following way : if $\phi \in \Delta(\mu)$,

$$w^* - \lim_{n\to\infty} |\phi|^n =: |\phi|^\infty$$

and

$$w^* - \lim_{n\to\infty} |\phi|^{1/n} =: |\phi|^0$$

are idempotents on μ satisfying : $|\phi|^\infty \leq |\phi| \leq |\phi|^0$.

The idempotents lead to new decompositions of the algebra $M(\mathbf{T})$.

Definition 1.2. We call L an *L-space* of $M(\mathbf{T})$ when L is a closed subspace of $M(\mathbf{T})$, invariant under absolute continuity of measures :

$$\mu \in L \quad \text{and} \quad \nu \ll \mu \Longrightarrow \nu \in L.$$

(In short "L"-property)

If we define the "orthogonal" of L by

$$L^\perp = \{\mu \in M(\mathbf{T}), \quad |\mu| \perp |\nu| \quad \forall \nu \in L\}$$

then we have the decomposition [123]

$$M(\mathbf{T}) = L \oplus L^\perp$$

i.e. every $\mu \in M(\mathbf{T})$ is uniquely decomposed into a sum

$$\mu = \mu_L + \mu_{L^\perp}, \quad \mu_L \in L, \quad \mu_{L^\perp} \in L^\perp.$$

Definition 1.3. We call $\mathscr{L}$ an *L-ideal* (resp. *L-sub-algebra*) of $M(\mathbf{T})$ if $\mathscr{L}$ is an ideal (resp. sub-algebra) of $M(\mathbf{T})$ enjoying the L-property.

Proposition 1.4. *Let us fix an idempotent $h \in \Delta$. Then $hM(\mathbf{T}) = \{h\mu,\ \mu \in M(\mathbf{T})\}$ is an L-sub-algebra and $(hM(\mathbf{T}))^{\perp}$ is the L-ideal :*

$$\mathscr{L}_h = \{\mu \in M(\mathbf{T}), \quad h\mu = 0\}.$$

Proof. By the definition of $h\mu$ with $h \in \Delta$, and from (1.9) (iii), we have, for $\gamma \in \Gamma$,

$$\int \gamma\, d(h \cdot \mu * \nu) = \int_{\mathbf{T}}\int_{\mathbf{T}} \gamma(x+y) h_{\mu*\nu}(x+y)\, d\mu(x)\, d\nu(y)$$

$$= \int_{\mathbf{T}}\int_{\mathbf{T}} \gamma(x)\gamma(y) h_{\mu}(x) h_{\nu}(y)\, d\mu(x)\, d\nu(y) = \int \gamma\, d(h\mu) \int \gamma\, d(h\nu);$$

whence the identity $h \cdot \mu * \nu = h\mu * h\nu$. The L-property follows from the Radon-Nikodym property. Thus we have the decomposition :

$$M(\mathbf{T}) = hM(\mathbf{T}) \oplus (hM(\mathbf{T}))^{\perp}.$$

Now, if h is an idempotent, $h(1-h) = 0$ and

$$(hM(\mathbf{T}))^{\perp} = (1-h)M(\mathbf{T}) = \{\mu \in M(\mathbf{T}), \quad h\mu = 0\}.$$

The property of $\mathscr{L}_h$ is now clear. □

We get this way all such canonical decompositions : assume that $M(\mathbf{T}) = A \oplus A^{\perp}$ where A is an L-sub-algebra and $A^{\perp}$ an L-ideal of $M(\mathbf{T})$; the natural projection on A defines an idempotent of Δ by the formula :

$$h\mu = \mu_A \quad \text{if} \quad \mu = \mu_A + \mu_{A^{\perp}}.$$

The positive *discrete idempotent*, h_d, associated to the decomposition (1.19) has been shown to belong to $\overline{\Gamma}$, the closure of Γ in Δ for the Gelfand topology [123]. It is proved that h_d is the smallest element in $\overline{\Gamma}_+$, the set of positive elements of $\overline{\Gamma}$. In particular,

Proposition 1.5. *For every $\mu \in M(\mathbf{T})$, the μ-component of h_d is in $\overline{\Gamma}(\mu)$.*

Proof. We fix $\mu \in M(\mathbf{T})$, and we write in short h instead of $(h_d)_{\mu}$. We claim that one can find a sequence of integers (n_j), $n_j \to \infty$, only depending on μ, with the property :

$$\hat{\tau}(n_j) \to \int h\, d\tau = \int d\tau_d = \hat{\tau}(0),$$

for all $\tau \ll \mu$. Thanks to the Radon-Nykodym theorem, it is sufficient to get the limit against the characters and to find (n_j) such that, for all $m \in \mathbf{Z}$,

$$\lim_{j\to\infty}\int e^{in_jt}e^{imt}\,d\mu(t)=\int e^{imt}\,d\mu_d(t)=\widehat{\mu}_d(m).$$

But

$$\widehat{\mu}(n_j+m)=\widehat{\mu}_c(n_j+m)+\widehat{\mu}_d(n_j+m),$$

and we consider μ_c and μ_d separately. We choose (m_j) such that $e^{im_jt}\to 1$ in $L^1(\mu_d)$. Thus,

$$|\widehat{\mu}_d(m_j+m)-\widehat{\mu}_d(m)|\leq\int|e^{i(m+m_j)t}\to e^{imt}|\,d\mu_d(t)\to 0$$

for all $m\in\mathbf{Z}$. Now, from Wiener's lemma, $\lim_{N\to\infty}\frac{1}{N}\sum_{n=K+1}^{N+K}|\widehat{\mu}_c(n)|^2=0$ uniformly in K. We may extract a subsequence (n_j) from (m_j) in such a way that

$$\widehat{\mu}_c(n_j+m)\to 0,\quad\forall m\in\mathbf{Z}.$$

This proves that h is the weak*-limit of e^{in_jt} in $L^\infty(\mu)$. □

This property of the idempotent h_d is more or less equivalent to the Dunkl-Ramirez inequality [75]:

Proposition 1.6 (Dunkl-Ramirez). *If $\mu\in M(\mathbf{T})$ then*

$$||\widehat{\mu_d}||_\infty\leq||\hat{\mu}||_\infty \tag{1.21}$$

Proof. Since $\mu_d=h_d\mu$, we have

$$\widehat{\mu_d}(\Gamma)=\widehat{h_d\mu}(\Gamma)=\hat{\mu}(h_d\Gamma)\subset\hat{\mu}\left(\overline{\Gamma(\mu)}\right)\subset\overline{\hat{\mu}(\Gamma)}$$

whence (1.21). □

As an application, we quote the following consequences :

Corollary 1.2. *Let $\mu\in M(\mathbf{T})$. The measure μ is discrete if and only if $\hat{\mu}$ is an almost-periodic function on $\mathbf{Z}$.*

Proof. Recall that f defined on $\mathbf{R}$ is said to be almost-periodic if f is the uniform limit on $\mathbf{R}$ of a sequence of almost-periodic polynomials :

$$\forall\varepsilon>0,\ \exists P,\ P(x)=\sum_{\text{finite}}a_je^{i\lambda_jx},\quad||f-P||_\infty\leq\varepsilon,$$

where a_j,λ_j are real numbers.

▷ If μ is a discrete measure, μ is the limit in norm of finitely supported measures μ_n, and $\hat{\mu}$ is the uniform limit of the almost-periodic polynomials $\hat{\mu}_n=:P_n$.

◁ Suppose now that $\hat{\mu}$ is an almost-periodic function on $\mathbf{Z}$: there exists a sequence of discrete measures, (μ_p), such that

$$\lim_{p\to\infty} ||\hat{\mu} - \widehat{\mu_p}||_\infty = 0,$$

observing in the previous definition that P is also a Fourier transform $\widehat{\mu_p}$, with $\mu_p = \sum_{\text{finite}} a_j \delta_{\lambda_j}$. Applying now (1.21) we get

$$||\widehat{\mu_p} - \widehat{\mu_d}||_\infty \leq ||\widehat{\mu_p} - \hat{\mu}||_\infty \to 0$$

and $\widehat{\mu} = \widehat{\mu_d}$. □

(Note that the decomposition : $M(\mathbf{T}) = M_s(\mathbf{T}) \oplus L^1(\mathbf{T})$ is not of the canonical form, just a decomposition into a sum of L-spaces.)

Corollary 1.3. *Let $E \subset \mathbf{Z}$ be a set with arbitrary large gaps. If the Fourier spectrum of $\mu \in M(\mathbf{T})$ is included into E, then μ is continuous.*

Proof. We give a short proof of this fact, involving the discrete idempotent $h := h_d$. By the assumption on E, there exists a sequence of integers (n_j) such that $\lim_{j\to\infty} n_j = +\infty$ and

$$[n_j - j, n_j + j] \subset E^c.$$

Consider χ a cluster point of (γ_{n_j}) in $\overline{\Gamma}$; then $\chi\Gamma \subset \overline{E^c}$ and $\chi\overline{\Gamma} \subset \overline{E^c}$ (the closure of E^c in $\overline{\Gamma}$). Since h is the smallest element in $\overline{\Gamma}_+$, we must have $h|\chi|^2 = h$; it follows that, for every $\gamma \in \Gamma$,

$$\widehat{\mu_d}(\gamma) = \widehat{\mu}(h\gamma) = \widehat{\mu}(h|\chi|^2\gamma) = \widehat{\mu}(h\overline{\chi}\gamma \cdot \chi).$$

But $h\overline{\chi}\gamma \in \overline{\Gamma}$ and $h\overline{\chi}\gamma \cdot \chi \in \overline{E}^c$ so that $\widehat{\mu}(h\overline{\chi}\gamma \cdot \chi) = 0$ for every $\gamma \in \Gamma$. Thereby, $\mu_d = 0$ and the corollary is proved. □

1.5 Dirichlet Measures

In this last section, we study a class of measures related to some mixing property of dynamical systems (see chapter 3).

Definition 1.4. 1) A Borel set E of $\mathbf{T}$ is said to be a *Dirichlet set*, if there exists a sequence $(\gamma_j) \subset \Gamma$ such that $\gamma_j \to \infty$ in Γ, and $\gamma_j \to 1$ uniformly on E; in other words, there exists a sequence $(n_j) \subset \mathbf{Z}$ such that $|n_j| \to \infty$ and $\sup_{t\in E} |e^{in_j t} - 1| \to 0$.

2) We say that E is *weak-Dirichlet set*, if, for every positive measure $\mu \in M(\mathbf{T})$ supported by E, there exists a sequence $(\gamma_j) \subset \Gamma$ such that $\gamma_j \to \infty$ in Γ, and $\int |\gamma_j - 1|\, d\mu \to 0$.

There is a nice description of probability measures supported by some weak-Dirichlet set, in terms of idempotents.

Proposition 1.7. *Let μ be a probability measure on* $\mathbf{T}$. *The following assertions are equivalent :*

a) *μ is supported by a weak-Dirichlet set.*
b) $\limsup_{\gamma\to\infty}|\hat{\mu}(\gamma)| = 1$.
c) *1 is the limit in $L^1(\mu)$ of a sequence $(\gamma_j)\subset\Gamma$ which tends to infinity in Γ.*
d) *There exists an idempotent $h\in\overline{\Gamma}$, $h\neq 1$, such that $h\mu=\mu$.*

Proof. $\star$ Let $h\in\overline{\Gamma}$, $h\neq 1$, such that $h\mu=\mu$. In particular, we have

$$1=\hat{\mu}(h)\leq\limsup_{\gamma\to\infty}|\hat{\mu}(\gamma)|\leq 1$$

and d) implies b).

$\star$ Now suppose that b) holds. Then, there exists a sequence $(\gamma_j)\subset\Gamma$ which tends to infinity in Γ and satisfies

$$\limsup_{j\to\infty}|\hat{\mu}(\gamma_j)|=1.$$

Let χ be a cluster point of (γ_j) in $\overline{\Gamma}$. For a subsequence (γ_{j_k}) of (γ_j),

$$\lim_{k\to\infty}\hat{\mu}(\gamma_{j_k}\gamma)=\hat{\mu}(\chi\gamma)$$

for every $\gamma\in\Gamma$. It follows that χ is constant on μ : indeed, if we prove that

$$|\hat{\mu}(\gamma_{j_k}\gamma)-\hat{\mu}(\gamma_{j_k})\hat{\mu}(\gamma)|\to 0 \tag{1.22}$$

when $k\to\infty$, we get

$$\hat{\mu}(\chi\gamma)=\hat{\mu}(\chi)\hat{\mu}(\gamma),\quad\forall\gamma\in\Gamma$$

and the claim results from (1.10). We are thus left to compute

$$\int|\gamma_{j_k}\gamma-\hat{\mu}(\gamma_{j_k})\gamma|^2d\mu=1-|\hat{\mu}(\gamma_{j_k})|^2$$

which tends to 0 by construction of (γ_j). Now (1.22) is a consequence of the Schwarz inequality, and $\chi\mu=C$. Note that necessarily $|C|=1$.
By similar arguments, we see that

$$\int|\gamma_{j_k}-C|\,d\mu\to 0;$$

to finish, we may choose (γ_{j_k}) in such a way that $\gamma_{j_{k+1}}\overline{\gamma_{j_k}}$ tends to ∞ in Γ together with

$$\lim_{k\to\infty}\int|\gamma_{j_{k+1}}\overline{\gamma_{j_k}}-1|\,d\mu=\lim_{k\to\infty}\int|\gamma_{j_{k+1}}\overline{\gamma_{j_k}}-|C|^2|\,d\mu=0.$$

Hence c) is proved.

⋆ Assume c) and let k be an integer ≥ 1. From a sequence of characters satisfying c), we can extract a sub-sequence converging uniformly on a compact set E_k of μ-measure $\geq 1 - 1/k$ (Egorov's theorem). By using the diagonal process, we get a sub-sequence (γ_j) converging uniformly to 1 on each E_k, $k \geq 1$. At each step, we may choose $E_{k+1} \supset E_k$ so that $E = \cup_{k \geq 1} E_k$ is a borelian support of the probability measure μ.

Let ν be any probability measure supported by E. For every $\varepsilon > 0$, we see that E contains a set E_ε of ν-measure $\geq 1 - \varepsilon$, on which γ_j converges uniformly to 1 (by taking E_k, k large enough). This implies that $\lim_{j \to \infty} \int_E |\gamma_j - 1| \, d\nu = 0$ and E is a weak-Dirichlet set.

⋆ Finally, let E be a weak-Dirichlet set, μ a probability measure supported by E and (γ_j) the associated sequence or characters. If χ is any cluster point of (γ_j) in $\overline{\Gamma}$, we get $\chi\mu = \mu$ and

$$h = \lim_{n \to \infty} |\chi|^{2n}$$

is an idempotent of $\overline{\Gamma}$. Moreover $\chi \notin \Gamma$ (otherwise (γ_j) would be stationary) so that $h \neq 1$ and $h\mu = \mu$, which achieves the proof. □

Definition 1.5. 1) A positive measure is called a *Dirichlet measure* if the probability measure $\mu/||\mu||$ enjoys one of the equivalent properties of the proposition (1.7)

2) A complex measure μ is now called a Dirichlet measure if $|\mu|$ is one.

So we have

Proposition 1.8. *The set of Dirichlet measures in $M(\mathbf{T})$ is an L-space of $M(\mathbf{T})$, whose orthogonal, denoted by $\mathscr{L}_I$, is an L-ideal, characterized by*

$$\mu \in \mathscr{L}_I \Longleftrightarrow h\mu = 0 \quad \textit{for every idempotent } h \in \overline{\Gamma}, \quad h \neq 1$$

Remark 1.4. One can prove that $\mu \in \mathscr{L}_I$ if and only if $\limsup_{\gamma \to \infty} |\hat{\nu}(\gamma)| < 1$ for every probability measure $\nu \ll \mu$ [123]. Clearly

$$M_0(\mathbf{T}) \subset \mathscr{L}_I \subset M_c(\mathbf{T})$$

since continuous measures are characterized by $h_d\mu = 0$.

In the same work can be found examples and new characterizations of $M_c(\mathbf{T})$, $\mathscr{L}_I$, and others, with the help of arithmetical properties of the Fourier spectrum of measures; more precisely, those ideals are characterized by properties of the sets

$$E(\mu, \varepsilon) = \{n \in \mathbf{Z}, \ |\hat{\mu}(n)| > \varepsilon\},$$

$\varepsilon > 0$ and μ in the ideal.

Chapter 2
Spectral Theory of Unitary Operators

We wish to classify isometric operators on Hilbert spaces, up to unitary equivalence (or *spectral equivalence*). We introduce for this purpose different notions of the spectral theory of unitary operators, such as : spectral measure, maximal spectral type, spectral multiplicity, multiplicity function, etc.; we establish two versions of the spectral decomposition theorem for these operators, with our familiar notations. The definitions and results will be used in the next chapter where we focus on dynamical systems, and later, when we study the spectral properties of substitutive sequences.

2.1 Representation Theorem of Unitary Operators

Let U be a unitary operator on the separable Hilbert space H, endowed with the inner product $\langle \cdot, \cdot \rangle$. In chapter 3, H will be $L^2(X, \mathscr{B}, \mu)$ and $U = U_T$, the unitary operator associated to an automorphism of the probability space $(X, \mathscr{B}, \mu)$, defined by $U_T f(x) = f(Tx)$ if $f \in L^2(X, \mathscr{B}, \mu)$.

Recall that the *spectrum* of an operator A on H is the set $sp(A)$ of complex numbers z such that $A - zI$ is not invertible. If $A = U$ is a unitary operator, $sp(U)$ is a compact subset of the circle. But we need more to distinguish unitary operators.

2.1.1 Construction of Spectral Measures

For each $f \in H$, the sequence $(t_n)_{n \in \mathbf{Z}}$ where $t_n = \langle U^n f, f \rangle$ is positive definite since

$$\begin{aligned}\Sigma_{i,j} z_i \overline{z_j} t_{i-j} &= \Sigma_{i,j} z_i \overline{z_j} \langle U^{i-j} f, f \rangle \\ &= \Sigma_{i,j} \langle z_i U^i f, z_j U^j f \rangle \\ &= || \Sigma_i z_i U^i f ||_H^2 \geq 0.\end{aligned}$$

M. Queffélec, *Substitution Dynamical Systems – Spectral Analysis: Second Edition*, Lecture Notes in Mathematics 1294, DOI 10.1007/978-3-642-11212-6_2,
© Springer-Verlag Berlin Heidelberg 2010

By the Bochner theorem (section 1.1 §3), we can associate to the element $f \in H$ a positive measure on $\mathbf{T}$, denoted by σ_f, that we call the *spectral measure* of f; σ_f is characterized by its Fourier coefficients

$$\widehat{\sigma}_f(n) = \langle U^n f, f \rangle, \qquad n \in \mathbf{Z},$$

and its total mass is :

$$||\sigma_f|| = ||f||_H^2. \tag{2.1}$$

Now, if f and g are two elements of H, we consider

$$a_n = \langle U^n f, g \rangle, \qquad n \in \mathbf{Z},$$

so that $\langle U^n g, f \rangle = \overline{a_{-n}}$. The elementary identity

$$\begin{aligned} 4\langle U^n f, g \rangle &= \langle U^n (f+g), f+g \rangle - \langle U^n (f-g), f-g \rangle \\ &+ i\langle U^n (f+ig), f+ig \rangle - i\langle U^n (f-ig), f-ig \rangle \end{aligned} \tag{2.2}$$

proves that (a_n) is the Fourier transform of a complex measure on $\mathbf{T}$:

$$a_n = \langle U^n f, g \rangle =: \widehat{\sigma}_{f,g}(n), \qquad n \in \mathbf{Z},$$

where, by (2.2),

$$\sigma_{f,g} = \frac{1}{4}(\sigma_{f+g} - \sigma_{f-g} + i\sigma_{f+ig} - i\sigma_{f-ig}).$$

We have $\sigma_{f,f} = \sigma_f, \quad \sigma_{f,g} = \overline{\sigma_{g,f}}$ and

$$\sigma_{f+g} = \sigma_f + \sigma_g + \sigma_{f,g} + \sigma_{g,f} \tag{2.3}$$

Definition 2.1. The family of measures $(\sigma_{f,g})_{f,g \in H}$ is referred to as the *spectral family* of the operator U.

2.1.2 Properties of the Spectral Family

With the help of the spectral family, we shall give an integral representation of the operator U.

Definition 2.2. If $f \in H$, we write $[U, f]$ for the *cyclic subspace* generated by f, which is the closure in H of the linear span of $\{U^n f,\ n \in \mathbf{Z}\}$.

More generally, $[U, f_1, \ldots, f_k]$ will denote the cyclic subspace generated by $f_1, \ldots, f_k \in H$.

We have the simple but fundamental following propositions.

Proposition 2.1. *Let R be a trigonometric polynomial on* **T**, *then*

$$||R(U)f||_H = ||R||_{L^2(\sigma_f)}. \tag{2.4}$$

Proof. If $R(t) = \sum_k \hat{R}(k)e^{ikt}$, then

$$\begin{aligned}
||R(U)f||_H^2 &= \langle \sum_k \hat{R}(k)U^k f, \sum_j \hat{R}(j)U^j f\rangle \\
&= \sum_{j,k} \hat{R}(k)\overline{\hat{R}(j)}\langle U^{k-j}f, f\rangle \\
&= \sum_{j,k} \hat{R}(k)\overline{\hat{R}(j)}\hat{\sigma}_f(k-j) \\
&= \int_{\mathbf{T}} \big(\sum_{j,k} \hat{R}(k)\overline{\hat{R}(j)}e^{ikt}e^{-ijt}\big)\, d\sigma_f(t) \\
&= \int_{\mathbf{T}} |R(t)|^2\, d\sigma_f(t).
\end{aligned}$$

□

Proposition 2.2. *Let R be a trigonometric polynomial on* **T**, *then*

$$\sigma_{R(U)f,g} = R \cdot \sigma_{f,g} \quad \text{and} \quad \sigma_{R(U)f} = |R|^2 \sigma_f. \tag{2.5}$$

Proof. The second assertion is a direct consequence of the first one, and for this first identity, it suffices to write

$$\begin{aligned}
\widehat{\sigma}_{R(U)f,g}(k) &= \langle U^k R(U)f, g\rangle = \int_{\mathbf{T}} e^{ikt} R(t)\, d\sigma_{f,g}(t) \\
&= (R \cdot \sigma_{f,g})^\frown(k).
\end{aligned}$$

□

Proposition 2.3. *The map* $(f,g) \mapsto \sigma_{f,g}$ *from* $H \times H$ *into* $M(\mathbf{T})$ *is bilinear and continuous.*

Proof. We shall prove the continuity by showing the inequality $||\sigma_{f,g}|| \leq ||f||_H ||g||_H$, the bilinearity being obvious by construction.
For a trigonometric polynomial R on **T**, we have,

$$\begin{aligned}
\int_{\mathbf{T}} R(t)\, d\sigma_{f,g}(t) &= |\langle R(U)f, g\rangle| \qquad \text{by (2.5)} \\
&\leq ||R(U)f||_H ||g||_H \\
&= ||R||_{L^2(\sigma_f)} ||g||_H \qquad \text{by (2.4)} \\
&\leq ||R||_\infty ||f||_H ||g||_H
\end{aligned}$$

since $||\sigma_f|| = ||f||_H^2$. From this inequality, we derive the claimed one by taking the sup on the trigonometric polynomials R with norm $||R||_\infty \leq 1$. □

Proposition 2.4. *For all $f,g \in H$, the measure $\sigma_{f,g}$ is absolutely continuous with respect to σ_f and σ_g; more precisely,*

$$|\sigma_{f,g}|(B) \leq \sqrt{\sigma_f(B)}\sqrt{\sigma_g(B)}$$

for every Borel set B in $\mathbf{T}$.

Proof. Fix a Borel set $B \subset \mathbf{T}$; applying the Schwarz inequality to the positive bilinear form $(f,g) \mapsto \sigma_{f,g}(B)$, we obtain

$$|\sigma_{f,g}(B)| \leq \sqrt{\sigma_f(B)}\sqrt{\sigma_g(B)}.$$

Now, by definition of the total variation measure $|\sigma_{f,g}|$, for any fixed $\varepsilon > 0$ there exists a partition (B_n) of B such that

$$\sum_n |\sigma_{f,g}(B_n)| \geq |\sigma_{f,g}|(B) - \varepsilon.$$

One more application of the Schwarz inequality leads to

$$\begin{aligned}\Sigma_n |\sigma_{f,g}(B_n)| &\leq \Sigma_n \sqrt{\sigma_f(B_n)}\sqrt{\sigma_g(B_n)} \\ &\leq (\Sigma_n \sigma_f(B_n))^{1/2}(\Sigma_n \sigma_g(B_n))^{1/2},\end{aligned}$$

so that, finally, $|\sigma_{f,g}|(B) \leq \sqrt{\sigma_f(B)}\sqrt{\sigma_g(B)} + \varepsilon$, and the result follows. □

Remark 2.1. Actually, this property is equivalent to the previous one. But it can be slightly improved in the following way : If both of σ_f and σ_g are absolutely continuous with respect to a same positive measure ω, so is $\sigma_{f,g}$, and we have

$$\left|\frac{d\sigma_{f,g}}{d\omega}\right| \leq \sqrt{\frac{d\sigma_f}{d\omega}}\sqrt{\frac{d\sigma_g}{d\omega}}.$$

As a consequence of (2.3), note that we have

Corollary 2.1. *For $f,g \in H$,*

$$\sigma_{f+g} \ll \sigma_f + \sigma_g.$$

Also

Corollary 2.2. *If (f_n) converges to f in H, σ_{f_n} converges to σ_f in $M(\mathbf{T})$.*
More precisely, we have the inequalities

$$||\sigma_{f_n} - \sigma_f|| \leq ||\sigma_{f_n - f}|| + 2||\sigma_{f,f_n - f}|| \leq ||f - f_n||_H^2 + 2||f||_H||f_n - f||_H.$$

2.1.3 Spectral Representation Theorem

Identity (2.4) defines, by extension, an *isometry* W from $[U,f]$ onto $L^2(\sigma_f)$ with the following properties : $Wf = \mathbf{1}$ and, if $h \in [U,f]$ then

$$(i) \quad ||h||_H = ||Wh||_{L^2(\sigma_f)}$$

$$(ii) \quad e^{it}Wh(t) = WUh(t) \tag{2.6}$$

In other words

Theorem 2.1. *There exists a unitary equivalence W between U restricted to the cyclic subspace $[U,f]$ and V, the operator of multiplication by e^{it} on $L^2(\sigma_f)$, so that the following diagram commutes :*

$$\begin{array}{ccc} [U,f] & \xrightarrow{W} & L^2(\sigma_f) \\ U \downarrow & & \downarrow V \\ [U,f] & \xrightarrow[W]{} & L^2(\sigma_f) \end{array}$$

This theorem has many consequences. Note first the following :

Corollary 2.3. *If $f,g \in H$, then $U_{|[U,f]}$ is unitarily equivalent to $U_{|[U,g]}$ if and only if $\sigma_f \sim \sigma_g$.*

Also, we can extend identity (2.5) to the cyclic subspace $[U,f]$.

Corollary 2.4. *Let $f \in H$; then $h \in [U,f]$ if and only if $\sigma_h = |\phi|^2\sigma_f$ for some $\phi \in L^2(\sigma_f)$.*

Proof. By using the above theorem and (2.6), we get for $h \in [U,f]$ and $g \in H$

$$\begin{aligned} \widehat{\sigma}_{h,g}(n) &= \langle U^n h, g\rangle_H = \langle WU^n h, Wg\rangle_{L^2(\sigma_f)} \\ &= \langle e^{int}Wh, Wg\rangle_{L^2(\sigma_f)} = \int_{\mathbf{T}} e^{int}Wh\,\overline{Wg}\,d\sigma_f \\ &= \left(Wh\,\overline{Wg}\,\sigma_f\right)^\frown(n) \end{aligned}$$

so that $\sigma_{h,g} = Wh\,\overline{Wg}\,\sigma_f$; in particular $\sigma_h = |Wh|^2\sigma_f$.

Conversely, if $\sigma \ll \sigma_f$, we may write $\sigma = |\phi|^2\sigma_f$ with $\phi \in L^2(\sigma_f)$; since W is onto, there exists $h \in [U,f]$ such that $\phi = Wh$, and $\sigma = \sigma_h$ by reverse calculations. □

Note that when ϕ is a trigonometric polynomial R, $h = R(U)f$ by (2.5). In case $\phi \in L^2(\sigma_f)$, we denote this element h by $\phi(U)f$ and the corollary becomes :

Corollary 2.5. *For every $f \in H$ and $\phi \in L^2(\sigma_f)$, we can define an element $\phi(U)f \in [U,f]$, satisfying :*

$$\sigma_{\phi(U)f,g} = \phi \sigma_{f,g}, \quad \forall g \in H. \tag{2.7}$$

In particular, $\sigma_{\phi(U)f} = |\phi|^2 \sigma_f$.

We are now in a position to extend the *functional calculus* to $L^\infty(\mathbf{T})$.

Proposition 2.5. *If ϕ is a bounded Borel function on $\mathbf{T}$, the map $f \in H \mapsto \phi(U)f$ defines a linear operator on H, bounded by $||\phi||_\infty$.*

Proof. When $\phi \in L^\infty(\mathbf{T})$, according to the previous corollary, $\phi(U)f$ is quite well defined for each $f \in H$; as a consequence of (2.7) and thanks to the bilinearity property of $\sigma_{f,g}$, we check that

$$\sigma_{\phi(U)(f+g)-\phi(U)f-\phi(U)g} =$$

$$|\phi|^2(\sigma_{f+g} + \sigma_f + \sigma_g - \sigma_{f+g,f} - \sigma_{f,f+g} - \sigma_{f+g,g} - \sigma_{g,f+g} + \sigma_{f,g} + \sigma_{g,f}) = 0$$

so that

$$||\phi(U)(f+g) - \phi(U)f - \phi(U)g||_H^2 = ||\sigma_{\phi(U)(f+g)-\phi(U)f-\phi(U)g}|| = 0$$

In the same way,

$$||\phi(U)(\lambda f) - \lambda\phi(U)f||_H = ||\sigma_{\phi(U)(\lambda f)-\lambda\phi(U)f}|| = 0,$$

and $\phi(U)$ is linear. Finally,

$$||\phi(U)f||_H^2 = ||\sigma_{\phi(U)f}|| = |||\phi|^2\sigma_f|| \leq ||\phi||_\infty^2 ||f||_H^2$$

gives the claimed bound. □

We have proved, in passing, the so-called *spectral representation theorem* for unitary operators, namely :

Theorem 2.2 (Spectral representation theorem). *Let U be a unitary operator on the separable Hilbert space H. Then there exists a family of measures on $\mathbf{T}$, $(\sigma_{f,g})_{f,g\in H}$, such that, for every bounded Borel function ψ on $\mathbf{T}$, we have*

$$\langle \psi(U)f, g \rangle = \int \psi \, d\sigma_{f,g}. \tag{2.8}$$

2.1.4 Invariant Subspaces and Spectral Projectors

If the subspace H_0 of H is invariant under U, so is $H_0^\perp$, and, if P_0 is the orthogonal projection onto H_0, then $P_0 U = U P_0$. We would like to identify the operators commuting with U, and, in view of theorem 2.1, it is natural to begin with the description of operators commuting with a multiplication operator.

Proposition 2.6. *Let σ be a probability measure on $\mathbf{T}$, and V the operator of multiplication by e^{it} on $L^2(\sigma)$. Then*

1) The bounded operator Q commutes with V on $L^2(\sigma)$ if and only if Q is the operator of multiplication by some real function $\varphi \in L^\infty(\sigma)$.

2) The subspace M of $L^2(\sigma)$ is invariant under V if and only if $M = \mathbf{1}_B L^2(\sigma)$, for some Borel set B in $\mathbf{T}$ and $\mathbf{1}_B$ its indicator function.

Proof. The assertion 2) can be derived from 1) : if P_M is the orthogonal projection onto M, $P_M f = \varphi f$ with $\varphi \in L^\infty_{\mathbf{R}}(\sigma)$ by 1). But $P_M^2 = P_M$ and $\varphi^2 = \varphi$ whence $\varphi = \mathbf{1}_B$ for some Borel set B.

Let us prove the first assertion. By assumption, $QV^n = V^n Q$ so that, putting $\varphi = Q\mathbf{1} \in L^2(\sigma)$, we have $e^{int}\varphi(t) = Q(e^{int})$ for every $n \geq 0$. This identity extends readily to the trigonometric polynomials P :

$$Q(P(t)) = P(t)\varphi(t) \tag{2.9}$$

We must show that $\varphi \in L^\infty(\sigma)$. If φ is not zero, consider $a > 0$ such that the set $B = \{|\varphi| \geq a\}$ has positive measure. Integrating (2.9) on B, we get

$$\int_B |P\varphi|^2 \, d\sigma = \int_B |Q(P)|^2 \, d\sigma \leq C^2 ||P||^2_{L^2(\sigma)}$$

if $C := ||Q|| > 0$. Hence

$$a||\mathbf{1}_B P||_{L^2(\sigma)} \leq C||P||_{L^2(\sigma)}$$

for every trigonometric polynomial P; in particular for P_n where P_n tends to $\mathbf{1}_B$ in $L^2(\sigma)$. Taking the limit on n in the inequality $a||\mathbf{1}_B P_n||_{L^2(\sigma)} \leq C||P_n||_{L^2(\sigma)}$, we deduce that $a \leq C$ and $\varphi \in L^\infty(\sigma)$. □

This proposition gives an answer to the commutator problem when the space H is cyclic under U. In the general case, we shall need a decomposition of H into cyclic subspaces, which is the purpose of the next section. A first reduction arises from the *spectral projectors* :

Proposition 2.7. *Let B be a Borel set in $\mathbf{T}$; we denote by $\mathbf{1}_B(U)$ the associated operator (proposition 2.5). Then, $\mathbf{1}_B(U)$ is the orthogonal projection onto the following subspace H_B of H :*

$$H_B = \{f \in H, \ \sigma_f(B^c) = 0\}.$$

These projectors are called *spectral projectors* and, from now on, denoted by E_B, $B \in \mathscr{B}_{\mathbf{T}}$.

Proof. Thanks to the absolute continuity relation $\sigma_{f+g} \ll \sigma_f + \sigma_g$, H_B is clearly a subspace of H. It is closed in H since $||\sigma_{f_n} - \sigma_f||$ tends to zero as soon as f_n tends to f in H (corollary 2.2). If we put $P = E_B$, we have $||P|| \leq 1$ (proposition 2.5) and it remains to prove that $P^2 = P$. By (2.7), for $f, g \in H$,

$$\sigma_{Pf,g} = \mathbf{1}_B \sigma_{f,g}$$

and $\sigma_{Pf,g}$ is supported on B. This implies

$$\begin{aligned} \langle P(Pf), g\rangle &= (\mathbf{1}_B \sigma_{Pf,g})\widehat{\ }(0) = (\mathbf{1}_B \sigma_{f,g})\widehat{\ }(0) && (2.10)\\ &= \sigma_{f,g}(B) && (2.11)\\ &= \langle Pf, g\rangle && (2.12)\end{aligned}$$

and $P^2 = P$. Thus P is the orthogonal projection onto H_B. □

The following corollary will be useful later:

Corollary 2.6. *Let $h \in H$ and $\sigma_h = \mu + \nu$ be a decomposition of σ_f into mutually singular positive measures. Then there exist f and g in $[U, h]$ such that $\sigma_f = \mu$ and $\sigma_g = \nu$.*

Proof. By mutual singularity of μ and ν, there exists a Borel set $B \subset \mathbf{T}$ such that $\mathbf{1}_B \mu = \mu$ and $\mathbf{1}_{B^c} \nu = \nu$. If we set $f = E_B\, h$ and $g = E_{B^c}\, h$, then $f, g \in [U, h]$ with

$$\sigma_f = \mathbf{1}_B \sigma_h = \mathbf{1}_B(\mu + \nu) = \mu,$$

and

$$\sigma_g = \mathbf{1}_{B^c} \sigma_h = \mathbf{1}_{B^c}(\mu + \nu) = \nu.$$

□

Turning back to our problem, we can show :

Proposition 2.8. *Q commutes with U if and only if Q commutes with all the spectral projectors.*

Proof. If $f, g \in H$ and if B is a Borel set in $\mathbf{T}$,

$$\langle QE_B f, g\rangle = \langle E_B f, Q^* g\rangle = \sigma_{f,Q^*g}(B)$$

while

$$\langle E_B Qf, g\rangle = \sigma_{Qf,g}(B).$$

But

$$\widehat{\sigma}_{Qf,g}(n) = \langle U^n Qf, g\rangle = \langle QU^n f, g\rangle = \widehat{\sigma}_{f,Q^*g}(n)$$

for all n, f, g if and only if Q commutes with U. This gives the result. □

2.2 Operators with Simple Spectrum

Before establishing the spectral decomposition theorem in the general case, we dwell upon the cyclic case, from the operator point of view.

Definition 2.3. The unitary operator U is said to have *simple spectrum* if there exists $h \in H$ such that $[U,h] = H$.

In this case, H is isometrically isomorphic to $L^2(\sigma_h)$ and U is unitarily equivalent to the multiplication operator V defined on $L^2(\sigma_h)$ by $(V\phi)(t) = e^{it}\phi(t)$ (theorem 2.1).

We begin with the following general remark, obvious from $\widehat{\sigma}_{f,g}(n) = \langle U^n f, g\rangle$:

Lemma 2.1. *For $f,g \in H$,*

$$\sigma_{f,g} = 0 \Longleftrightarrow [U,f] \perp [U,g]. \tag{2.13}$$

and we say, in this case, that f and g are U-orthogonal.

So we always have : $\sigma_f \perp \sigma_g \Longrightarrow \sigma_{f,g} = 0 \Longleftrightarrow [U,f] \perp [U,g]$; but the reverse implication may be false and this is related to multiplicity (see section 2.4).

2.2.1 Basic Examples

1. Consider $H = L^2(\mathbf{T},m) =: L^2(\mathbf{T})$ and U the unitary operator associated to the rotation $R_\theta : x \mapsto x+\theta \mod 2\pi$, θ/π irrational. We claim that U *has simple spectrum*. In other words, we can exhibit a function $f \in H$, for which the linear span of $R_\theta^n f$, $n \in \mathbf{Z}$, is dense in H. It is well-known that the candidates are exactly those functions f whose Fourier transform never vanishes. Let us sketch the proof rapidly : suppose that $\phi \in H$ is orthogonal to the $R_\theta^n f$, $n \in \mathbf{Z}$; this means that
$$(\phi * \check{f})(n\theta) = 0 \quad \forall n \in \mathbf{Z},$$
where $\check{f}(x) = \overline{f(-x)}$. By continuity of the $L^2 * L^2$- functions, necessarily $\phi * \check{f} = 0$ and $\hat{\phi} \cdot \overline{\hat{f}} = 0$. This implies $\phi = 0$ unless $\hat{f}$ vanishes somewhere, whence the claim. (We shall construct explicitly the isometry W between H and $L^2(\sigma_f)$ in subsection 2.5.2)
2. Consider now $H = L^2([0,1])$ and V the pointwise multiplication by ϕ, where ϕ is any continuous, one-to-one and unimodular function on $[0,1]$. Once more, V has simple spectrum : indeed, the sub-algebra of $C([0,1])$ generated by ϕ is dense because ϕ is one-to-one (Stone-Weierstrass theorem), so that $H = [V,f]$ where $f = 1$. We deduce that $\hat{\sigma}_f(n) = \int \phi^n \, dm$, or equivalently, σ_f is the *pull-back* of m under ϕ.
3. Let us go back to the previous example, but with $\phi(x) = e^{2ix}$; ϕ is π-periodic. A new application of the Stone-Weierstrass theorem shows that $[V, \mathbf{1}_{[0,\pi]}] =$

$L^2([0,\pi],m)$ and $[V,\mathbf{1}_{[\pi,2\pi]}] = L^2([\pi,2\pi],m)$; hence H admits the decomposition $H = [V,f] \oplus [V,g]$ where f and g are V-orthogonal functions with $\sigma_f = \sigma_g = m/2$. V has no longer simple spectrum and we say that the multiplicity of V is equal to 2 (see next subsection 2.4.2).

The following characterization is in fact a consequence of a result which will be proved later.

Proposition 2.9. *The following properties are equivalent :*
(a) U has simple spectrum.
(b) For every $f,g \in H$, $\sigma_{f,g} = 0 \Longrightarrow \sigma_f \perp \sigma_g$.

Proof. Suppose that (a) holds and $H = [U,h]$. If W denotes the natural isometry of H onto $L^2(\sigma_h)$, we have

$$\sigma_f = |Wf|^2\sigma_h,\ \sigma_g = |Wg|^2\sigma_h,\ \sigma_{f,g} = Wf\cdot\overline{Wg}\,\sigma_h.$$

If $\sigma_{f,g} = 0$, then $Wf\cdot\overline{Wg} = 0$ σ_h- a.e. and the supports of Wf and Wg are disjoint in $L^2(\sigma_h)$; hence $\sigma_f \perp \sigma_g$.

Under assumption (b) we have to prove that U has simple spectrum. If this is not the case, for any $h \in H$, there would exist $g \in H$ such that $[U,g] \perp [U,h]$, so that, by (b), $\sigma_h \perp \sigma_g$. But this would contradict the following result ensuring the existence of *maximal elements* : *There exists $h \in H$ such that*

$$\sigma_g \ll \sigma_h \quad \textit{for every } g \in H.$$

This is a simplified version of the forthcoming lemma 2.4 □

2.2.2 Simple Lebesgue Spectrum

Definition 2.4. U has *simple Lebesgue spectrum* on H if $H = [U,h]$ for some $h \in H$ with $\sigma_h \sim m$.

Proposition 2.10. *U has simple Lebesgue spectrum on H if and only if there exists $h \in H$ such that $(U^n h)_{n\in\mathbf{Z}}$ forms an orthonormal basis of H.*

We shall prove an apparently stronger result: *Given $f \in H$, the following assertions are equivalent:*
a) There exists $h \in [U,f]$ such that $(U^n h)_{n\in\mathbf{Z}}$ forms an orthonormal basis of $[U,f]$.
b) $\sigma_f \sim m$.

Proof. If $(U^n h)_{n\in\mathbf{Z}}$ is complete in $[U,f]$, then $[U,h] = [U,f]$ and $\sigma_h \sim \sigma_f$ (corollary 2.4). In addition, $(U^n h)$ is an orthonormal system if and only if $||h|| = 1$ and $\langle U^n h, h\rangle = 0$ for all $n \neq 0$. This is exactly the property $\sigma_h = m$. Whence the first implication a)$\Longrightarrow$ b).

Conversely, suppose that $\sigma_f \sim m$. Then, $m = \varphi\sigma_f$ with $\varphi \in L^1(\sigma_f)$, $\varphi \geq 0$; putting $\varphi = |\phi|^2$, $\phi \in L^2(\sigma_f)$ and we get $\sigma_h := \sigma_{\phi(U)f} = m$. It follows that $(U^n h)$ is an orthonormal system with, clearly, $h \in [U,f]$. But $\sigma_f \sim \sigma_h = m$ and $[U,f] = [U,h]$. This proves that $(U^n h)_{n\in\mathbf{Z}}$ forms an orthonormal basis of $[U,f]$. □

Via theorem 2.1, this property means that H is isometrically isomorphic to $L^2(\mathbf{T},m)$ and that U is unitarily equivalent to the operator V of multiplication by e^{it}, on $L^2(\mathbf{T},m)$. The isometry W of H onto $L^2(\mathbf{T},m)$ is defined by setting $W(U^n h) = e^{int}$ if $(U^n h)$ is an orthonormal basis of H.

2.3 Spectral Decomposition Theorem and Maximal Spectral Type

Let H and H' be separable Hilbert spaces. Recall that unitary operators U on H and U' on H' are said to be *spectrally equivalent* or just *conjugate* if there exists an isometric isomorphism $W : H \to H'$ such that $WU = U'W$. We already achieved the description of a unitary operator (up to equivalence) in the cyclic case (spectral representation theorem 2.1), and now, we consider the general case.

Definition 2.5. The class of measures equivalent to a fixed measure $\mu \in M(\mathbf{T})$ is called the *type* of μ and will be denoted by $[\mu]$.

Theorem 2.3 (First formulation of the spectral decomposition theorem). *Let U be a unitary operator on a separable Hilbert space H. There exists a (possibly finite) sequence $(h_n)_{n\geq 1}$ of elements of H, such that*

(a) $H = \oplus_{n=1}^{\infty}[U,h_n]$ and $[U,h_i] \perp [U,h_j]$ for $i \neq j$.
(b) $\sigma_{h_1} \gg \sigma_{h_2} \gg \cdots$

and for any other sequence $(h'_n)_{n\geq 1}$ of elements of H satisfying (a) and (b), we have $\sigma_{h_i} \sim \sigma_{h'_i}$ for each $i \geq 1$.

Remark 2.2. This formulation means that U is unitarily equivalent to the operator V, defined on the space $\oplus_{n=1}^{\infty} L^2(\mathbf{T},\sigma_{h_n})$ by

$$V(f_1,f_2,\ldots)(t_1,t_2,\ldots) = (e^{it_1}f_1(t_1),e^{it_2}f_2(t_2),\ldots), \quad t_i \in \mathbf{T},$$

the sequence $([\sigma_n])$ being uniquely determined by U.

Proof. We shall construct inductively a sequence satisfying (a); then, we shall modify it to realize (b) in addition.

1. **Lemma 2.2.** *There exists a sequence (e_n) of elements of H such that the cyclic subspaces $H_j = [U,e_j]$ satisfy (a).*

   Proof. We start with an orthonormal basis (ε_j) of the separable space H, and we put $e_1 = \varepsilon_1$, $H_1 = [U,e_1]$. Suppose $e_1,\ldots,e_j$ have been so constructed that the spaces $H_1,\dots,H_j$ are orthogonal and $\varepsilon_1,\dots,\varepsilon_j$ belong to $\oplus_{i\le j}H_i$. We now proceed this way : suppose that n_j is the first index n for which ε_n is not in this sum, if such exists. Hence, $n_j \ge j+1$, and we put

$$e_{j+1} = \varepsilon_{n_j} - P_{\oplus_{i\le j}H_i}(\varepsilon_{n_j}),$$

(where, in general, P_M is the orthogonal projection onto the closed subspace M). Clearly, $H_{j+1} = [U, e_{j+1}]$ is orthogonal to $\oplus_{i\le j}H_i$. If ε_{j+1} is already in $H_1 \oplus \cdots \oplus H_j$, we are done; otherwise, $n_j = j+1$ and therefore $\varepsilon_{j+1} = e_{j+1} + P_{\oplus_{i\le j}H_i}(\varepsilon_{j+1})$ belongs to $H_1 \oplus \cdots \oplus H_{j+1}$. The lemma follows by induction on j. □

We write $H = \oplus^{\perp}_{n\ge1}[U, e_n]$ for the orthogonal sum. Note that the so-constructed sequence (e_n) satisfy $\sigma_{e_i,e_j} = 0$ for $i \ne j$, and H_n is isomorphic to $L^2(\sigma_{e_n})$.

2. **Lemma 2.3.** *If $H = \oplus^{\perp}_{n\ge1}[U, e_n]$ and if the positive measure σ is such that $\sigma_{e_n} \ll \sigma$ for every n, then $\sigma_x \ll \sigma$ for every $x \in H$.*
(σ is maximal *for H, in relation to the absolute continuity property of measures.)*

Proof. Every $x \in H$ can be decomposed into a sum $\sum_{n\ge1} x_n$ where $x_n \in [U, e_n]$, and the sequence of measures $(\sigma_{\sum_{n\le N} x_n})_N$ converges in norm to σ_x (corollary 2.2). In addition, $\sigma_{\sum_{n\le N} x_n} = \sum_{n\le N} \sigma_{x_n}$ since $\sigma_{x_i,x_j} = 0$ for $i \ne j$, and $\sigma_{x_n} \ll \sigma_{e_n} \ll \sigma$ for every n; thus $\sum_{n\le N} \sigma_{x_n} \ll \sigma$ for every N, and $\sigma_x \ll \sigma$ for every $x \in H$. □

3. Combining these two lemmas, we shall construct maximal spectral measures.

Lemma 2.4. *For every $e \in H$, there exists $h \in H$ such that*

$$e \in [U, h] \quad \text{and} \quad \sigma_x \ll \sigma_h \quad \forall x \in H.$$

Proof. Given $e \in H$, we construct, according to the first lemma, an orthonormal basis (e_n) such that

$$H = \oplus^{\perp}_{n\ge1}[U, e_n], \quad e_1 = e.$$

We put $f_1 = e_1$ and $g_1 = 0$. For $n \ge 2$, we can choose $f_n, g_n \in [U, e_n]$ such that :

$$f_n + g_n = e_n, \quad \sigma_{f_n} \perp \sigma_e \quad \text{and} \quad \sigma_{g_n} \ll \sigma_e;$$

indeed, remembering that the σ_{e_n} are not necessarily mutually singular, we decompose $\sigma_{e_n} = \mu + \nu$, with $\mu \perp \sigma_e$ and $\nu \ll \sigma_e$, and corollary 2.6 provides both functions f_n and g_n. We claim that

$$h = \sum_{n\ge1} \frac{f_n}{||f_n||2^{n-1}} = e + \sum_{n\ge2} \frac{f_n}{||f_n||2^{n-1}} = e + f$$

will do the job. First of all, $\sigma_e \ll \sigma_h$ and $e \in [U, h]$ (corollary 2.4); then, $\sigma_e \perp \sigma_f$ so that $\sigma_h = \sigma_e + \sigma_f$ and for $n \ge 2$, $\sigma_{e_n} = \sigma_{f_n} + \sigma_{g_n} \ll \sigma_f + \sigma_e = \sigma_h$. We conclude with the help of the second lemma. □

4. We are now in a position to finish the proof of the theorem. Assume, as before, $H = \oplus_{n\geq 1}^{\perp}[U,e_n]$ where (e_n) is an orthonormal basis of H. Applying lemma 2.4 to e_1, we get $h_1 \in H$ such that $e_1 \in [U,h_1]$ and σ_{h_1} is maximal for H. We consider $H_1 = [U,h_1]$; $H_1^{\perp}$ is invariant under U. Let k be the first index $n \geq 2$ for which $e_n \notin H_1$; then
$$e'_k = e_k - P_{H_1}(e_k)$$
is in $H_1^{\perp}$, and $[U,e_1,\ldots,e_{k-1},e_k] = [U,e_1,\ldots,e_{k-1},e'_k]$. So we choose $h_2 \in H_1^{\perp}$ such that $e'_k \in [U,h_2]$ and σ_{h_2} is maximal for $H_1^{\perp}$. Clearly, $\sigma_{h_2} \ll \sigma_{h_1}$, and we put $H_2 = [U,h_2]$. We repeat this process indefinitely to obtain $\sigma_{h_{n+1}} \ll \sigma_{h_n}$ for all $n \geq 1$ and $H = \oplus_{n\geq 1}^{\perp} H_n = \oplus_{n\geq 1}^{\perp}[U,h_n]$.
5. **Unicity** : Suppose that $H = \oplus_{n\geq 1}^{\perp}[U,h'_n]$ is another decomposition of H with property (b). By maximality of both, $\sigma_{h_1} \sim \sigma_{h'_1}$; hence, the spaces $L^2(\sigma_{h_1})$ and $L^2(\sigma_{h'_1})$ are isomorphic, and there exists an isometry conjugating the multiplication operators V on $L^2(\sigma_{h_1})$ and V' on $L^2(\sigma_{h'_1})$ (take the operator of multiplication by $\varphi^{1/2}$ if $\sigma_{h_1} = \varphi\sigma_{h'_1}$). Now, by theorem 2.1, the restrictions $U_{|[U,h_1]}$ and $U_{|[U,h'_1]}$ are conjugate too (we write in short $U_{|[U,h_1]} \simeq U_{|[U,h'_1]}$). The unicity will follow easily by induction if we prove the following ultimate lemma :

Lemma 2.5. *Fix $f,g \in H$. If $U_{|[U,f]} \simeq U_{|[U,g]}$, then $U_{|[U,f]^{\perp}} \simeq U_{|[U,g]^{\perp}}$.*

Proof. First of all, we may assume that $H = \overline{[U,f]+[U,g]}$ since $U_{|\overline{[U,f]+[U,g]}} \simeq U_{|\overline{[U,f]+[U,g]}}$ obviously.
By corollary 2.6, we can find g_0 and g_1 in $[U,g]$ such that
$$\sigma_g = \sigma_{g_0} + \sigma_{g_1}, \quad \sigma_{g_0} \ll \sigma_f \quad \text{and} \quad \sigma_{g_1} \perp \sigma_f.$$
Thanks to the canonical isometry W, we see that $[U,g]$ is exactly $[U,g_0] \oplus^{\perp} [U,g_1]$.

Next, we decompose σ_f. If $\sigma_f = \mu + \nu$ with $\mu \ll \sigma_{g_0}$ and $\nu \perp \sigma_{g_0}$, necessarily $\mu \sim \sigma_{g_0}$ since $\sigma_{g_0} \ll \sigma_f$ must be $\ll \mu$. This implies that $\mu = |\phi|^2\sigma_{g_0} = \sigma_{\phi(U)g_0}$ for some $\phi \in L^2(\sigma_{g_0})$. Thus $g'_0 = \phi(U)g_0$ is such that : $\sigma_{g'_0} \leq \sigma_f$ and $[U,g'_0] = [U,g_0]$. We may suppose already $\sigma_{g_0} \leq \sigma_f$. Whence the decomposition
$$\sigma_f = \sigma_{g_0} + \sigma_{f_0}, \quad \sigma_{f_0} \perp \sigma_{f_0} \quad \text{and} \quad f_0 \in [U,f].$$
Moreover $[U,f] = [U,g_0] \oplus^{\perp} [U,f_0]$ and finally we obtain a new decomposition for H :
$$H = [U,f_0] \oplus^{\perp} [U,g_0] \oplus^{\perp} [U,g_1].$$
Turning back to the hypothesis, we proceed now by equivalence :
$$U_{|[U,f]} \simeq U_{|[U,g]} \Longleftrightarrow \sigma_f \sim \sigma_g$$

(corollary 2.3), and, thanks to our decomposition,

$$\sigma_f \sim \sigma_g \Longleftrightarrow \sigma_{f_0} \sim \sigma_{g_1};$$

but this means $U\big|_{[U,f_0]} \simeq U\big|_{[U,g_1]}$, which was to be proved. □

The proof of the theorem is complete.

□

We can now precise some notions. Keeping in mind the notations of the theorem,

Definition 2.6. $[\sigma_{h_1}]$ is called the *maximal spectral type* of the operator U acting on H, and, from now on, denoted by $[\sigma_{\max}]$.

U possesses a *discrete–continuous–singular–absolutely continuous*–or *Lebesgue spectrum* if the measure $\sigma_{\max}$ is discrete–continuous–singular–absolutely continuous (with respect to the Lebesgue measure)– or equivalent to the Lebesgue measure (respectively).

U has a *simple spectrum* if $0 = \sigma_{h_2} = \sigma_{h_2} = \cdots$, and U has a *countable spectrum* if all the σ_{h_j} are $\neq 0$.

Obviously, from the construction of the σ_{h_j}, we have the following characterization :

Corollary 2.7. $\sigma_{\max}$ *is characterized, up to equivalence, by the following properties*

(i) $\sigma_f \ll \sigma_{\max}$ *for every* $f \in H$,
(ii) If $\sigma \in M(\mathbf{T})$ *satisfies* $0 \leq \sigma \ll \sigma_{\max}$, *then* $\sigma = \sigma_f$ *for some* $f \in H$ *(in particular* $\sigma_{\max} = \sigma_{f_0}$ *for some* $f_0 \in H$*).*

The next corollary has already been noticed (subsection 2.2.1)

Corollary 2.8. *U possesses a simple spectrum if and only if, for every pair* $f, g \in H$, $\sigma_{f,g} = 0$ *implies* $\sigma_f \perp \sigma_g$.

We can now exhibit, using the decomposition, a candidate h with $[U,h] = H$: if $H = \oplus_{n\geq 1}^{\perp}[U,e_n]$, any $h = \sum_{n\geq 1} a_n e_n$, where $a_n \neq 0$ for all n, will work.

2.4 Spectral Decomposition Theorem and Spectral Multiplicity

A mathematical object, related to Hilbert space operator theory, is called a *spectral invariant* if it is preserved under unitary equivalence. A system of spectral invariants is said to be *complete* if it characterizes unitary operators up to equivalence.

The unicity assertion in the last theorem means that the spectral types $[\sigma_{h_n}]$ constitute a complete system of spectral invariants; actually, in view to describing U up to equivalence, we need only know the maximal spectral type $[\sigma_{\max}]$ and the *multiplicity function* that we define now.

2.4.1 Multiplicity Function

We will write more briefly σ for $\sigma_{\max}$. According to the proof of lemma 2.5, each h_n may be replaced by an f_n such that $[U, f_n] = [U, h_n]$ and $\sigma_{f_n} = \mathbf{1}_{B_n}\sigma$, where (B_n) is a non-increasing sequence of Borel sets. This leads to a re-formulation of theorem 2.3 :

There exist a sequence (f_n) of elements of H and a non-increasing sequence of Borel sets (B_n) in $\mathbf{T}$ such that

$$\sigma_{f_n} = \mathbf{1}_{B_n}\sigma \quad \text{and} \quad H = \oplus^{\perp}_{n\geq 1}[U, f_n].$$

Definition 2.7. The function $M = \sum_{n\geq 1} \mathbf{1}_{B_n}$ is called the *multiplicity function.*

Note that M takes its values in $\mathbf{N} \cup \{\infty\}$ and

$$M(t) = \sup\{n \geq 1,\ t \in B_n\} \quad \sigma - a.e.$$

Remark 2.3. The following notation may be convenient. For a separable Hilbert space with an orthonormal basis $(e_n)_{n\geq 1}$, one denotes by H_m the subspace of H generated by the m first vectors $e_1, \ldots, e_m$, $H_\infty = H$. If τ is some positive measure on $\mathbf{T}$, $L^2(\tau, H)$ is the space of square-integrable vector-valued functions : $\{\varphi, \int_{\mathbf{T}} ||\varphi(t)||^2\, d\tau(t) < \infty\}$. Let now $t \mapsto m(t)$ with $m(t) \in \mathbf{N} \cup \{\infty\}$: then,

$$\int^{\oplus} H_{m(t)}\, d\tau(t) = \{\varphi \in L^2(\tau, H), \quad \varphi(t) \in H_{m(t)}\}.$$

With those notations, the spectral decomposition theorem says: *There exists an isomorphism of Hilbert spaces: $W : H \to \int^{\oplus} H_{M(t)}\, d\sigma(t)$ conjugating U with the operator V of multiplication by e^{it}.*

This allows us to put the finishing touches on the commutator problem.

Proposition 2.11. *Let σ be the maximal spectral type of the unitary operator U on H. Then the operator Q commutes with U on H if and only if Q is, up to equivalence, a multiplication operator on $L^2(\sigma, H)$, i.e.*

$$(QF)(t) = Q(t)F(t)$$

where, for σ-a.e. t, $Q(t)$ is an operator on $H_{M(t)}$ with $t \mapsto Q(t)$ weakly measurable and $t \mapsto ||Q(t)||$ in $L^\infty(\sigma)$.

For the purpose of determinating the multiplicity function, a new formulation of the spectral decomposition theorem will be needed.

Theorem 2.4 (Second formulation of the spectral decomposition theorem). *Let U be a unitary operator on a separable Hilbert space H. For $n \in \mathbf{N} \cup \{\infty\}$, one can find a Borel set A_n in $\mathbf{T}$, and a finite sequence $(h_n^{(k)})_{1\leq k\leq n}$ of elements of H such that*

(a) *(A_n) is a partition of* $\mathbf{T}$.
(b) $H = \oplus_{n\geq 1} \oplus_{1\leq k\leq n} [U, h_n^{(k)}]$, *and* $[U, h_n^{(k)}] \perp [U, h_{n'}^{(k')}]$ *for* $(n,k) \neq (n',k')$.
(c) $\sigma_{h_n^{(k)}} = \sigma^{(n)}$ *for* $1 \leq k \leq n$, *and* $\sigma^{(n)}(A_n^c) = 0$.

Moreover, for any other sequence $(h_n'^{(k)})_{n\geq 1}$ *of elements of* H *satisfying (a), (b) and (c), we have* $\sigma_{h_n^{(k)}} \sim \sigma_{h_n'^{(k)}}$, *for each* (n,k).

Proof. Suppose that (a), (b) and (c) hold; if we set

$$h_n = \sum_{k\geq n} h_n^{(k)},$$

the elements h_n fulfill conditions (a) and (b) of theorem 2.3.

Conversely, let us assume that there exist a sequence (f_n) of elements of H and a non-increasing sequence of Borel sets (B_n) in $\mathbf{T}$ such that

$$\sigma_{f_n} = \mathbf{1}_{B_n}\sigma \quad \text{and} \quad H = \oplus_{n\geq 1}^{\perp}[U, f_n].$$

We set

$$A_n = B_n \backslash B_{n+1}, \quad A_\infty = \cap_{n\geq 1} B_n;$$

(A_n) is a Borel partition of $\mathbf{T}$ whence (a).

Now, for $1 \leq k \leq n$, $\mathbf{1}_{A_n}\sigma_{f_k}$ is the spectral measure of some element in $[U, f_k]$ that we call $h_n^{(k)}$. Since $B_k \supset B_n$,

$$\sigma_{h_n^{(k)}} = \mathbf{1}_{B_n\backslash B_{n+1}} \mathbf{1}_{B_k}\sigma = \mathbf{1}_{B_n\backslash B_{n+1}}\sigma$$

and $\sigma_{h_n^{(k)}}$ does not depend on k, $1 \leq k \leq n$; so we put $\sigma^{(n)} = \mathbf{1}_{A_n}\sigma$ and (c) holds.

Clearly, for each $k \geq 1$,

$$[U, f_k] = \oplus_{n\geq k}^{\perp}[U, h_n^{(k)}].$$

It follows that

$$H = \oplus_{n\geq 1}^{\perp}[U, f_n] = \oplus_{n\geq 1}^{\perp} \oplus_{1\leq k\leq n}^{\perp} [U, h_n^{(k)}],$$

and the theorem is proved. □

The multiplicity function is now more readable : with the same notations,

Corollary 2.9. *The multiplicity function* M *is defined on* $\mathbf{T}$ *by the relation :*

$$M(t) = n \quad \text{if} \quad t \in A_n.$$

2.4.2 Global Multiplicity

We start with the following notions and notations :

Definition 2.8. $[\sigma^{(n)}]$ is referred to as the *spectral type of multiplicity n*.

One says that U has *finite multiplicity* on H if $\sigma^{(n)} = 0$ for $n > n_0$; if $\sigma^{(n)} = 0$ for $n \neq n_0$, i.e. M is constant, one says that U has a *homogeneous spectrum of multiplicity* n_0.

Definition 2.9. The essential supremum of the function M is called the *spectral multiplicity* of U on H or *global multiplicity*, and denoted by $m(U)$.

Hence, U has a simple spectrum if $m(U) = 1$.

Corollary 2.10. *The global multiplicity of U is also the cardinality of the maximal set of U-orthogonal elements in H with identical spectral measures.*

Proof. If $g_1, \ldots, g_p$ are p elements of H, U-orthogonal and such that $\sigma_{g_1} = \cdots = \sigma_{g_p} = \tau$, we shall prove that $m(U) \geq p$. We decompose for that,

$$H = H(\tau) \oplus H(\tau)^{\perp}$$

where $H(\tau) = \{f \in H,\ \sigma_f \ll \tau\}$. U has $[\tau]$ as its maximal spectral type on $H(\tau)$. $H(\tau)$, in turn, admits the decomposition

$$H(\tau) = [U, g_1] \oplus^{\perp} \cdots \oplus^{\perp} [U, g_p] \oplus^{\perp} H'$$

H' being U-invariant too. Applying now theorem 2.3 to H' and U we get:

$$H(\tau) = [U, g_1] \oplus^{\perp} \cdots \oplus^{\perp} [U, g_p] \oplus^{\perp}_{n \geq 1} [U, h_n]$$

with $\tau \gg \sigma_{h_1} \gg \sigma_{h_2} \gg \cdots$. We deduce that $U_{|H(\tau)}$ has multiplicity $\geq p$, and the corollary follows. □

This notion of global multiplicity turns out to be very important : in the homogeneous case, the operator is determined, up to equivalence, by the pair of invariants $\sigma_{\max}$ and $m(U)$. The following results deal with estimations of $m(U)$ and will be used in a forthcoming section.

Proposition 2.12. *Assume (H_n) to be a non-decreasing sequence of U-invariant subspaces of H, such that $H = \overline{\cup_{n \geq 1} H_n}$. Let r be fixed. If the spectral multiplicity of U on H_n is $\leq r$ for all n, so is the global multiplicity.*

Proof. If the global multiplicity $m(U)$ is at least $r+1$, by corollary 2.10 we can find $r+1$ U-orthogonal vectors $f_1, \ldots, f_{r+1}$ with identical spectral measure σ. We have to estimate the multiplicity of U on H_n.

Denote by P_n the orthogonal projection onto H_n. Since $H = \overline{\cup_{n\geq 1} H_n}$ and by the continuity property of the map $f \mapsto \sigma_f$ (corollary 2.2), for $\varepsilon > 0$ we can choose $n(\varepsilon)$ such that

$$||\sigma_{P_n f_j} - \sigma|| \leq \varepsilon, \quad \text{for every } 1 \leq j \leq r+1, \text{ and } n \geq n(\varepsilon).$$

But $\sigma_{P_n f_j} \leq \sigma$, and making ε small enough, one can find $\tau \neq 0$ such that

$$\tau \ll \sigma_{P_n f_j},\ 1 \leq j \leq r+1,\ n = n(\varepsilon).$$

Now, for each $j = 1, \ldots, r+1$, we choose $g_j \in [U, P_n f_j] \subset [U, f_j]$, with $\sigma_{g_j} = \tau$. It follows that $g_1, \ldots, g_{r+1}$ are U-orthogonal elements in H_n with identical spectral measures. Thus, the multiplicity of U on H_n is $\geq r+1$, which was to be proved. □

This proposition will be convenient in aim of establishing the simplicity of the spectrum. The next one (referred to as Chacon's theorem) provides a lower bound for the multiplicity [54, 106].

Proposition 2.13. *The global multiplicity $m(U)$ is $\geq m$ if and only if there exists an orthonormal family $f_1, \ldots, f_m$ in H such that, for any cyclic subspace H_0, we have :*

$$\sum_{i=1}^{m} d(f_i, H_0)^2 \geq m-1 \tag{2.14}$$

($d(.,H_0)$ denoting the distance to H_0)

Proof. ▷ If the global multiplicity $m(U) \geq m$ we can find, as above, m U-orthogonal vectors $f_1, \ldots, f_m$ with $\sigma_{f_j} = \sigma$, for $j = 1, \ldots, m$. Without loss of generality, we assume $||f_j|| = 1$ for all j so that σ is a probability measure. We claim that the vectors (f_j) do the job. We are led to prove the following : if $h \in H$ is fixed,

$$S := \sum_{i=1}^{m} ||f_i - g_i||^2 \geq m-1$$

for all $g_1, \ldots, g_m$ in the cyclic space $[U, h]$, and more simply, for all $g_1, \ldots, g_m$ in the linear span of the $U^n h$, $n \in \mathbf{Z}$.
We thus consider, for $i = 1, \ldots, m$, $g_i = P_i(U)h$ where P_i is a polynomial in $\mathbf{C}[X, X^{-1}]$ and we decompose h into the form

$$h = \sum_{j=1}^{m} \langle h, f_j \rangle f_j + h'$$

with $h' \perp [U, f_j]$ for $j = 1, \ldots, m$. We deduce

$$S = \sum_{i=1}^{m} \left(||f_i - \sum_{j=1}^{m} \langle h, f_j \rangle P_i(U) f_j||^2 + ||P_i(U)h'||^2 \right)$$

$$\geq \sum_{i=1}^{m} ||f_i - \sum_{j=1}^{m} \langle h, f_j \rangle P_i(U) f_j||^2$$
$$= \sum_{i=1}^{m} ||f_i - \alpha_i P_i(U) f_i||^2 + \sum_{\substack{1 \leq i,j \leq m \\ j \neq i}} |\alpha_j|^2 ||P_i(U) f_j||^2$$

by U-orthogonality, where $\alpha_j = \langle h, f_j \rangle$. Using now the isometric isomorphism between each cyclic space $[U, f_i]$ and $L^2(\sigma)$, we obtain

$$S \geq \int_{\mathbf{T}} \Big(\sum_{i=1}^{m} |1 - \alpha_i P_i(t)|^2 + \sum_{\substack{1 \leq i,j \leq m \\ j \neq i}} |\alpha_j|^2 |P_i(t)|^2 \Big) \, d\sigma(t)$$

$$= \int_{\mathbf{T}} \sum_{i=1}^{m} \Big(|1 - \alpha_i P_i(t)|^2 + |P_i(t)|^2 (\sum_{\substack{1 \leq j \leq m \\ j \neq i}} |\alpha_j|^2) \Big) \, d\sigma(t)$$

where $P(t)$ as usual denotes the associated trigonometric polynomial. We end the proof by establishing that $Q(t) := \sum_{i=1}^{m} \Big(|1 - \alpha_i P_i(t)|^2 + |P_i(t)|^2 (\sum_{\substack{1 \leq j \leq m \\ j \neq i}} |\alpha_j|^2) \Big) - m + 1$ is a positive polynomial.

Indeed, putting $a = \Big(\sum_{i=1}^{m} |\alpha_i|^2 \Big)^{1/2}$ and expanding the sum in Q, we readily get via the Cauchy-Schwarz inequality

$$Q(t) = 1 + \sum_{i=1}^{m} |\alpha_i|^2 |P_i(t)|^2 - 2 \sum_{i=1}^{m} \Re(\overline{\alpha_i} P_i(t)) + \sum_{i=1}^{m} |P_i(t)|^2 (a^2 - |\alpha_i|^2)$$
$$= 1 + a^2 \sum_{i=1}^{m} |P_i(t)|^2 - 2 \sum_{i=1}^{m} \Re(\overline{\alpha_i} P_i(t))$$
$$\geq 1 + a^2 \sum_{i=1}^{m} |P_i(t)|^2 - 2a \Big(\sum_{i=1}^{m} |P_i(t)|^2 \Big)^{1/2} \geq 0.$$

The necessity of the condition is proved.

$\triangleleft$ Let $f_1, \ldots, f_m$ be an orthonormal family in H such that, for any cyclic subspace H_0, we have : $\sum_{i=1}^{m} d(f_i, H_0)^2 \geq m - 1$. If P_0 is the orthogonal projection on H_0,

$$d(f_i, H_0)^2 = ||f_i||^2 - ||P_0 f_i||^2 = 1 - ||P_0 f_i||^2$$

and

$$\sum_{i=1}^{m} ||P_0 f_i||^2 \leq 1.$$

Taking $H_0 = [U, f_j]$, we deduce that $f_i \perp [U, f_j]$ for $i \neq j$. We shall prove that the spectral measures $\sigma_j := \sigma_{f_j}$ are equivalent. Otherwise, we may assume that $\sigma_1 \not\sim \sigma_2$; let A be a Borel set satisfying

$$\sigma_1(A) > 0 \quad \text{and} \quad \sigma_2(A) = 0.$$

We shall get a contradiction by considering $H_0 = [U,g]$ with

$$g = E_A f_1 + f_2.$$

Indeed, $E_A g = E_A f_1$ so that $E_A f_1$ and f_2 are in H_0, and

$$\sum_{i=1}^{m} ||P_0 f_i||^2 \geq ||f_2||^2 + ||P_0 f_1||^2 \geq 1 + ||E_A f_1||^2 \geq 1 + \sigma_1(A) > 1,$$

which is impossible. This proves the sufficiency of condition (2.11). □

2.5 Eigenvalues and Discrete Spectrum

In this section, U is an unitary operator on a separable Hilbert space.

2.5.1 *Eigenvalues*

Recall that, U being unitary, the spectrum of U is a compact subset of the circle.

Definition 2.10. The complex number $e^{i\lambda}$, $\lambda \in \mathbf{T}$, is an *eigenvalue* of U if $Uf = e^{i\lambda} f$ for some $f \in H$, $f \neq 0$, which is an *eigenvector* corresponding to $e^{i\lambda}$.

We establish now the connection between the eigenvalues of U and the maximal spectral type $[\sigma_{\max}]$.

Lemma 2.6. *If $\lambda \in \mathbf{T}$, the operator $\mathbf{1}_{\{\lambda\}}(U) =: E_\lambda$ is the orthogonal projection onto the eigen-subspace associated to the eigenvalue $e^{i\lambda}$.*

Proof. By the definition-proposition 2.7, $H_\lambda := E_\lambda H$ is the subspace of $h \in H$ whose spectral measure σ_h is supported by $\{\lambda\}$. We shall now identify H_λ and the eigen-subspace associated with $e^{i\lambda}$.

Suppose that $Uh = e^{i\lambda} h$; then $\hat{\sigma}_h(n) = e^{in\lambda} ||h||^2$, $n \in \mathbf{Z}$, and $\sigma_h = ||h||^2 \delta_\lambda$.

Conversely, suppose there is $h \in H$ with $\sigma_h = \alpha \delta_\lambda$; if R is a trigonometric polynomial,

$$||R(U)h||_H = ||R||_{L^2(\sigma_h)} = \alpha R(\lambda).$$

Choose $R(t) = e^{it} - e^{i\lambda}$; then $||R(U)h||_H = ||Uh - e^{i\lambda} h||_H = 0$ and h is an eigenvector corresponding to $e^{i\lambda}$. □

As a consequence, with the same notations,

Proposition 2.14. *$e^{i\lambda}$ is an eigenvalue of U if and only if $\sigma_{\max}\{\lambda\} \neq 0$.*

Proof. This means : $e^{i\lambda}$ is an eigenvalue as soon as one can find $f \in H$ with $\sigma_f\{\lambda\} \neq 0$. But, by (2.8)

$$\sigma_f\{\lambda\} = \langle E_\lambda f, f\rangle = ||E_\lambda f||^2, \quad \forall f \in H. \tag{2.15}$$

It follows that

$$f \in H_\lambda^\perp \Longleftrightarrow E_\lambda f = 0 \Longleftrightarrow \sigma_f\{\lambda\} = 0,$$

whence the claim. □

Remark 2.4. We are able to precise the link between the spectral set $sp(U)$ and the maximal spectral type. Let $h \in H$ be such that $\sigma_{\max} = \sigma_h$. If $e^{i\lambda} \notin sp(U)$, there exists a positive constant C such that

$$||Uf - e^{i\lambda} f|| \geq C||f|| \quad \text{for every } f \in H.$$

Fix $\varepsilon > 0$ and consider the function $\mathbf{1}_{]\lambda-\varepsilon,\lambda+\varepsilon[} \in L^2(\sigma_h)$. If g_ε is the associated element in $[U,h] \simeq L^2(\sigma_h)$, then we have

$$\sigma_{g_\varepsilon} = \mathbf{1}_{]\lambda-\varepsilon,\lambda+\varepsilon[}\sigma_h,$$

$$||g_\varepsilon||^2 = \sigma_h(]\lambda-\varepsilon,\lambda+\varepsilon[),$$

and

$$||Ug_\varepsilon - e^{i\lambda} g_\varepsilon||^2 = \int_{]\lambda-\varepsilon,\lambda+\varepsilon[} |e^{it} - e^{i\lambda}|^2 d\sigma_h(t)$$

$$\leq \varepsilon^2 \sigma_h(]\lambda-\varepsilon,\lambda+\varepsilon[);$$

Since $||Ug_\varepsilon - e^{i\lambda} g_\varepsilon||^2 \geq C^2||g_\varepsilon||^2$, this imposes $C \leq \varepsilon$ for every positive ε and a contradiction, unless $g_\varepsilon = 0$ and $\sigma_{\max}(]\lambda-\varepsilon,\lambda+\varepsilon[) = 0$.

For any $f \in H$, the equation $Ug - e^{i\lambda} g = f$ has a solution $g \in H$ if and only if one can solve the equation $e^{it}\varphi - e^{i\lambda}\varphi = \mathbf{1}$ in $L^2(\sigma_f)$, and this is possible if $1/(e^{it} - e^{i\lambda})$ belongs to $L^2(\sigma_f)$. Suppose now that $e^{i\lambda}$ is avoiding the topological support of σ_h; in this case, $1/(e^{it} - e^{i\lambda})$ is bounded σ_f-a.e. and consequently, $e^{i\lambda} \notin sp(U)$.

The link between the spectrum of the unitary operator U and its maximal spectral type is now clear : $sp(U)$ is exactly the *topological support* of the measure $\sigma_{\max}$.

2.5.2 Discrete Spectral Measures

To any decomposition of $\sigma_{\max}$ into a sum of mutually singular measures, there corresponds a decomposition of H into a sum of U-invariant orthogonal closed subspaces: If $H(\tau) = \{f \in H,\ \sigma_f \ll \tau\}$,

$$H = H(\sigma_1) \oplus^{\perp} H(\sigma_2)$$

for any decomposition $\sigma_{\max} = \sigma_1 + \sigma_2$ with $\sigma_1 \perp \sigma_2$.

In particular, we introduce the following definition :

Definition 2.11. H_d is the closed subspace of elements in H with *discrete spectral measure*, and H_c, the closed subspace of elements in H with *continuous spectral measure*; also

$$H = H_d \oplus^{\perp} H_c.$$

We have an almost obvious description of H_d :

Proposition 2.15. *H_d is the closure of the linear span of eigenvectors of U.*

Proof. Let $f \in H$; σ_f is discrete if and only if $||\sigma_f|| = \sum_\lambda \sigma_f\{\lambda\} = ||f||_H^2$; but, using (2.15),

$$\sum_\lambda \sigma_f\{\lambda\} = \sum_\lambda ||E_\lambda f||^2,$$

so that $f \in H_d$ if and only if

$$||f||_H^2 = \sum_\lambda ||E_\lambda f||^2.$$

By the converse of the Parseval identity, this condition means exactly that $f = \sum_\lambda E_\lambda f$ which was to be proved. □

Definition 2.12. U is said to have *discrete (or pure-point) spectrum* when $\sigma_{\max}$ is a (purely) discrete measure.

If $H_d \neq \{0\}$ (resp. $H_c \neq \{0\}$), U is said to possess a *discrete component* (resp. a *continuous component*).

If U possesses a discrete as well as a continuous component, U has a *mixed spectrum*.

Whence this simple remark.

Corollary 2.11. *U has a discrete spectrum if and only if the linear span of eigenvectors of U is dense in H.*

We conclude this subsection by a useful observation made by Solomyak [230].

Proposition 2.16. *Let U be a unitary operator on a Hilbert space H. Suppose that there exists a total set $\mathscr{F} \subset H$, and a linearly recurrent sequence of integers (q_n) such that* $\lim_{n\to\infty} q_n = \infty$ *and*

$$\sum_{n\geq 1} ||U^{q_n} f - f||^2 < \infty, \ \textit{for every} \ \ f \in \mathscr{F}.$$

Then U has purely discrete spectrum.

Proof. Since

$$||U^{q_n}f - f||^2 = \int_{\mathbf{T}} \left|e^{itq_n} - 1\right|^2 d\sigma_f(t)$$

for any $f \in H$, the assumption in the proposition implies that

$$\sum_{n\geq 1} \sigma_f(\{t,\ \left|e^{itq_n} - 1\right| \geq 2^{-\ell}\}) < \infty, \text{ for every } f \in \mathscr{F} \text{ and every } \ell > 0.$$

By the Borel-Cantelli lemma, $\sigma_f(\mathbf{T}\backslash\{t,\ e^{itq_n} \to 1\}) = 0$ for those f and, $\mathscr{F}$ being total, one deduces that $\sigma_{max}(\mathbf{T}\backslash\{t,\ e^{itq_n} \to 1\}) = 0$. But, the sequence (q_n) satisfies a linear integral recurrence relation so that the set $\{t,\ e^{itq_n} \to 1\}$ is countable and $\sigma_{\max}$ is a discrete measure. This is the claim. □

2.5.3 Basic Examples

1. *Irrational rotation* : we turn back to the operator U defined on $H = L^2(\mathbf{T})$ by

$$Uf(t) = f(t+\theta), \quad \theta/\pi \notin \mathbf{Q},$$

with simple spectrum as observed in 2.2.1. We claim that : *U has discrete spectrum and* $\sigma_{\max} \sim \sum_{k\in\mathbf{Z}} 2^{-|k|}\delta_{k\theta}$.

Indeed, for each $k \in \mathbf{Z}$, e^{ikt} is obviously an eigenvector of U with eigenvalue $e^{ik\theta}$, and the family $(e^{ikt})_{k\in\mathbf{Z}}$ is complete in H; whence the discrete spectrum.

Let now $f \in H$ be such that $\hat{f}(k) \neq 0$ for every $k \in \mathbf{Z}$; as we already observed in subsection 2.2.1, $\sigma_f \sim \sigma_{\max}$ and $H = [U,f] \simeq L^2(\sigma_f)$; moreover, $\sigma_f\{k\theta\} = ||E_{k\theta}f||^2$ so that

$$\sigma_f = \sum_{k\in\mathbf{Z}} |\hat{f}(k)|^2 \delta_{k\theta}.$$

We wish to determine an isometry W conjugating U on $L^2(\mathbf{T},m)$ and V on $L^2(\sigma_f)$, such that $Wf = \mathbf{1}$.

If $g \in L^2(\mathbf{T},m)$, there exists $(a_n) \in \ell^2(\mathbf{Z})$ such that

$$g = \sum_{n\in\mathbf{Z}} a_n U^n f$$

and, if $k \in \mathbf{Z}$,

$$\hat{g}(k) = \sum_{n\in\mathbf{Z}} a_n (U^n f)\hat{}\,(k) = \sum_{n\in\mathbf{Z}} a_n e^{ink\theta} \hat{f}(k)$$

We shall see that the formula $Wg = h$ with $h(k\theta) = \hat{g}(k)/\hat{f}(k)$ for all $k \in \mathbf{Z}$, gives the result; indeed, such an h satisfies

$$\int |h|^2\, d\sigma = \sum_{k\in\mathbf{Z}} |h(k\theta)|^2 |\hat{f}(k)|^2 = \sum_{k\in\mathbf{Z}} |\hat{g}(k)|^2 < \infty,$$

and W is an isometry from $L^2(\mathbf{T},m)$ into $L^2(\sigma_f)$. Also, W is clearly invertible, and $W(U^n f) = h_n$ with $h_n(t) = e^{int}$ so that, finally, the conjugation identity

$$WU = e^{it}W,$$

valid on $\{U^n f, n \in \mathbf{Z}\}$, extends to $L^2(\mathbf{T},m) = [U,f]$.

2. Let $H = L^2(\mathbf{T})$ once more, and U be the unitary operator defined on $f \in H$ by

$$Uf(t) = \phi(t)f(t+\theta), \quad \theta \in [0,2\pi],$$

ϕ being unimodular. We claim that :
$\sigma_{\max} \sim \sum_{k\in\mathbf{Z}} 2^{-|k|}\sigma * \delta_{k\theta}$, *where* σ *is the spectral measure of the function* $\mathbf{1}$.
If $f \in H$, $U^n f(t) = \phi^{(n)}(t)f(t+n\theta)$, where the *cocycle* $(\phi^{(n)})$ is defined by

$$\phi^{(n)}(t) = \begin{cases} \phi(t)\phi(t+\theta)\cdots\phi(t+(n-1)\theta) & \text{if } n \geq 1, \\ \overline{\phi^{(-n)}(t+n\theta)} & \text{if } n \leq -1, \\ 1 & \text{if } n = 0; \end{cases}$$

since

$$\widehat{\sigma}_f(n) = \int_{\mathbf{T}} \phi^{(n)}(x)f(t+n\theta)\overline{f(t)}\,dm(t),$$

we see that

$$\sigma_{fe^{it}} = \sigma_f * \delta_\theta. \tag{2.16}$$

It follows that the measure $\nu := \sum_{k\in\mathbf{Z}} 2^{-|k|}\sigma * \delta_{k\theta}$ is $\ll \sigma_{\max}$.

Conversely, if $f(t) = e^{ikt}$, $\sigma_f = \sigma * \delta_{k\theta}$ by (2.16) and $\sigma_f \ll \nu$ for every trigonometric polynomial f. Approximating any $f \in H$ by a trigonometric polynomial and invoking the continuity of the map $f \mapsto \sigma_f$ (corollary 2.2), we get $\sigma_f \ll \nu$ for all $f \in H$; whence $\sigma_{\max} \sim \nu$.

3. Let U be the operator defined by $Uf(t) = f(2t)$ on $H = L^2(\mathbf{T})$; then U has infinite multiplicity. Actually, for every finite set $f_1, f_2, \ldots, f_p$ in H, $[U,(f_j)_{1\leq j\leq p}]^\perp$ is infinite dimensional.

2.6 Application to Ergodic Sequences

1. By using the isometry machinery, we can now easily prove the following classical theorem due to von Neumann :

Theorem 2.5 (von Neumann). *If U is an isometry of the separable Hilbert space H, then, for all $f \in H$,*

$$\frac{1}{N}\sum_{n=0}^{N-1} U^n f \to Pf,$$

where P is the orthogonal projection onto the subspace of U-invariant vectors.

Proof. Suppose first that U is an unitary operator.

For any fixed $f \in H$, the sequence $(M_N f)_{N\geq 1}$, with $M_N = \frac{1}{N}\sum_{n=0}^{N-1} U^n$, stays inside the cyclic subspace $[U,f]$; so, we are reduced to study the limit of the sequence of trigonometric polynomials $\frac{1}{N}\sum_{n=0}^{N-1} e^{int}$ in $L^2(\sigma_f)$. But it is easily seen that $\lim_{N\to\infty} \frac{1}{N}\sum_{n=0}^{N-1} e^{int} = \mathbf{1}_{\{0\}}$, and even, $\lim_{N\to\infty} \frac{1}{N}\sum_{n=K}^{N+K-1} e^{int} = \mathbf{1}_{\{0\}}$ uniformly with respect to K. We conclude that $\lim_{N\to\infty} \frac{1}{N}\sum_{n=K}^{N+K-1} e^{int} = \mathbf{1}_{\{0\}}$ in $L^2(\sigma_f)$ by the dominated convergence theorem, and $\frac{1}{N}\sum_{n=K}^{N+K-1} U^n f \to E_0 f$ in H uniformly in K. In particular, $M_N f \to Pf$, for all $f \in H$.

When the isometry U is no longer an unitary operator, the spectral family, associated to U, is defined in this way : if $f \in H$, we put

$$\gamma_n = \begin{cases} \langle U^n f, f\rangle & \text{if } n \geq 0 \\ \langle f, U^{-n} f\rangle & \text{if } n < 0 \end{cases}$$

Since $\gamma_n = \overline{\gamma_{-n}}$ for $n < 0$, $(\gamma_n)_{n\in\mathbf{Z}}$ is still a positive definite sequence and the spectral measure σ_f is well defined by

$$\widehat{\sigma}_f(n) = \gamma_n, \quad n \in \mathbf{Z}.$$

We easily check that the fundamental identity (2.4) : $||R(U)f||_H = ||R||_{L^2(\sigma_f)}$, remains valid for trigonometric polynomials on $\mathbf{T}$ with *nonnegative frequences* : $R(t) = \sum_{k\geq 0} a_k e^{ikt}$. This is sufficient to conclude in this case. □

2. **Coboundaries**. Consider an isometry U of the separable Hilbert space H and $g \in H$. Then g is a *U-coboundary* if there exists $f \in H$ such that $g = Uf - f$. Note that in this case, $(U^n g)$ is a bounded sequence in H; in fact the converse is true and holds in a more general context [167].

Theorem 2.6. *Let U be an isometry of the separable Hilbert space H. Then $g \in H$ is a U-coboundary if and only if $(U^n g)$ is a bounded sequence in H.*

Proof. The weak closure of convex hull of the $U^n g, n \geq 0$, is a weak compact set K, which is stable under the continuous map $\varphi : f \mapsto Uf - g$. By the Markov-Kakutani theorem, φ has a fixed point in K : there exists $f \in K$ such that $\varphi(f) = f$ or $Uf - f = g$. □

3. The following result may be viewed as a consequence of von Neumann's theorem.

Proposition 2.17. *Let U be an isometry of the separable Hilbert space H. Then, for every $f \in H$ and every $\varepsilon > 0$, the set of non-negative integers*

$$\{n \in \mathbf{N}, \quad |\widehat{\sigma}_f(n)| \geq \sigma_f\{0\} - \varepsilon\}$$

is relatively dense.

We recall that a set of integers is said to be *relatively dense* or *syndetic* if it has bounded gaps.

Proof. Observe first that $\sigma_f\{0\} = ||\sigma_{Pf}|| = ||Pf||^2 = \langle Pf, f\rangle$ with $P = E_0$. Now

$$\langle Pf, f\rangle = \lim_{N\to\infty} \frac{1}{N} \sum_{n=K}^{N+K-1} \langle U^n f, f\rangle$$

uniformly in K. Let us fix $\varepsilon > 0$; there exists $N > 0$ such that

$$|\frac{1}{N} \sum_{n=K}^{N+K-1} \langle U^n f, f\rangle - \langle Pf, f\rangle| \leq \varepsilon,$$

hence

$$\frac{1}{N} \sum_{n=K}^{N+K-1} |\widehat{\sigma}_f(n)| \geq \sigma_f\{0\} - \varepsilon$$

for all $K \geq 0$. This means that $\{K, \ldots, K+N-1\}$ contains at least one integer n for which $|\widehat{\sigma}_f(n)| \geq \sigma_f\{0\} - \varepsilon$, whatever $K \geq 0$, and the property is established. □

Remark 2.5. Actually, this property is shared by any positive measure $\sigma \in M(\mathbf{T})$.

4. More generally, we consider sequences (U^{k_n}) for some increasing sequence of non-negative integers (k_n).

Proposition 2.18. *The two assertions (a) and (b) are equivalent :*
(a) The sequence $(\frac{1}{N}\sum_{n<N} e^{ik_n\alpha})$ *converges for every* $\alpha \in \mathbf{R}$.
(b) The sequence $(\frac{1}{N}\sum_{n<N} U^{k_n} f)$ *converges for every* $f \in H$ *and every isometry* U *of the separable Hilbert space* H.

Proof. In the same way,

$$||\frac{1}{N}\sum_{n<N} U^{k_n} f - \frac{1}{N'}\sum_{n<N'} U^{k_n} f||_H^2 = ||\frac{1}{N}\sum_{n<N} e^{ik_n\alpha} - \frac{1}{N'}\sum_{n<N'} e^{ik_n\alpha}||_{L^2(\sigma_f)}^2$$

whence (a) implies (b) via the dominated convergence theorem.

Conversely, for each α we apply (b) with $H = L^2(\mathbf{T}, m)$, $Uf = f \circ R_\alpha$ and $f(t) = e^{it}$ to get (a). □

The limit need not be a projection in the general case for it may fail to be invariant under U.

After Eberlein [79], Blum and Eisenberg [30], Blum and Reich [31], Bellow and Losert [22] and others (with a non unified terminology), we shall say that a sequence of integers (k_n) fulfilling one of these equivalent conditions is *weakly ergodic* (for reasons which will be clear later).

It is said to be *ergodic* if $\frac{1}{N}\sum_{n<N} e^{ik_n\alpha}$ tends to 0 for all $\alpha \neq 0$. For these last sequences, we thus have "a" von Neumann theorem.

We develop now two examples of weakly ergodic sequences which in fact are well-known.

Proposition 2.19. *The set of s-powers :* $\mathscr{S} = \{n^s,\ n \in \mathbf{N}\}$, *with* s *integer* ≥ 2, *and the set of primes* $\mathscr{P}$, *are weakly ergodic sequences (but not ergodic ones).*

Proof. We introduce the usual number theoretic notation : $e(x) = e^{2i\pi x}$.
$*$ We shall prove that the limit of

$$p_N(x) = \frac{1}{N}\sum_{n=1}^{N} e(n^s x)$$

exists for all $x \in \mathbf{R}$ as $N \to \infty$; more precisely

$$\frac{1}{N}\sum_{n=1}^{N} e(n^s x) \to \begin{cases} 0 & \text{if } x \notin \mathbf{Q} \\ \frac{1}{q}\sum_{r=1}^{q} e^{2i\pi r^s \frac{a}{q}} & \text{if } x = \frac{a}{q} \end{cases}$$

When $x \notin \mathbf{Q}$, the sequence $(n^s x)$ is uniformly distributed modulo one and the limit is zero (see chapter 4 and [153]).

Suppose now that $x = a/q$ with $(a,q) = 1$. Then

$$\sum_{n=1}^{N} e(n^s x) = \sum_{r=0}^{q-1}\Big(\sum_{\substack{n \leq N \\ n \equiv r(q)}} e(n^s \frac{a}{q})\Big) \sim \frac{N}{q}\sum_{r=0}^{q-1} e(r^s \frac{a}{q}),$$

whence the expected limit.

For $s = 2$, we recognize in the sum on r, the classical Gauss sum whose behaviour is well-known [112]:

$$S_q(a) = \sum_{r=0}^{q-1} e(r^2 \frac{a}{q}) = O(\sqrt{q});$$

It follows that

$$\lim_{N\to\infty} \frac{1}{N}\sum_{n=1}^{N} e(n^2 x) = \frac{S_q(a)}{q} = O(1/\sqrt{q})$$

as $q \to \infty$ which may be useful.

$*$ We denote by $p_1 < p_2 < \cdots$ the prime numbers and we will prove that the limit of the sequence

$$q_N(x) = \frac{1}{N}\sum_{n=1}^{N} e(p_n x)$$

exists for all $x \in \mathbf{R}$ as $N \to \infty$. We refer to [81] and [112] for results in analytic number theory.

When $x \notin \mathbf{Q}$, $q_N(x)$ tend to zero according to a deep theorem of Vinogradov.

Suppose now that $x = a/q$ with $(a,q) = 1$. Then

$$\sum_{n=1}^{N} e(p_n x) = \sum_{\substack{(r,q)=1 \\ r \bmod q}} \Big(\sum_{\substack{p \le p_N \\ p \equiv r(q)}} e(p\frac{a}{q}) \Big),$$

$$= \sum_{\substack{(r,q)=1 \\ r \bmod q}} e(r\frac{a}{q}) \Big(\sum_{\substack{p \le p_N \\ p \equiv r(q)}} 1 \Big).$$

Dirichlet's theorem on arithmetic progressions asserts that prime numbers are fairly distributed in the $\varphi(q)$ arithmetic progressions : $q\mathbf{N} + r$, $(r,q) = 1$ (Here φ is the Euler function). This results in

$$\frac{1}{\pi(p_N)} \sum_{\substack{p \le p_N \\ p \equiv r(q)}} 1 \sim \frac{1}{\varphi(q)}, \quad \text{for prime } (r,q)$$

where $\pi(x)$ is the number of prime numbers less or equal to x. One deduce from this estimate that

$$\frac{1}{N} \sum_{n=1}^{N} e(p_n x) = \sum_{\substack{(r,q)=1 \\ r \bmod q}} e(r\frac{a}{q}) \Big(\frac{1}{N} \sum_{\substack{p \le p_N \\ p \equiv r(q)}} 1 \Big) \sim \frac{1}{\varphi(q)} \sum_{\substack{(r,q)=1 \\ r \bmod q}} e(r\frac{a}{q}).$$

But we recognize in $\Sigma_q := \sum_{\substack{(r,q)=1 \\ r \bmod q}} e(r\frac{a}{q})$ the classical Ramanujan sum whose value is well-known : in fact $\Sigma_q = \mu(q)$, where μ is the Möbius function. So we have proved

$$\lim_{N \to \infty} \frac{1}{N} \sum_{n=1}^{N} e(p_n \frac{a}{q}) = \frac{\mu(q)}{\varphi(q)}$$

for all a/q, $(a,q) = 1$. □

Remark 2.6. a. Note that general lacunary sequences fail to be ergodic : this is a consequence of a deep result on bad distribution, due to Erdòs and Taylor.

b. Random ergodic sequences of zero density have been constructed [36] (see also [161]). Later, in [33], the authors exhibit deterministic and natural such sequences; for example, the sequence $(k_n) = ([\alpha n^\beta])$ with β irrational, $\beta > 1$, and $\alpha \neq 0$, is ergodic.

Chapter 3
Spectral Theory of Dynamical Systems

In this chapter, we deal with dynamical systems to which we apply the foregoing results. In particular, we give a spectral characterization of the different mixing properties (weak, mild, strong). All the results are well-known, and we omit the classical proofs for which we refer to [61, 93, 111, 139, 145, 151, 193, 197, 200, 241] or others. We close this chapter by an overview on group extensions over an ergodic rotation.

3.1 Notations and Definitions

1. Let $(X, \mathscr{B}, \mu)$ be a measure space, where the measure μ is finite, and let T be a measurable map from X onto X (μ-a.e.), which *preserves the measure* μ, which means that
$$\mu(T^{-1}B) = \mu(B), \quad \forall B \in \mathscr{B}.$$
We call T an *endomorphism* of X and for such a *measure-preserving transformation*, we say as well : "μ is T-invariant" when emphasizing on the measure. $(X, \mathscr{B}, \mu, T)$ is called a *dynamical system* : it denotes the action on $(X, \mathscr{B}, \mu)$ of the countable semi-group $(T^n)_{n \in \mathbf{N}}$. Without any restriction, the measure μ will be assumed to have mass one i.e. $(X, \mathscr{B}, \mu)$ will be a probability space.
2. When T is in addition a one-to-one, bi-measurable map of X onto X, we say that T is an *automorphism* of X, in which case the operator $U := U_T$, defined on $H = L^2(X, \mu)$ by
$$U_T f(x) = f(Tx),$$
is a unitary operator. Otherwise, U_T is an isometry of H. Throughout this chapter, we suppose that $L^2(X, \mu)$ is separable.
3. We agree that the automorphism T has the spectral property $(\mathscr{P})$ if the operator U_T does have it. Note that such an operator U_T has always a discrete component in its spectrum : indeed, the constants are U_T-invariant functions, thus eigenvectors corresponding to the eigenvalue 1. The orthogonal of the constants in $L^2(X, \mu)$ will be denoted by $\mathbf{1}^\perp$. Observe that
$$\mathbf{1}^\perp = \{f \in L^2(X, \mu), \int f \, d\mu = 0\}.$$

M. Queffélec, *Substitution Dynamical Systems – Spectral Analysis: Second Edition*,
Lecture Notes in Mathematics 1294, DOI 10.1007/978-3-642-11212-6_3,
© Springer-Verlag Berlin Heidelberg 2010

To the maximal spectral type of U_T on $L^2(X,\mu)$ (mostly denoted by σ without [ ]), we shall prefer the so-called *reduced* maximal spectral type, namely the maximal spectral type of U_T on $\mathbf{1}^\perp$ and we shall use the notation : σ_0.

4. From now on, we simply say that a measurable function f on $(X,\mathscr{B})$ is T-*invariant* if it is invariant under U_T, that is :

$$f(Tx) = f(x), \quad \text{for every } x \in X.$$

A measurable set $B \in \mathscr{B}$ in turn is said to be T-*invariant* if the function $\mathbf{1}_B$ is, in other words $T^{-1}B = B$. The class of all T-invariant sets in $\mathscr{B}$ is a sub-σ-algebra of $\mathscr{B}$ denoted by $\mathscr{I}$.

5. Two dynamical systems are said to be *metrically isomorphic* if there exists an isomorphism between the probability spaces, exchanging the transformations, more precisely : $(X,\mathscr{B},\mu,T)$ and $(Y,\mathscr{C},\nu,S)$ are metrically isomorphic if there exist invariant sets $X_1 \in \mathscr{B}$, $Y_1 \in \mathscr{C}$, $\mu(X_1) = 1$, $\nu(Y_1) = 1$, and a one-to-one map ϕ from X_1 onto Y_1 such that the following diagram commutes

$$\begin{array}{ccc} X_1 \subset X & \xrightarrow{\phi} & Y_1 \subset Y \\ T\downarrow & & \downarrow S \\ X_1 \subset X & \xrightarrow[\phi]{} & Y_1 \subset Y \end{array}$$

i.e.

$$\phi \circ T = S \circ \phi$$

and

$$\mu(\phi^{-1}(B)) = \nu(B), \quad \text{for} \quad B \in \mathscr{C} \cap Y_1.$$

6. If the homomorphism ϕ is only onto in the previous definition, $(Y,\mathscr{C},\nu,S)$ is called a *factor* of the system $(X,\mathscr{B},\mu,T)$, and $(X,\mathscr{B},\mu,T)$ an *extension* of $(Y,\mathscr{C},\nu,S)$. When $\mathscr{B}'$ is a T-invariant sub-σ-algebra of $\mathscr{B}$, $(X,\mathscr{B}',\mu,T)$ is a factor of $(X,\mathscr{B},\mu,T)$.

7. Two dynamical systems are said to be *spectrally isomorphic* if the associated operators are unitarily equivalent. It is clear that metrically isomorphic systems have the same spectral invariants (but the reverse implication may be false) : with same notations, it is easy to check that, for any $f \in L^2(X,\mu)$, we have $f \circ \phi^{-1} \in L^2(Y,\nu)$ and

$$\sigma_f = \sigma_{f \circ \phi^{-1}}.$$

8. To avoid pathologies, the probability space $(X,\mathscr{B},\mu)$ will thereafter be assumed to be a *Lebesgue space* i.e. isomorphic as measure space to the standard Lebesgue space $([0,1],\mathscr{B}_{[0,1]},\lambda)$.

3.2 Ergodic Dynamical Systems

In the measurable context the following invariance property is more natural : B is μ-almost invariant under T if $T^{-1}B \subset B$ up to a μ-negligible set; this condition is obviously equivalent to the next $\mu(T^{-1}B\Delta B) = 0$. The following property of the dynamical system is, in some sense, a metric version of connexity.

Definition 3.1. The system $(X, \mathscr{B}, \mu, T)$ is said to be *ergodic* if $\mu(B) = 0$ or 1 for any μ-almost invariant B (and we use again the notation $B \in \mathscr{I}$).

This property is equivalent to the following : *The T-invariant L^2-functions are reduced to the constants.* Actually, for $f \in L^2(X, \mu)$, the weaker hypothesis $f \circ T \leq f \ \mu - a.e.$ already implies that f is constant $\mu - a.e.$.
According to whether we focus on the transformation T, or on the measure μ, we shall say more briefly : "T is μ-ergodic", or, "μ is T-ergodic". We deduce the following characterization :

Proposition 3.1. *The dynamical system $(X, \mathscr{B}, \mu, T)$ is ergodic if and only if* 1 *is a simple eigenvalue for U_T.*

If T is ergodic, each eigenvalue has multiplicity one, each eigenfunction has a constant modulus, and the eigenvalues form a countable subgroup of $\mathbf{U}$.

3.2.1 First Examples

1. *Rotations*. The transformation $R_\theta : x \mapsto x + \theta$ on $\mathbf{T}$ is an automorphism of $(\mathbf{T}, \mathscr{B}_\mathbf{T}, m)$. The system $(\mathbf{T}, \mathscr{B}_\mathbf{T}, m, R_\theta)$ is ergodic for all θ, $\theta/\pi \notin \mathbf{Q}$. We shall refer to this system as to the "irrational rotation" in short.
   More generally, consider G an abelian compact *monothetic group*, i.e. $\mathbf{Z}g$ is dense in G for some $g \in G$. Recall that a compact and connected abelian group is monothetic. We denote by $R : x \mapsto x + g$ the rotation by g, and by m_G the Haar measure on G, which is the *unique* R-invariant probability measure on G. Then, the system $(G, \mathscr{B}_G, m_G, R)$ is ergodic.
   Indeed, suppose that B is a R-invariant Borel set in G with $0 < m_G(B) < 1$; then $\mu = \mathbf{1}_B \cdot m_G / m_G(B)$ would be a R-invariant probability measure, different from m_G.
   We shall refer to this system as to an "ergodic rotation" on G.
2. *q-adic transformation*. Let q be any integer ≥ 2. The transformation $S_q : x \mapsto qx \mod 1$ on $[0,1]$ is an endomorphism of $([0,1], \mathscr{B}_{[0,1]}, m)$, and the system $([0,1], \mathscr{B}_{[0,1]}, m, S_q)$ is ergodic too, as it can be checked on the Fourier decomposition of $f \in L^2([0,1])$. This system is metrically isomorphic to the *q-shift*, shortened terminology for the following symbolic system :

$$(\{0, 1, \ldots, q-1\}^\mathbf{N}, \mathscr{B}, \mu, T)$$

where T is the unilateral shift on $\{0,1,\ldots,q-1\}^{\mathbf{N}}$, the infinite sequences taking their values in the alphabet $\{0,1,\ldots,q-1\}$, and where μ is the infinite convolution of the same discrete measure $\sum_{j=0}^{q-1}\delta_j/q$.

By abusing the notation, we shall speak of "q-shift" in both cases.

3. *Product.* Consider the direct product of two dynamical systems :

$$(X_1,\mathscr{B}_1,\mu_1,T_1)\times(X_2,\mathscr{B}_2,\mu_2,T_2)=(X_1\times X_2,\mathscr{B}_1\otimes\mathscr{B}_2,\mu_1\otimes\mu_2,T)$$

where $T(x_1,x_2)=(T_1(x_1),T_2(x_2))$. A direct product of two ergodic dynamical systems need not be ergodic : suppose for example that there exist $t\in\mathbf{R}$, $t\neq 0$, $f_1\in L^2(X_1,\mu_1)$ and $f_2\in L^2(X_2,\mu_2)$ such that

$$f_1(T_1x_1)=e^{it}f_1(x_1)\quad\text{and}\quad f_2(T_2x_2)=e^{it}f_2(x_2)$$

where f_1,f_2 are non-constant unimodular eigenfunctions. Obviously $f_1\otimes\bar{f}_2$ is a T-invariant function in $\mathbf{1}^{\perp}$.

4. *Group extension.* Let $(X,\mathscr{B},\mu,T)$ be a dynamical system, G a metric compact abelian group (with additive notations) equipped with its Haar measure m_G and $\varphi: X\to G$ a measurable map. We define the transformation T_φ on the product space $(X\times G,\mathscr{B}\otimes\mathscr{B}_G,\mu\otimes m_G)$ by

$$T_\varphi(x,z)=(Tx,z+\varphi(x)).$$

It is easy to check that T_φ is an endomorphism of the product space. The system $(X\times G,\mathscr{B}\otimes\mathscr{B}_G,\mu\otimes m_G,T_\varphi)$ is called a *group extension* of the initial one. H. Anzai [16] has studied the ergodicity of this new system.

Theorem 3.1 (Anzai). *Assume the system $(X,\mathscr{B},\mu,T)$ to be ergodic. Then, T_φ is not ergodic if and only if there exist $\gamma\in\widehat{G}$, the dual group of G, $\gamma\neq 1$ and f a measurable map : $X\to\mathbf{U}$ such that*

$$\gamma(\varphi(x))=f(x)\cdot f(Tx)^{-1}\quad \mu-a.e. \tag{3.1}$$

Proof. We just sketch the proof. If $\gamma\in\widehat{G}$, $\gamma\neq 1$, satisfies (3.1), consider $h(x,z)=f(x)\gamma(z)$. It is easy to check that h is T_φ-invariant. In addition, $\int_G\gamma\,dm_G=0$ implies $\int h\,d\mu\otimes m_G=0$ in turn. Thus h cannot be constant unless this constant would be zero, which is incompatible with $|h|=1$.

Conversely, suppose T_φ non-ergodic and let $h: X\times G\to\mathbf{C}$ be a non-constant invariant function. By a straightforward calculation, we see that

$$f_\gamma: x\mapsto\int_G h(x,z)\bar{\gamma}(z)\,dm_G(z)$$

has a T-invariant, thus constant, modulus.

If $f_\gamma = 0$ for every γ, h would be independent of z by Fourier unicity, and consequently, h would be a T-invariant function. But T is supposed to be ergodic and h a non-constant function which is impossible.

Therefore, there exists $\gamma \in \widehat{G}$ for which $|f_\gamma| =: C_\gamma$ is a non-zero constant, and the function $f = C_\gamma^{-1} f_\gamma$ realizes (3.1). □

Remark 3.1. This example is a particular case of what we usually call a *skew product* : $(X, \mathscr{B}, \mu, T)$ is a dynamical system and $(S_x)_{x\in X}$ is a family of endomorphisms of another measure space $(Y, \mathscr{C}, \nu)$. If $(x,y) \mapsto S_x y$ is measurable from $(X \times Y, \mathscr{B} \times \mathscr{C})$ into $(Y, \mathscr{C})$, the transformation $(x,y) \in X \times Y \mapsto (Tx, S_x y)$ is an endomorphism of the product measure space, called a skew product (see [223]).

3.2.2 Main Theorems

The forthcoming major theorems play a key role in the theory, as well in the understanding of measure-preserving transformations as in various applications, that one proposes to show later. Comments on history, detailed proofs and complements will be found in the references. Certain of these theorems are called "ergodic" though ergodicity is not presently assumed.

$(X, \mathscr{B}, \mu, T)$ is a dynamical system with a probability measure μ. The first result we mention is the *recurrence theorem of Poincaré* which asserts that, given $B \in \mathscr{B}$ of positive measure, $T^n x$ visits B infinitely often for μ-a.e. $x \in B$.

Theorem 3.2 (Recurrence theorem of Poincaré). *Let $(X, \mathscr{B}, \mu, T)$ be a dynamical system and $B \in \mathscr{B}$. Then*

$$\mu(B \cap (\cap_{N=0}^{\infty} \cup_{n=N}^{\infty} T^{-n}B)) = \mu(B).$$

A quantitive improvement has been obtained by Khintchin in the ergodic case.

Theorem 3.3 (Recurrence theorem of Khintchin). *Let $(X, \mathscr{B}, \mu, T)$ be an ergodic dynamical system. For every $B \in \mathscr{B}$ and every $\varepsilon > 0$, the set*

$$\{n \in \mathbf{N},\ \mu(T^{-n}B \cap B) \geq \mu(B)^2 - \varepsilon\}$$

is relatively dense.

Proof. This is an immediate consequence of the proposition 2.17 applied with $f = \mathbf{1}_B$: indeed, for all $n \geq 0$,

$$\widehat{\sigma}_{\mathbf{1}_B}(n) = < \mathbf{1}_B \circ T^n, \mathbf{1}_B > = \mu(T^{-n}B \cap B)$$

while

$$\sigma_{\mathbf{1}_B}\{0\} = \mu(B)^2,$$

since, by ergodicity, $\sigma_f\{0\} = ||E_0 f||^2 = \left|\int f\, d\mu\right|^2$ for any $f \in L^2(X, \mu)$. □

The formulation of von Neumann's theorem (theorem 2.5) in the dynamical context goes as follows :

Theorem 3.4 (von Neumann mean ergodic theorem). *Let $(X,\mathscr{B},\mu,T)$ be a dynamical system. Then, for every $f \in L^2(X,\mu)$,*

$$\left\| \frac{1}{N} \sum_{n=0}^{N-1} f \circ T^n - E^{\mathscr{I}}(f) \right\|_2 \to \text{o},$$

where $E^{\mathscr{I}}(f)$ is the conditional expectation of f, given the σ-algebra $\mathscr{I}$ of T-invariant sets in $\mathscr{B}$. If, in addition, the system is ergodic, $\mathscr{I}$ is trivial and $E^{\mathscr{I}}(f) = \int f\, d\mu$.

We deduce from this theorem several characterizations of ergodicity in terms of spectral measures.

Proposition 3.2. *$(X,\mathscr{B},\mu,T)$ is ergodic if and only if one of the following equivalent properties holds :*
(i) $\sigma_f\{0\} = 0$ for every $f \in \mathbf{1}^{\perp}$;
(ii) The reduced maximal spectral type has no point mass in 0;
(iii) For all $A, B \in \mathscr{B}$,

$$\lim_{N\to\infty} \frac{1}{N} \sum_{n<N} \mu(T^{-n}A \cap B) = \mu(A)\cdot\mu(B) \tag{3.2}$$

which may be written : $\sigma_{\mathbf{1}_A,\mathbf{1}_B}\{0\} = \mu(A)\cdot\mu(B)$;
(iv) For all $A \in \mathscr{B}$,

$$\lim_{N\to\infty} \frac{1}{N} \sum_{n<N} \mu(T^{-n}A \cap A) = \mu(A)^2$$

(v) For all $f, g \in L^2(X,\mu)$,

$$\lim_{N\to\infty} \frac{1}{N} \sum_{n<N} \int f \circ T^{-n} \cdot \bar{g}\, d\mu = \int f\, d\mu \cdot \int \bar{g}\, d\mu; \tag{3.3}$$

(vi) For all $f \in L^2(X,\mu)$

$$\lim_{N\to\infty} \frac{1}{N} \sum_{n<N} \int f \circ T^{-n} \cdot \bar{f}\, d\mu = \left| \int f\, d\mu \right|^2.$$

Proof. These equivalences have a flavour of "déjà vu", and we will just give a direct proof of the characterization (iii). Setting $B = A$ in (3.2) where A is a T-invariant set, we get $\mu(A) = \mu(A)^2$ and $\mu(A) = 0$ or 1; the system is ergodic. Conversely, we apply von Neumann's theorem to $f = \mathbf{1}_A$ and the result follows by taking the inner product with $\mathbf{1}_B$. □

The famous next theorem is a pointwise version of the mean ergodic theorem, which contains the recurrence theorem of Poincaré as a particular case; its proof usually requires an additional maximal inequality (for another approach see [145]).

Theorem 3.5 (Birkhoff's ergodic theorem). *Let $(X,\mathscr{B},\mu,T)$ be a dynamical system. Then, for every $f \in L^1(X,\mu)$,*

$$\frac{1}{N}\sum_{n=0}^{N-1} f\circ T^n \to E^{\mathscr{I}}(f) \quad \mu - a.e.$$

If, in addition, the system is ergodic, the a.e. limit is $\int f\, d\mu$.

Remark 3.2. Specifying $f = \mathbf{1}_A$ where $\mu(A) > 0$, we get that almost all points of X (and not only those already in A) are visiting A infinitely often under T; moreover, the quantitative information is noteworthy : for such $x \in X$, the frequency of n with $T^n x \in A$ is $\mu(A) > 0$.

Remark 3.3. As a direct consequence of the spectral isomorphism, we already noticed in section 2.6 the correspondence between trigonometric sums behaviour and convergence in norm of ergodic means. We reformulate the result for dynamical systems :

Proposition 3.3. *Let (k_n) be an increasing sequence of integers. The following assertions are equivalent :*

1) The sequence $(\frac{1}{N}\sum_{n<N} e^{ik_n\alpha})$ converges for every $\alpha \in \mathbf{T}$.

2) For any dynamical system $(X,\mathscr{B},\mu,T)$ and for any $f \in L^2(X,\mu)$, the Birkhoff's means $(\frac{1}{N}\sum_{n<N} f\circ T^{k_n})$ converge in $L^2(X,\mu)$.

We derived in this way the norm-convergence of means of the form $\frac{1}{N}\sum_{n<N} U^{k_n} f$, for sublacunary sequences (k_n) ; a dynamical formulation of proposition 2.19 now is :

Corollary 3.1. *The Birkhoff means $\frac{1}{N}\sum_{n<N} f\circ T^{n^s}$ with s integer ≥ 1, or $\frac{1}{N}\sum_{n<N} f\circ T^{p_n}$ with $p_n \in \mathscr{P}$, the set of primes, converge in $L^2(X,\mu)$, for every $f \in L^2(X,\mu)$.*

In order to get pointwise convergence (instead of L^2-one) one generally needs a maximal inequality as for the Birkhoff ergodic theorem.
$\star$ In case of sequences (k_n) with a positive density, the following can easily be deduced from the standard maximal inequality.

Proposition 3.4. *Assume that* $\inf(k/n_k) = \delta > 0$*; then, for any dynamical system $(X,\mathscr{B},\mu,T)$, for any $f \in L^1(X,\mu)$ and any $\lambda > 0$,*

$$\mu\{x \in \mathbf{T},\ \sup_K \Big|\frac{1}{K}\sum_{k<K} f\circ T^{n_k}(x)\Big| > \lambda\} \leq \frac{||f||_1}{\delta\lambda}.$$

⋆ The case of zero density sequences is much more involved. The improvement of corollary 3.1 as pointwise statements has been performed by J. Bourgain [35, 36].

One will return to these ergodic theorems along subsequences in the next chapter.

The ultimate result we quote now, sheds a different light on measure-preserving transformations by revealing a geometric aspect.

Proposition 3.5 (Rokhlin's lemma). *Let $(X,\mathscr{B},\mu,T)$ be an ergodic dynamical system. Then, for every $\varepsilon > 0$, for every $n \geq 1$, one can find $F \in \mathscr{B}$ such that*

$$F, TF, \ldots, T^{n-1}F \quad \text{are disjoint}$$

and

$$\mu(\cup_{j<n} T^j F) \geq 1 - \varepsilon.$$

(We call such a sequence a *Rokhlin ε-tower* of height n and base F.)

Note the immediate consequence on the spectral set : *for an ergodic transformation T on $(X,\mathscr{B},\mu)$, the spectrum $sp(U_T)$ contains the whole circle.* Recall that $\lambda \in \mathbf{C}$ is an *approximate eigenvalue* of the operator U on H if there exists a sequence of normalized $f_n \in H$ such that $Uf_n - \lambda f_n \to 0$. We shall see that any $e^{iu} \in \mathbf{T}$ is an approximated eigenvalue of U_T on $L^2(X,\mu)$: let us fix $\varepsilon > 0$, $n \geq 1$ and a Rokhlin ε-tower of height n and base $F \in \mathscr{B}$. We define $f \in L^2(X,\mu)$ by putting

$$f = e^{iku} \quad \text{on} \quad T^k F, \quad 0 \leq k \leq n-1$$

and

$$f = 1 \quad \text{on} \quad E = \left(\cup_{0 \leq k \leq n-1} T^k F\right)^c.$$

Then,

$$f(Tx) = e^{iu} f(x) \quad \text{on} \quad \cup_{0 \leq k \leq n-2} T^k F$$

so that

$$||f \circ T - e^{iu} f||^2_{L^2(X,\mu)} = \int_{T^{n-1}F} |f \circ T - e^{iu} f|^2 d\mu + \int_E |f \circ T - e^{iu} f|^2 d\mu$$

and

$$||f \circ T - e^{iu} f||^2_{L^2(X,\mu)} \leq 2/n + \varepsilon;$$

whence the claim. Therefore, either T is invertible and $sp(U_T) = \mathbf{T}$, or $sp(U_T) = \mathbf{D}$ the closed unit disc.

3.2.3 Quasi-Invariant Systems

A whole book [1] has been devoted to those systems and we just quote a few observations.

Definition 3.2. Let T be a one-to-one, bi-measurable transformation of the probability space $(X,\mathscr{B},\mu)$. We say that the system $(X,\mathscr{B},\mu,T)$ is *quasi-invariant* (or μ is T-quasi-invariant) if $T(\mu)$ and μ are equivalent measures, where $T(\mu)$ is the pullback of μ under T.

We say that the system is ergodic if $\mu(A)=0$ or 1 for any T-invariant set A in $\mathscr{B}$.

For such a system, we consider the operator U on $L^2(X,\mu)$ defined by

$$Uf = r\cdot f\circ T$$

where r^2 is the density $\dfrac{dT^{-1}(\mu)}{d\mu}\in L^\infty(\mu)$; thus we still have a unitary operator :

$$\int |Uf|^2\,d\mu = \int_X r^2(x)|f(Tx)|^2\,d\mu(x) = \int_X |f|^2(Tx)\,d(T^{-1}\mu)(x) = \int |f|^2\,d\mu.$$

But, generally, U may fail to have eigenfunctions, which would be solutions of the equation $e^{i\alpha}f(x)=f(Tx)r(x)$: If such an f exists, $r^2(x)|f(Tx)|^2=|f(x)|^2$ and $\nu=|f|^2\mu$ is a T-invariant measure; thus, except when there exists a T-invariant measure $\nu\sim\mu$, U has no eigenfunction.

However, $L^\infty(\mu)$ is invariant under T and $e^{i\alpha}$ is said to be an *L^∞-eigenvalue* if there exists $f\in L^\infty(\mu)$, $f\neq 0$, such that

$$f\circ T = e^{i\alpha}f,\ \text{on } X.$$

J.F. Méla proved the following [182]:

Proposition 3.6. *If $(X,\mathscr{B},\mu,T)$ is an ergodic quasi-invariant system with $(X,\mathscr{B},\mu)$ a non-atomic Lebesgue-space, then the set of all L^∞-eigenvalues is a Borel (not necessarily countable) proper subgroup Λ of* **T***, with the following property :*

Let ν be a probability measure supported by Λ; if σ is a probability measure on **T** *such that*

$$\|\gamma_{n_j}-1\|_{L^1(\nu)}\to 0 \Longrightarrow \|\gamma_{n_j}-1\|_{L^1(\sigma)}\to 0,$$

then σ also is supported by Λ.

More generally, a subgroup H of **T** enjoying this property, will be called a *saturated subgroup*.

Suppose that there exists a probability measure ν supported by H, satisfying $\limsup_{n\to\infty}|\hat{\nu}(n)|<1$. Then, the only sequences (n_j) for which $\|\gamma_{n_j}-1\|_{L^1(\nu)}\to 0$ are the trivial ones since n_j must be zero for j large enough, and $\|\gamma_{n_j}-1\|_{L^1(\sigma)}\to 0$ for any σ obviously. H must be the whole circle if saturated.

From now on we suppose that H is a *Dirichlet subgroup*, which means that every probability measure μ supported by H is a Dirichlet measure (section 1.5) :

$$\limsup_{n\to\infty}|\hat{\mu}(n)|=1.$$

Corollary 3.2. *Let $(X,\mathscr{B},\mu,T)$ be an ergodic quasi-invariant system with $(X,\mathscr{B},\mu)$ a non-atomic Lebesgue-space, and Λ be the set of all L^∞-eigenvalues. Then Λ is a saturated Dirichlet subgroup.*

Examples and counter-examples A very interesting class of saturated Dirichlet subgroups is obtained in this way [123]:

Definition 3.3. Let μ a probability measure on $\mathbf{T}$. We call the *translation group of* μ and we denote by $H(\mu)$ the subgroup of $x \in \mathbf{T}$ such that $\mu * \delta_x \sim \mu$.

If $\mu \sim m$, the Lebesgue measure on $\mathbf{T}$, then $H(\mu) = \mathbf{T}$, and $H(\mu)$ is countable if μ is discrete.

Proposition 3.7. *Let ρ be the classical Riesz product*

$$\rho = \prod_{k=1}^{\infty}(1+\cos n_k x)$$

where the sequence of positive integers (n_k) is dissociate. Then

$$H(\rho) = \{\alpha \in \mathbf{T}, \sum_k ||n_k\alpha||^2 < \infty\},$$

where $||\cdot||$ is the distance to the nearest integer.

Proof. For any $\alpha \in \mathbf{T}$, the measure $\delta_\alpha * \rho$ is still a dissociate Riesz product namely

$$\delta_\alpha * \rho = \prod_{k=1}^{\infty}\left(1+\cos n_k(x-\alpha)\right)$$

A generalization, due to F. Parreau, of the equivalence criterion for Riesz products (proposition 1.2) gives the description of $H(\rho)$ [196]. □

But there exist probability measures σ such that $\delta_\alpha * \sigma \perp \sigma$ for every non-zero α so that $H(\sigma) = \{0\}$ [123].

Definition 3.4. A subgroup H of $\mathbf{T}$ is said to be a *translation group* if $H = H(\mu)$ for some probability measure μ on $\mathbf{T}$.

Proposition 3.8. *A translation group is a saturated subgroup. And $H = H(\mu)$ is a saturated Dirichlet subgroup if μ is not absolutely continuous.*

By considering $H(\rho)$ with $\rho = \prod_{k=1}^{\infty}(1+a_k\cos n_k x)$, we get the following.

Corollary 3.3. *Let*

$$H = \{x \in \mathbf{R}, \sum_{j\geq 1} a_j(1-\cos(n_j x)) < \infty\}$$

where $n_{j+1}/n_j \geq 3$ and $a_j \geq 0$ with $\sum_{j\geq 1} a_j^2 = \infty$. Then H is a saturated Dirichlet subgroup.

One can show that a translation group is the group of L^∞-eigenvalues of some quasi-invariant dynamical system. What about the reverse assertion?

Conjecture 3.1 (Aaronson). The group of L^∞-eigenvalues of some quasi-invariant dynamical system is a translation group.

3.3 Mixing Properties

We focus, in this section, on mixing properties which are spectral invariants and also may appear as new measuring instruments regarding rigidity and randomness of the dynamical system $(X, \mathscr{B}, \mu, T)$.

3.3.1 Weak and Strong Mixing Properties

Definition 3.5. A dynamical system is said to possess *weak mixing property*, or, merely, the system itself (or T or μ) is *weakly mixing* if, for all $A, B \in \mathscr{B}$,

$$\lim_{N\to\infty} \frac{1}{N} \sum_{n<N} |\mu(T^{-n}A \cap B) - \mu(A) \cdot \mu(B)| = 0 \tag{3.4}$$

Definition 3.6. A dynamical system is said to possess *strong mixing property*, or, merely, the system itself (or T or μ) is *strongly mixing* if, for all $A, B \in \mathscr{B}$,

$$\lim_{N\to\infty} \mu(T^{-n}A \cap B) = \mu(A) \cdot \mu(B) \tag{3.5}$$

Strong mixing implies weak mixing since, for a bounded sequence of real numbers (a_n), $\frac{1}{N} \sum_{n<N} |a_n| \to 0$ means $\lim_{n \notin R} a_n = 0$ for some sequence R of integers of density zero. Now, in turn, weak mixing implies ergodicity obviously, in view of (3.2) (see [130]).

Basic Examples

1. The irrational rotation R_θ on $\mathbf{T}$ is never strongly mixing : indeed let μ be any probability measure on $\mathbf{T}$ and $A \subset [0, \pi]$; let now (n_j) be such that $n_j\theta \to \pi$ mod 2π; if $\varepsilon > 0$ is given,
$$\mu(T^{-n_j}A \cap A) \geq \mu(A) - \varepsilon$$
for j large enough, and one cannot have $\lim_{j\to\infty} \mu(T^{-n_j}A \cap A) = \mu(A)^2$ if $\mu(A)^2 < \mu(A)$.
2. Consider the auto-similar Riesz product $\prod_{n\geq 0}(1 + r\cos q^n t)$, where $0 < |r| \leq 1$ and $q \geq 2$, and the q-adic transformation S_q on $\mathbf{T}$. Observe first that the measure ρ is preserved by the transformation $S := S_q$. Indeed, as a consequence of the

particularized identity (1.8) with constant coefficients, we get that $\hat{\rho}(qn) = \hat{\rho}(n)$. This leads to

$$\int f \circ S \, d\mu = \int f \, d\mu$$

for every trigonometric polynomial f; the S-invariance of ρ follows from an approximation argument. We say, in short, that ρ is *q-invariant*.

We now claim that the dynamical system $(\mathbf{T}, \mathscr{B}_{\mathbf{T}}, \rho, S)$ is strongly mixing.

We already saw (1.11) that

$$\lim_{n\to\infty} \hat{\rho}(aq^n + b) = \hat{\rho}(a)\hat{\rho}(b), \quad \forall a, b \in \mathbf{Z}.$$

Thus

$$\lim_{n\to\infty} \int f \circ S^n \cdot \bar{g} \, d\mu = \int f \, d\mu \cdot \int \bar{d} \, d\mu$$

for every trigonometric polynomials f, g, then, with the same approximation argument, for $f = \mathbf{1}_A$, $g = \mathbf{1}_B$, $A, B \in \mathscr{B}_{\mathbf{T}}$.

In the same way, the generalized Riesz product $\prod_{n\geq 0} P(q^n t)$ is strongly mixing with respect to S_q (section 1.3). For another example of such measures see [144]. This property turns out to be fundamental and we simply say in the sequel that such a measure is q-strongly mixing.

We shall now give spectral characterizations of these mixing properties. We suppose $U := U_T$ to be a unitary operator and we refer to chapter 1 for the notations.

Proposition 3.9. *Let $(X, \mathscr{B}, \mu, T)$ be a dynamical system.*

a) T is weakly mixing if and only if $\sigma_f \in M_c(\mathbf{T})$ for every $f \in \mathbf{1}^{\perp}$; this means that the reduced maximal spectral type σ_0 a is continuous *measure.*

b) T is strongly mixing if and only if $\sigma_f \in M_0(\mathbf{T})$ for every $f \in \mathbf{1}^{\perp}$; this means that the reduced maximal spectral type σ_0 is a Rajchmann-measure.

This proposition can easily be derived from the following lemmas.

Lemma 3.1. *T is weakly mixing if and only if one of the equivalent properties is satisfied :*

a) For all $f, g \in L^2(X, \mu)$,

$$\lim_{N\to\infty} \frac{1}{N} \sum_{n<N} \left| \int f \circ T^n \cdot \bar{g} \, d\mu - \int f \, d\mu \cdot \int \bar{g} \, d\mu \right| = 0;$$

b) For all $f \in L^2(X, \mu)$,

$$\lim_{N\to\infty} \frac{1}{N} \sum_{n<N} \left| \int f \circ T^n \cdot \bar{f} \, d\mu - \int f \, d\mu \cdot \int \bar{f} \, d\mu \right| = 0;$$

c) For all $f,g \in L^2(X,\mu)$,

$$\lim_{N\to\infty} \frac{1}{N} \sum_{n<N} \left| \int f\circ T^n \cdot \bar{g}\, d\mu - \int f\, d\mu \cdot \int \bar{g}\, d\mu \right|^2 = 0.$$

The equivalence between a) and c) in the lemma is a consequence of the following remark : for a bounded sequence of real numbers (a_n),

$$\frac{1}{N} \sum_{n<N} |a_n| \to 0 \Longleftrightarrow \frac{1}{N} \sum_{n<N} |a_n|^2 \to 0.$$

Lemma 3.2. *T is strongly mixing if and only if one of the equivalent properties is satisfied :*

a) For all $f,g \in L^2(X,\mu)$,

$$\lim_{N\to\infty} \int f\circ T^n \cdot \bar{g}\, d\mu = \int f\, d\mu \cdot \int \bar{g}\, d\mu;$$

b) For all $f \in L^2(X,\mu)$,

$$\lim_{N\to\infty} \int f\circ T^n \cdot \bar{f}\, d\mu = \int f\, d\mu \cdot \int \bar{f}\, d\mu.$$

Proof. The proposition a) is just a reinterpretation of Wiener's lemma (lemma 1.1) and the definition of $M_0(\mathbf{T})$ yields the proposition b). □

The following equivalent spectral characterization of weak mixing will be useful.

Proposition 3.10. *Let $(X,\mathscr{B},\mu,T)$ be a dynamical system. Then T is weakly mixing if and only if the eigenvalue $\lambda = 1$ has multiplicity one and is the unique eigenvalue.*

We observed in subsection 3.2.1 that the direct product of two ergodic dynamical systems may fail to be ergodic. It is quite clear with the following :

Proposition 3.11. *Let $(X,\mathscr{B},\mu,T)$ be a dynamical system with T invertible. The system is weakly mixing if and only if its product with any ergodic system is still ergodic.*

In particular a weakly mixing automorphism is *totally ergodic* (with ergodic powers) and the product of two weakly mixing dynamical systems is still weakly mixing.

Proof. ▷ Suppose that $(X,\mathscr{B},\mu,T)$ is weakly mixing and let $(Y,\mathscr{C},\nu,S)$ be any ergodic dynamical system. According to the characterization (3.2), we have to take account of rectangles only and we must prove : for every $A,B \in \mathscr{B}$, $C,D \in \mathscr{C}$,

$$\lim_{N\to\infty} \frac{1}{N} \sum_{n<N} (\mu\otimes\nu)[(T\times S)^{-n}(A\times C)\cap(B\times D)] = \mu(A)\mu(B)\nu(C)\nu(D).$$

This can be deduced from the following general fact on bounded sequences of real numbers : let (a_n) and (b_n) be two bounded and non-negative sequences such that

$$\lim_{N\to\infty}\frac{1}{N}\sum_{n<N}|b_n-b|=0,\quad \text{and}\quad \lim_{N\to\infty}\frac{1}{N}\sum_{n<N}a_n=a;$$

then $\lim_{N\to\infty}\frac{1}{N}\sum_{n<N}a_nb_n=ab.$

$\triangleleft$ Conversely, we assume that the transformation $S=T\times T$ on $(X\times X,\mathscr{B}\otimes\mathscr{B},\mu\otimes\mu)$ is ergodic and aim to derive the weak mixing property for T. We will make use of the last proposition. Let $e^{i\alpha}$ be an eigenvalue of T and $f\in L^2(X,\mu)$ a corresponding eigenfunction. Consider F defined on $X\times X$ by

$$F(x,y)=f(x)\overline{f(y)}.$$

F is obviously S-invariant, hence constant by assumption on S. So is f and $e^{i\alpha}$ must be one. This proves the expected property. □

3.3.2 *Mild Mixing Property*

A dynamical system $(Y,\mathscr{C},\nu,S)$, where the measure ν is no longer a probability measure but is σ-finite only, is said to be *ergodic* if, for every S-invariant set A, we have either $\nu(A)=0$ or $\nu(A^c)=0$.

H. Furstenberg and B. Weiss describe in [97] the dynamical systems $(X,\mathscr{B},\mu,T)$, whose cartesian product with any ergodic, σ-finite, dynamical system $(Y,\mathscr{C},\nu,S)$ is still ergodic. These systems are called *mildly mixing* systems, and it is rather easy to give a spectral characterization of them.

Let $(X,\mathscr{B},\mu,T)$ be a dynamical system, with $(X,\mathscr{B},\mu)$ a probability space, and $U:=U_T$.

Definition 3.7. The transformation T is said to be *rigid* provided there exists a sequence (n_k) going to infinity, such that

$$\lim_{k\to\infty}U^{n_k}f=f\quad\text{in}\quad L^2(X,\mu)\text{ for all }f\in L^2(X,\mu).$$

By extension, we shall say that $f\in L^2(X,\mu)$ itself is *rigid* if

$$f=\lim_{k\to\infty}U^{n_k}f\quad\text{in}\quad L^2(X,\mu)$$

or, equivalently, if

$$\widehat{\sigma}_f(n_k)\to||f||^2=||\sigma_f||,$$

for some sequence (n_k) going to infinity.

Thus, clearly in view of section 1.5 :

Definition 3.8. A function $f \in L^2(X,\mu)$ is *rigid* if σ_f is a Dirichlet measure.

We rather adopt the following equivalent definition for a mildly mixing system.

Definition 3.9. The dynamical system $(X,\mathscr{B},\mu,T)$ is *mildly mixing* if the only rigid functions are the constants, in other words, if it has no rigid factor.

Mild mixing implies weak mixing and is itself implied by strong mixing. This will be quite clear with the following characterization, which also relates to section 1.5.

Proposition 3.12. *The dynamical system $(X,\mathscr{B},\mu,T)$ is mildly mixing if and only if, for every $f \in \mathbf{1}^\perp$, σ_f belongs to $\mathscr{L}_I$, the L-ideal orthogonal to the Dirichlet measures.*

Proof. $\triangleleft$ Suppose that σ_f belong to $\mathscr{L}_I$ whenever $f \in \mathbf{1}^\perp$ and let $g \in L^2(X,\mu)$ be a rigid function; we have to prove that g is constant. But, since σ_g is a Dirichlet measure and according to proposition 1.7, there exists an idempotent $h \in \overline{\Gamma}\backslash\Gamma$ annihiling $\mathscr{L}_I$ and satisfying

$$h\sigma_g = \sigma_g.$$

Consider now $f = g - \int g \, d\mu$ so that $f \in \mathbf{1}^\perp$. By our assumption,

$$0 = h\sigma_f = h(\sigma_g - |\int g \, d\mu|^2\delta_0)$$

and

$$\sigma_g = |\int g \, d\mu|^2\delta_0.$$

This means that g is a constant $C = \int g \, d\mu$.

$\triangleright$ In the opposite direction, we have to prove, under the mild mixing property, that, for $f \in \mathbf{1}^\perp$,

$$h\sigma_f = 0 \quad \text{for every idempotent } h \in \overline{\Gamma}, \quad h \neq 1$$

(proposition 1.8). But, for any such idempotent h,

$$h\sigma_f = \sigma_{h(U)f}$$

where the operator $P := h(U)$ is an orthogonal projection since $h^2 = h$. Putting $g = Pf$, it follows that

$$h\sigma_g = h(h\sigma_f) = \sigma_g;$$

so, g is rigid and must be constant. But remember that f is in $\mathbf{1}^\perp$, so that $0 = \langle g, f\rangle = \langle Pf, f\rangle = ||Pf||^2$ and $g = 0$. The proposition is proved. $\square$

Remark 3.4. We noticed, in the foregoing, that the operator $h(U)$, where h is an idempotent in $\overline{\Gamma}\backslash\Gamma$, is an orthogonal projection. We can say more : actually

$h(U)$ is a conditional expectation $E^{\mathscr{A}}$ given a T-invariant sub-σ-algebra $\mathscr{A}$.

Proof. Let σ be the maximal spectral type of U. If γ_{n_j} tends to h (or h_σ) in $\overline{\Gamma}_\sigma$, for every $f,g \in L^2(X,\mu)$, for every $m \in \mathbf{Z}$, $(h\sigma_{f,g})\hat{}(m) = \lim_{j\to\infty} \widehat{\sigma}_{f,g}(m+n_j)$ so that, taking $m = 0$,

$$< h(U)f,g >= \lim_{j\to\infty} \int f \circ T^{n_j} \cdot \overline{g}\, d\mu. \tag{3.6}$$

It is then clear that $h(U)f \geq 0$ if $f \geq 0$ and $h(U)1 = 1$. This proves that $h(U)$ is a conditional expectation $E^{\mathscr{A}}$ for some sub-σ-algebra $\mathscr{A}$ to be specified.

If $A \in \mathscr{A}$, $h(U)\mathbf{1}_A = \mathbf{1}_A$, and

$$\begin{aligned} h(U)\mathbf{1}_{T^{-1}A} &= h(U)(\mathbf{1}_A \circ T) = h(U) \cdot U\mathbf{1}_A \\ &= Uh(U)\mathbf{1}_A \\ &= \mathbf{1}_A \circ T \end{aligned}$$

so that $T^{-1}A \in \mathscr{A}$ if $A \in \mathscr{A}$ already and $\mathscr{A}$ is T-invariant. □

Later, we shall give an explicit description of this σ-algebra for the discrete idempotent h_d.

Question 3.1. What can we say about the reciprocal? If $\mathscr{A}$ is T-invariant, does $E^{\mathscr{A}}$ respect the cyclic spaces, and, in this case, is it possible to find $h \in \overline{\Gamma}$ such that $E^{\mathscr{A}} = h(U)$?

Now, referring to [182], we prove the following formulation of Furstenberg and Weiss' result [97].

Theorem 3.6. *Let $(X,\mathscr{B},\mu,T)$ a non-atomic dynamical system, where T is an ergodic automorphism. The two assertions are equivalent :*

a) $(X,\mathscr{B},\mu,T)$ is mildly mixing;

b) For any invertible, quasi-invariant and ergodic transformation S on the probability space $(Y,\mathscr{C},\nu)$, $T \times S$ is ergodic on $(X \times Y, \mathscr{B}\otimes\mathscr{C}, \mu\otimes\nu)$.

Proof. ▷ Suppose first that $T \times S$ is not ergodic on the product space : there exists a non-constant measurable function f on $X \times Y$ which is $T \times S$-invariant, but $|f| = 1$. From the identity

$$f(Tx,y) = f(x,S^{-1}y),$$

we get a measurable family $(f_y)_{y\in Y}$ of functions defined on (X,μ) and satisfying

$$f_y(Tx) = f_{S^{-1}y}(x). \tag{3.7}$$

Fix $\varepsilon > 0$, and consider, for all $y \in Y$

$$A_y = \{z \in Y, \int |f_y - f_z|^2\, d\mu < \varepsilon\}.$$

For almost all $y \in Y$, $\nu(A_y) > 0$; we choose such a generic y_0 and we denote by A the set A_{y_0}. By ergodicity,

$$\nu(\cup_{n \in \mathbf{Z}} S^{-n} A) = 1,$$

and ν-almost all $y \in Y$ may be written $S^n z$ for some $z \in A$. But $\int |f_{y_0} - f_z|^2 \, d\mu < \varepsilon$ and, using (3.7) and the T-invariance of μ, we get that

$$\int |f_{y_0} - f_{S^{-n}y}|^2 \, d\mu = \int |f_{y_0} \circ T^n - f_y|^2 \, d\mu < \varepsilon.$$

If $f = f_{y_0}$, for ν-almost all $y \in Y$, we deduce the existence of a sequence (n_j) going to infinity, such that

$$\lim_{j \to \infty} \int |f_y - f \circ T^{n_j}|^2 \, d\mu = 0,$$

which yields, after expansion,

$$\lim_{j \to \infty} \widehat{\sigma}_{f,f_y}(n_j) + \widehat{\sigma}_{f_y,f}(n_j) = 2.$$

The measures σ_f and σ_{f_y} being probability measures, this implies that (γ_{n_j}) converges to a modulus one character in $\overline{\Gamma}(\sigma_f)$. Refining the sequence (n_j) if necessary, $\gamma_{n_j} \overline{\gamma_{n_{j+1}}}$ converges to 1 in $\overline{\Gamma}(\sigma_f)$ and σ_f is a Dirichlet measure. The system $(X, \mathscr{B}, \mu, T)$ fails to be mildly mixing.

$\triangleleft$ Conversely, if the non-atomic system $(X, \mathscr{B}, \mu, T)$ is not mildly mixing, there exists some $f \in L^2(X, \mu)$ for which σ_f is a Dirichlet continuous measure. We shall exhibit a quasi-invariant dynamical system $(Y, \mathscr{C}, \nu, S)$ such that $T \times S$ fails to be ergodic on the product.

Put $\sigma = \sigma_f$, and let us choose $Y = \overline{\Gamma_1}(\sigma)$, the weak-star closure of Γ in the unit ball of $L^\infty(\sigma)$. Y is a metric compact set, and S, defined on $y \in Y$ by

$$S(y) = \gamma \cdot y, \quad \text{where} \quad \gamma(\lambda) = e^{i\lambda}, \; \lambda \in \mathbf{T},$$

is an homeomorphism of Y. There exists a sequence (n_j) such that $S^{n_j}(1) \to 1$ (we say that 1 is a *recurrent point* for S) : indeed, σ being a Dirichlet measure, such a sequence can be found so that

$$\lim_{j \to \infty} \gamma_{n_j} = 1 \quad \text{in} \quad L^1(\sigma),$$

which provides the claim. It then follows from [142] that we can construct a continuous probability measure ν on Y satisfying $S\nu \sim \nu$.

We now construct a non-constant $T \times S$-invariant function on $X \times Y$. If γ_{k_i} tends to y in Y, then, for every $f \in L^2(X, \mu)$, $U^{k_i} f$ converges in $L^2(X, \mu)$ to a limit function that we denote by $U^y f$. Observe that $Sy = \lim_{i \to \infty} \gamma \cdot \gamma_{k_i}$ so that

$$U^{Sy} f = \lim_{i \to \infty} U U^{k_i} f = U U^y f.$$

We consider the non-constant function g defined on $X \times Y$ by

$$g(x,y) = (U^y)^{-1} f(x),$$

and we check that g is $T \times S$-invariant since

$$\begin{aligned} g(Tx, Sy) &= (U^{Sy})^{-1} f(Tx) = (UU^y)^{-1} f(Tx) \\ &= (U^y)^{-1} U^{-1} f(Tx) \\ &= g(x,y). \end{aligned}$$

It follows that $(X, \mathscr{B}, \mu, T)$ is not mildly mixing and the theorem is proved. □

3.3.3 Multiple Mixing Properties

Another concept of mixing, emphasized by the famous H. Furstenberg 's ergodic-theoretic proof of Szemerédi's theorem [92], actually has been introduced by Rokhlin in 1949. We just do a quick review of the notions and we choose to develop one pioneer result .

Definition 3.10. An endomorphism T on the probability space $(X, \mathscr{B}, \mu)$ is said to be *weakly mixing of order* 3 if, for all $A, B, C \in \mathscr{B}$,

$$\lim_{N \to \infty} \frac{1}{N} \sum_{n<N} |\mu(A \cap T^{-\ell_n} B \cap T^{-k_n} C) - \mu(A)\mu(B)\mu(C)| = 0$$

whenever $(\ell_n), (k_n)$ satisfy $\ell_n \to \infty$ and $k_n - \ell_n \to \infty$.

Definition 3.11. An endomorphism T on the probability space $(X, \mathscr{B}, \mu)$ is said to be *strongly mixing of order* 3 if, for all $A, B, C \in \mathscr{B}$,

$$\lim_{N \to \infty} \mu(A \cap T^{-\ell_n} B \cap T^{-k_n} C) = \mu(A)\mu(B)\mu(C)$$

whenever $(\ell_n), (k_n)$ satisfy $\ell_n \to \infty$ and $k_n - \ell_n \to \infty$.

Similarly, we define multiple mixing properties of higher orders and we speak of "n-fold mixing systems" more generally.

Two-fold mixing is the same as mixing and r-fold mixing implies two-fold mixing for every $r > 1$. The question of whether two-fold mixing implies r-fold mixing for every $r > 1$ was raised by Rokhlin [216] and in its ergodic version of Szemeredi's theorem, Furstenberg proved in passing that weak mixing implies weak mixing of all orders [92]. We give a detailed proof of this fundamental theorem.

Theorem 3.7 (Furstenberg). *If $(X,\mathscr{B},\mu,T)$ is a weakly mixing system, then, for every $A_1,\ldots,A_k \in \mathscr{B}$ and $k \geq 1$,*

$$\lim \frac{1}{N}\sum_{1}^{N} |\mu(A_1 \cap T^{-n}A_2 \cap \ldots \cap T^{-(k-1)n}A_k) - \mu(A_1)\mu(A_2)\ldots\mu(A_k)| = 0$$

Remark 3.5. Observe that the weaker property

$$\lim_{N\to\infty} \frac{1}{N} \sum_{n<N} \mu(A \cap T^{-n}B \cap T^{-2n}C) = \mu(A)\mu(B)\mu(C) \text{ for all } A,B,C \in \mathscr{B}$$

already implies the weak mixing property. Actually T must be ergodic and we shall see that T has no other eigenvalue than 1. Suppose that there exists ϕ such that $\phi(Tx) = e^{i\lambda}\phi(x)$ with $|\phi| = 1$ and $e^{i\lambda} \neq 1$. In particular,

$$\int \phi(Tx)d\mu(x) = \int \phi(x)d\mu(x) = e^{i\lambda}\int \phi(x)d\mu(x),$$

and necessarily $\int \phi(x)d\mu(x) = 0$. Formulated with measurable bounded functions instead of measurable sets, the assumption says that

$$\lim_{N\to\infty} \frac{1}{N} \sum_{n<N} \int f.g\circ T^n.h\circ T^{2n}\, d\mu = \int f d\mu \int g d\mu \int h d\mu.$$

Now, by choosing $f = h = \phi$ and $g = \phi^{-2}$, we get

$$f(x)g(T^n x)h(T^{2n}x) = e^{2ni\lambda}\phi^2(x)e^{-2ni\lambda}\bar{\phi}^2(x) = 1$$

while the right integrated term vanishes, whence a contradiction.

The Furstenberg theorem can easily be deduced from the following functional version, with Tf denoting $f \circ T$ in short.

Theorem 3.8. *If $(X,\mathscr{B},\mu,T)$ is a weakly mixing system, then, for every $f_1,\ldots,f_k \in L^\infty(X,\mu)$ and $k \geq 1$,*

$$\left\|\frac{1}{N}\sum_{1}^{N} T^n f_1 \cdots T^{kn} f_k - \int f_1 d\mu \cdots \int f_k d\mu\right\|_2 \to 0.$$

Specifying $f_j = \mathbf{1}_{A_j}$ in this version, we get that

$$\lim_{N\to\infty} \frac{1}{N}\sum_{1}^{N} \mu(A_1 \cap T^{-n}A_2 \cap \cdots \cap T^{-(k-1)n}A_k) = \mu(A_1)\mu(A_2)\ldots\mu(A_k); \tag{3.8}$$

but the system is assumed to be weakly mixing and so is its cartesian product : $(X \times X, \mathscr{B} \otimes \mathscr{B}, \mu \otimes \mu, T \times T)$; applying (3.8) to the sets $A_j \times A_j$, $1 \leq j \leq k$, we obtain

$$\frac{1}{N} \sum_1^N [\mu(A_1 \cap T^{-n} A_2 \cap \cdots \cap T^{-(k-1)n} A_k)]^2 \to [\mu(A_1)\mu(A_2) \cdots \mu(A_k)]^2,$$

since $(A \times A) \cap (B \times B) = (A \cap B) \times (A \cap B)$. Furstenberg's theorem follows from a classical property of Cesaro means.

Proof. The functions may be assumed to be real. The proof goes by induction on $k \geq 1$: for $k = 1$, this is the mean ergodic theorem of von Neumann. For $k \geq 2$, consider

$$M_k(N) = \|\frac{1}{N} \sum_1^N T^n f_1 \cdots T^{kn} f_k\|_2$$

where $\int f_j \, d\mu = 0$, $1 \leq j \leq k$. We are left to prove that

$$\lim_N M_k(N) = 0$$

assuming the limit holds for $k-1$ functions. This will be a consequence of the useful van der Corput lemma (for example [206]).

Lemma 3.3 (van der Corput). *Let (u_n) be a bounded sequence in a Hilbert space H. Let*

$$\gamma_m = \limsup_N \frac{1}{N} \sum_{n=1}^N < u_n, u_{n+m} > \quad \text{for } m \geq 0.$$

If

$$\lim_{M \to \infty} \frac{1}{M} \sum_{m=1}^M |\gamma_m| = 0, \quad \textit{then} \quad \lim_{N \to \infty} \|\frac{1}{N} \sum_{n=1}^N u_n\| = 0.$$

For our purpose, we put $u_n = T^n f_1 \cdots T^{kn} f_k$ so that

$$\begin{aligned}
\gamma_m &= \limsup_N \tfrac{1}{N} \textstyle\sum_1^N \int T^n f_1 \cdots T^{kn} f_k \cdot T^{n+m} f_1 \cdots T^{k(n+m)} f_k \, d\mu \\
&= \limsup_N \tfrac{1}{N} \textstyle\sum_1^N \int (f_1.T^m f_1) \circ T^n \cdots (f_k.T^{km} f_k) \circ T^{kn} \, d\mu \\
&= \limsup_N \tfrac{1}{N} \textstyle\sum_1^N \int (f_1.T^m f_1) \cdot T^n (f_2.T^m f_2) \cdots T^{(k-1)n} (f_k.T^{km} f_k) \, d\mu
\end{aligned}$$

by measure-preserving property again. Now, making use of the inductive assumption with $\frac{1}{N} \sum_1^N T^n (f_2.T^m f_2) \cdots T^{(k-1)n} (f_k.T^{km} f_k)$ leads to

$$\begin{aligned}
\gamma_m &= \int f_1.T^m f_1 \, d\mu \cdots \int f_k.T^{km} f_k \, d\mu \\
&= \widehat{\sigma}_{f_1}(m) \cdots \widehat{\sigma}_{f_k}(km),
\end{aligned}$$

hence

$$\frac{1}{M}\sum_{m=1}^{M}|\gamma_m| \leq \frac{C}{M}\sum_{m=1}^{M}|\widehat{\sigma}_{f_k}(km)|.$$

But in turn, T^k is weakly mixing and $\frac{1}{M}\sum_{m=1}^{M}|\widehat{\sigma}_{f_k}(km)| \to 0$ as $M \to \infty$ by Wiener's criterion (chapter 1), which ends the proof. □

Remark 3.6. Partial solutions have been obtained for the multiple strong mixing but the general problem remains open [120, 131].

3.4 Discrete Ergodic Systems

In section 2.5, we gave a characterization of unitary operators with discrete spectrum; thereby, the system $(X,\mathcal{B},\mu,T)$ (or merely T) is said to have *discrete spectrum*, if there exists an orthonormal basis of $L^2(X,\mu)$ consisting of eigenfunctions of T. We studied, as an example, the irrational rotation R_θ, $\theta/\pi \notin \mathbf{Q}$ and proved that $(\mathbf{T},\mathcal{B}_\mathbf{T},m,R_\theta)$ is a discrete system. More generally, if G is a compact abelian monothetic group, the characters of G are eigenfunctions for the ergodic rotation R_g, where $\overline{\mathbf{Z}g} = G$, and generate $L^2(G,m_G)$; thus $(G,\mathcal{B}_G,m_G,R_g)$ is discrete too.

3.4.1 Von Neumann's Theorem

We shall see that the ergodic rotation is in some sense the prototype of discrete systems and, that, for such systems, the metric isomorphism and spectral isomorphism properties are equivalent [111, 240]. We start with an elementary remark.

Proposition 3.13. *An ergodic discrete system always possesses a simple spectrum.*

Proof. If the conclusion does not hold, according to corollary 2.10, there exist two U- orthogonal elements $f,g \in L^2(X,\mu)$ such that $\sigma_f \not\perp \sigma_g$. Since the system is assumed to be discrete, so are the measures σ_f and σ_g and, there must be some $\lambda \in \mathbf{T}$ for which both $\sigma_f\{\lambda\}$ and $\sigma_g\{\lambda\}$ are positive; if $P := E_\lambda$ denotes the orthogonal projection onto the eigensubspace associated to the eigenvalue $e^{i\lambda}$, the non-zero functions $F = Pf$ and $G = Pg$ are eigenfunctions in $[U,f]$ and $[U,g]$ respectively, corresponding to the same eigenvalue. But $[U,f] \perp [U,g]$ and F,G cannot be proportional. This is in contradiction with ergodicity and the result follows. □

Theorem 3.9 (von Neumann). *Let $(X,\mathcal{B},\mu,T)$ be an ergodic dynamical system. This system is discrete if and only if there exist a compact abelian metrizable group G and an ergodic rotation R on G such that $(X,\mathcal{B},\mu,T)$ and $(G,\mathcal{B}_G,m_G,R)$ are metrically isomorphic.*

Proof. $\triangleleft$ The condition is clearly sufficient.

$\triangleright$ Suppose now that $(X,\mathscr{B},\mu,T)$ is a discrete ergodic system with a countable group $\Lambda \subset \mathbf{U}$ of eigenvalues. We set $G=\widehat{\Lambda}$, the dual group of Λ endowed with the discrete topology, and, as such, G is a compact metrizable group. Define $g_0 \in G$ by $g_0(\lambda)=\lambda$ for all $\lambda \in \Lambda$, and $R := R_{g_0}$, the rotation on $G : g \mapsto g+g_0$.

We shall determine the group of eigenvalues of R.

By the Pontryagin duality theorem [218], $\widehat{G} \simeq \Lambda$ and we can consider, for each $\lambda \in \Lambda$, the function $\chi_\lambda \in L^2(G,m_G)$ defined by $\chi_\lambda(g)=g(\lambda)$. The family $\{\chi_\lambda,\ \lambda \in \Lambda\}$ is an orthonormal basis of $L^2(G,m_G)$; moreover

$$\chi_\lambda(Rg)=\chi_\lambda(g+g_0)=g(\lambda)g_0(\lambda)=\lambda\chi_\lambda(g)$$

and each χ_λ is an eigenfunction. It follows that Λ is the spectrum of R.

We now show that the system $(G,\mathscr{B}_G,m_G,R)$ is ergodic; actually, proving the density of $\mathbf{Z}g_0$ in G will be sufficient (see subsection 3.2.1). Assume that one can find $\lambda \in \Lambda$ with $\chi_\lambda(ng_0) := (g_0(\lambda))^n = 1$ for every $n \in \mathbf{Z}$; then $g_0(\lambda)=1$ and λ must be one. This proves that $\overline{\mathbf{Z}g_0}=G$ and G is monothetic; consequently, R is ergodic with m_G as the unique R-invariant measure.

To end the proof of the theorem, we are left to establish the following lemma

Lemma 3.4. *Two ergodic and discrete systems with the same eigenvalues are metrically isomorphic.*

Proof. Observe first that $(X,\mathscr{B},\mu,T)$ being a discrete system, we can construct an orthonormal basis (f_λ) of unimodular functions on X, such that, for some subset X' of X of full measure, we have

$$f_{\lambda\mu}=f_\lambda f_\mu \quad \text{and} \quad f_\lambda \circ T=\lambda f_\lambda \quad \text{on} \quad X'.$$

Indeed, for each $\lambda \in \Lambda$, we choose a measurable unimodular function h_λ on X in such a way that $h_\lambda \circ T=\lambda h_\lambda$ so that, thanks to ergodicity, $h_{\lambda\mu}=C_{\lambda,\mu}\, h_\lambda h_\mu$ *a.e.* for some constant $C_{\lambda\mu}$. Λ being countable, there exist a subset $X' \subset X$ of full measure, on which $h_{\lambda\mu}=C_{\lambda,\mu}\, h_\lambda h_\mu$. Fix now $x_0 \in X'$; the functions $f_\lambda = h_\lambda/h_\lambda(x_0)$ do the job.

Suppose now that $(X,\mathscr{B},\mu,T)$ and $(Y,\mathscr{C},\nu,S)$ are two ergodic discrete systems with the same group Λ of eigenvalues, and consider $(f_\lambda)_{\lambda\in\Lambda}$ and $(g_\lambda)_{\lambda\in\Lambda}$ two orthonormal bases of eigenfunctions in $L^2(X,\mu)$ and $L^2(Y,\nu)$ respectively (writing X instead of X', Y instead of Y'). From the Fourier-Plancherel isomorphism between $L^2(X,\mu)$ or $L^2(Y,\nu)$ and $\ell^2(\Lambda)$, arises an isomorphism $V : L^2(X,\mu) \to L^2(Y,\nu)$ with the following multiplicative property

$$V(\alpha\beta)=V(\alpha)\cdot V(\beta)$$

for every $\alpha,\beta \in L^2(X,\mu)$ such that $\alpha\beta \in L^2(X,\mu)$.

We shall use and admit the following result : *For Lebesgue spaces, every σ-algebra-isomorphism derives from an isomorphism of the spaces themselves* [111].

So we just need, using V, to construct an isomorphism ϕ between the σ-algebras $\mathscr{B}$ and $\mathscr{C}$; this will be made via indicator functions :

For $C \in \mathscr{C}$, $V(\mathbf{1}_C^2) = V(\mathbf{1}_C)^2 = V(\mathbf{1}_C)$ so that $V(\mathbf{1}_C) = \mathbf{1}_B$ for some $B \in \mathscr{B}$. Now, we define ϕ on $\mathscr{C}$ by putting $\phi(C) = B$. Note that ϕ is well defined, invertible as V and

$$\mu(C) = ||V(\mathbf{1}_C)||^2 = ||\mathbf{1}_{\phi(C)}||^2 = \mu(\phi(C))$$

by the isometry property of V. It follows that $\phi(Y) = X$ (μ-a.e.). It remains to check that ϕ preserves complementation, and countable unions. Fix $C \in \mathscr{C}$. Since $V(\mathbf{1}) = \mathbf{1}$,

$$\begin{aligned}\mathbf{1}_{\phi(C^c)} &= V(\mathbf{1}_{C^c}) = V(\mathbf{1} - \mathbf{1}_C) = \mathbf{1} - V(\mathbf{1}_C)\\ &= \mathbf{1} - \mathbf{1}_{\phi(C)} = \mathbf{1}_{\phi(C)^c}.\end{aligned}$$

Now, applying V to the identity $\mathbf{1}_{C\cup D} = \mathbf{1}_C + \mathbf{1}_D - \mathbf{1}_{C\cap D}$ where $C, D \in \mathscr{C}$, we get

$$\mathbf{1}_{\phi(C\cup D)} = \mathbf{1}_{\phi(C)} + \mathbf{1}_{\phi(D)} - \mathbf{1}_{\phi(C)}\mathbf{1}_{\phi(D)} = \mathbf{1}_{\phi(C)\cup\phi(D)}.$$

Finally, if $C_n \in \mathscr{C}$ for every n, $\mathbf{1}_{\cup_{i=1}^n C_i}$ tends to $\mathbf{1}_{\cup_{i=1}^\infty C_i}$ in $L^2(Y, \nu)$ by the monotone convergence theorem; the continuity of V ensures that

$$\mathbf{1}_{\cup_{i=1}^n \phi(C_i)} \text{ tends to } \mathbf{1}_{\cup_{i=1}^\infty \phi(C_i)}$$

and the lemma follows. □

The proof of the theorem is complete. □

Remark 3.7. This lemma raises an interesting question : for which other systems does spectral isomorphism imply metric isomorphism? This is true for some Gaussian systems [89, 237, 238]. Kwiatkowski investigates the problem for dynamical systems associated to *generalized Morse sequences* [154, 155], see also [162]; a complete answer has been given for certain substitution dynamical systems [64].

3.4.2 The Kronecker Factor

If $e^{i\lambda} \in \mathbf{U}$ is an eigenvalue of the ergodic dynamical system $(X, \mathscr{B}, \mu, T)$, the system admits a discrete factor, isomorphic to $(\mathbf{T}, \mathscr{B}_\mathbf{T}, m, R_\lambda)$, and the corresponding eigenfunction $f_\lambda : X \to \mathbf{U}$ realizes the isomorphism since $f_\lambda \circ T = R_\lambda \circ f_\lambda$.

We aim to determine the maximal discrete factor of the initial system $(X, \mathscr{B}, \mu, T)$.

Let $\mathscr{F} \subset L^2(X, \mathscr{B}, \mu)$ be a class of functions, and $\mathscr{B}_\mathscr{F}$ the sub-σ-algebra generated by $\mathscr{F}$; then $\mathscr{F} \subset L^2(X, \mathscr{B}_\mathscr{F}, \mu) \subset L^2(X, \mathscr{B}, \mu)$. If $\mathscr{F}$ is a closed sublattice of $L^2(X, \mathscr{B}, \mu)$ containing the constants, then $\mathscr{F} = L^2(X, \mathscr{B}', \mu)$ for some sub-σ-algebra $\mathscr{B}'$; actually

$$\mathscr{B}' = \{B \in \mathscr{B},\ \mathbf{1}_B \in \mathscr{F}\} \subset \mathscr{B}_\mathscr{F}.$$

(By a "sublattice", we mean a self-adjoint subspace $\mathscr{F}$, containing $\max(f,g)$ for every pair of real functions $f,g \in \mathscr{F}$).

Recall that for a unitary operator U acting on the Hilbert space H, H_d is the closed subspace generated by the eigenfunctions of U (proposition 2.15).

Proposition 3.14. *Let $(X,\mathscr{B},\mu,T)$ be a dynamical system and $H = L^2(X,\mathscr{B},\mu)$; then there exists a sub-σ-algebra $\mathscr{B}_d$ of $\mathscr{B}$, namely the sub-σ-algebra of $\mathscr{B}$ generated by eigenfunctions, such that*

$$H_d = L^2(X,\mathscr{B}_d,\mu).$$

Moreover, if E_d is the orthogonal projection onto H_d, then, for every $f \in H$,

$$E_d f = E^{\mathscr{B}_d}(f) = h_d(U)f$$

where h_d is the discrete idempotent in $\overline{\Gamma}$.

Proof. Note that the eigenfunctions of U_T are in fact in $L^\infty(X,\mu)$ so that H_d is a closed self-adjoint sub-algebra of $L^2(X,\mathscr{B},\mu)$. Moreover $1 \in H_d$, and $|f| \in H_d$ if f is already in H_d; therefore H_d is a closed sublattice of $L^2(X,\mathscr{B},\mu)$. We denote by $\mathscr{B}_d$ the sub-σ-algebra of $\mathscr{B}$ such that $H_d = L^2(X,\mathscr{B}_d,\mu)$ and E_d the orthogonal projection onto H_d. Clearly,

$$B \in \mathscr{B}_d \Longleftrightarrow \sigma_{\mathbf{1}_B} \in M_d(\mathbf{T})$$

since H_d is also $\{f \in H,\ \sigma_f \in M_d(\mathbf{T})\}$ (proposition 2.15), and $\mathscr{B}_d$ is the sub-σ-algebra of $\mathscr{B}$ generated by the eigenfunctions.

We already observed in subsection 3.3.2 that $h_d(U) = E^{\mathscr{A}}$ for some sub-σ-algebra of $\mathscr{B}$. We wish to prove that $\mathscr{A} = \mathscr{B}_d$. By definition of $A \in \mathscr{A}$, $h_d(U)\mathbf{1}_A = \mathbf{1}_A$ so that $\sigma_{\mathbf{1}_A} = h_d\sigma_{\mathbf{1}_A} \in M_d(\mathbf{T})$ which proves that $A \in \mathscr{B}_d$. Conversely, if $\sigma_{\mathbf{1}_B}$ is discrete so is $\sigma_{\mathbf{1}_B,g}$ for every $g \in L^2(X,\mu)$, then $h_d\sigma_{\mathbf{1}_B,g} = \sigma_{\mathbf{1}_B,g}$. This implies that $\sigma_{h_d(U)\mathbf{1}_B,g} = \sigma_{\mathbf{1}_B,g}$ and $< h_d(U)\mathbf{1}_B, g > = < \mathbf{1}_B, g >$ for every $g \in L^2(X,\mu)$. Thus $h_d(U)\mathbf{1}_B = \mathbf{1}_B$, and $B \in \mathscr{A}$. □

Moreover, the Dunkl-Ramirez property (proposition 1.21) shows that

$$||E_d f||^2 \leq \limsup_{n \to +\infty} |\widehat{\sigma}_f(n)|.$$

Von Neumann's theorem 3.9 admits a slight generalization under the following form

Theorem 3.10. *Let $(X,\mathscr{B},\mu,T)$ an ergodic dynamical system. There exist an ergodic rotation R and a homomorphism $\phi : (X,\mathscr{B},\mu,T) \to (G,\mathscr{B}_G,m_G,R)$ such that, $\mathscr{B}_d = \phi^{-1}(\mathscr{B}_G)$. In other words, $L^2(X,\mathscr{B}_d,\mu)$ factors through ϕ.*

Definition 3.12. The discrete system $(X,\mathscr{B}_d,\mu,T)$ is called the *Kronecker factor* of the initial one and it is the maximal discrete factor.

Therefore, every ergodic dynamical system contains an ergodic rotation as a factor, possibly reduced to constants and may be viewed as an extension over its Kronecker factor (see subsection 3.6.1).

Proposition 3.15. *Let* $(X, \mathscr{B}, \mu, T)$ *be an ergodic dynamical system. Then the maximal spectral type* σ *is* quasi-invariant *under the translations by* Λ*, the group of eigenvalues of* T *:*

$$\delta_\lambda * \sigma \sim \sigma \quad \textit{for every eigenvalue } e^{i\lambda} \text{ of } T. \tag{3.9}$$

Proof. For each eigenvalue $e^{i\lambda}$ of T, let us choose f_λ the corresponding eigenfunction normalized by $|f_\lambda| = 1$. We obtain, for every $f \in L^2(X, \mu)$,

$$\begin{aligned} U(f_\lambda f) = (f_\lambda f) \circ T &= (f_\lambda \circ T)(f \circ T) \\ &= e^{i\lambda} f_\lambda \cdot f \circ T \\ &= e^{i\lambda} f_\lambda \cdot U f. \end{aligned}$$

We deduce that the spectral family satisfies the following equation :

$$\sigma_{f_\lambda f} = \delta_\lambda * \sigma_f, \quad \text{for any } e^{i\lambda} \in \Lambda,\ f \in L^2(X, \mu). \tag{3.10}$$

Indeed,

$$\begin{aligned} \sigma_{f_\lambda f}\widehat{\ }(n) &= \langle U^n f_\lambda f, f_\lambda f \rangle \\ &= e^{in\lambda} \langle f_\lambda \cdot U^n f, f_\lambda f \rangle \\ &= e^{in\lambda} \widehat{\sigma}_f(n) \quad \text{since } |f_\lambda| = 1, \end{aligned}$$

whence the equation (3.10). The proposition follows from the definition of the maximal spectral type and the relations $\sigma_{f_\lambda f} \ll \sigma$, $\sigma_f \ll \sigma$. □

3.5 Purity Law and D-Ergodicity

A class of measures is said to verify a *purity law* with respect to a decomposition in L-spaces : $M(\mathbf{T}) = L \oplus L^\perp$, if, whenever μ in the class, either $\mu \in L$ or $\mu \in L^\perp$ (section 1.4). We already encountered a purity law with Riesz products in section 1.1. A first purity result has been proved in 1935 by Jessen and Wintner [127], concerning infinite convolution products of Bernoulli measures; they are either discrete or absolutely continuous. But, later, another type of dichotomy result has appeared in [130], involving pairs of measures in some class : any two measures must be either equivalent or mutually singular. These properties are related to ergodicity and this has been emphasized by Brown and Moran in [47].

3.5.1 Extremal Properties of Ergodic Probabilities

We start with classical dichotomy properties of ergodic probabilities. Quasi-invariant measures have already been considered in subsection 3.2.3

Proposition 3.16. *Suppose that $(X,\mathscr{B},\mu_1)$ and $(X,\mathscr{B},\mu_2)$ are probability spaces, and let T be a transformation on X.*

a) If both μ_1 and μ_2 are T-invariant and T-ergodic, then either $\mu_1=\mu_2$ or $\mu_1 \perp \mu_2$.

b) If T is invertible, and if μ_1 and μ_2 are T-quasi-invariant and T-ergodic, then either $\mu_1 \sim \mu_2$ or $\mu_1 \perp \mu_2$.

Proof. a) If $\mu_1 \neq \mu_2$, we wish to construct a T-invariant set $A \in \mathscr{I}$ realizing

$$\mu_1(A)=1 \quad \text{and} \quad \mu_2(A)=0;$$

this will establish the mutual singularity of μ_1 and μ_2. By our assumption, there exists $B \in \mathscr{B}$ such that $\mu_1(B) \neq \mu_2(B)$; applying twice Birkhoff's theorem to the function $f=\mathbf{1}_B$ with the measures μ_1 and μ_2, we get

$$\frac{1}{N}\sum_{n<N} f(T^n x) \to \mu_i(B) \quad \mu_i - a.e., \qquad \text{for } i=1,2 \tag{3.11}$$

Let A be the set of $x \in X$ for which the limit in (3.11) exists μ_1-a.e. and is equal to $\mu_1(B)$; then, $\mu_1(A)=1$ and $\mu_2(A)=0$ since $\mu_2(B)\neq\mu_1(B)$.

b) If now $\mu_1 \not\prec \mu_2$, there exists a set $A \in \mathscr{B}$ such that $\mu_1(A)=0$ and $\mu_2(A)>0$. Consider $B=\cup_{n\in\mathbf{Z}}T^{-n}A$. B is a T-invariant set which contains A so that, $\mu_2(B)$ being $\geq \mu_2(A)>0$, it must be equal to 1 by ergodicity. On the other hand, $T(\mu_1)$ being equivalent to μ_1, $\mu_1(T^{-n}A)=0$ for all $n\in\mathbf{Z}$ and $\mu_1(B)=0$. The measures are thus mutually singular. (Note that μ_1 needs not be ergodic). □

We denote by $\mathscr{P}_T$, the set of T-invariant probability measures on $(X,\mathscr{B})$, which is a convex subset of the set of all probability measures on $(X,\mathscr{B})$.

Corollary 3.4. *$\mu \in \mathscr{P}_T$ is T-ergodic if and only if μ is an extreme point of the convex set $\mathscr{P}_T$.*

Proof. Suppose that μ is T-ergodic and let $\mu=\alpha\mu_1+\beta\mu_2$ be a convex decomposition of μ, where μ_1,μ_2 belong to $\mathscr{P}_T$, $\alpha>0$, $\beta>0$ and $\alpha+\beta=1$. If $A\in\mathscr{I}$, $\mu(A)=0$ or 1 and therefore, $\mu_1(A)=\mu_2(A)=0$ or 1. This proves that μ_1 and μ_2 are T-ergodic too. We shall see that $\mu_1=\mu_2$; if not, using the previous proposition, $\mu_1\perp\mu_2$ necessarily and we can find $A\in\mathscr{I}$ with $\mu_1(A)=1$ and $\mu_2(A)=0$; this entails $\mu(A)=\alpha$ with $0<\alpha<1$, in contradiction with the ergodicity of μ. Thereby $\mu_1=\mu_2$ and μ is extremal.

Conversely, suppose that the measure $\mu\in\mathscr{P}_T$ is not ergodic, and let $A\in\mathscr{I}$ be such that $\mu(A)=\alpha\neq 0$ or 1. We set $\mu_1=\frac{1}{\alpha}\mu\cdot\mathbf{1}_A$ and $\mu_2=\frac{1}{1-\alpha}\mu\cdot\mathbf{1}_{A^c}$. Since A is an invariant set, μ_1 and μ_2 are easily seen to belong to $\mathscr{P}_T$; now $\mu=\alpha\mu_1+(1-\alpha)\mu_2$ and μ is not extremal. □

As a consequence, if $\mathscr{P}_T$ consists of a singleton, this unique probability measure is automatically ergodic. This is the case when $T = R_\theta$, an irrational rotation on $\mathbf{T}$, the unique invariant probability measure being the Lebesgue measure m; more generally, when $T = R_g$, an ergodic rotation on a monothetic compact group with its Haar measure. Whence the property below, which will play an important role in the sequel :

Definition 3.13. The dynamical system $(X, \mathscr{B}, \mu, T)$ is said to be *uniquely ergodic* if μ is the unique T-invariant probability measure on $(X, \mathscr{B})$.

3.5.2 D-Ergodicity

We now focus on the probability space $(\mathbf{T}, \mathscr{B}_\mathbf{T}, m)$; instead of iterating some rotation R_θ on the circle (or translation modulo 2π), which involves the arithmetical nature of θ, G. Brown and W. Moran in [49] considered more general countable subgroups of $\mathbf{T}$ acting by translation on $\mathbf{T}$. This leads, in this context, to new dichotomy properties.

Let μ be a probability measure on $\mathbf{T}$ and D a countable subgroup of $\mathbf{T}$.

Definition 3.14. μ is said to be *D-quasi-invariant* if $\delta_d * \mu \ll \mu$ for every $d \in D$.

If $D = D(\alpha)$, the subgroup generated by some α with $\alpha/\pi \notin \mathbf{Q}$, such a measure is exactly what we called a R_α-quasi-invariant measure.

Definition 3.15. μ is said to be *D-ergodic* (without assuming the D-quasi-invariance of μ) if $\mu(A) = 0$ or 1 for every D-invariant set A (i.e. $A + d = A$ for every $d \in D$).

We have for such measures the analogue of the proposition 3.16 b) :

Proposition 3.17. *If μ_1 and μ_2 are two D-quasi-invariant and D-ergodic probability measures on $\mathbf{T}$, then either $\mu_1 \sim \mu_2$ or $\mu_1 \perp \mu_2$.*

Remark 3.8. A probability measure μ in an obvious way is $H(\mu)$-quasi-invariant, since $H(\mu)$ is precisely the subgroup of $x \in \mathbf{T}$ such that $\mu * \delta_x \sim \mu$! But $H(\mu)$ may happen to be reduced to $\{0\}$ (subsection 3.2.3).

With some D-ergodic probability measure μ on $\mathbf{T}$, we are sometimes led to associate a D-quasi-invariant probability measure by putting

$$\lambda = \sum_{d_n \in D} 2^{-n} \delta_{d_n} * \mu. \tag{3.12}$$

This measure λ is still D-ergodic, and, since obviously $\mu \ll \lambda$, μ inherits properties from λ.

The following purity law, with respect to L-spaces, is due to Brown and Moran [49].

Definition 3.16. An L-space L of $M(\mathbf{T})$, is said to be *D-invariant*, if $\mu * \delta_d \in L$ for every $d \in D$ and every $\mu \in L$. Notice that $L^{\perp}$ is then D-invariant too.

Proposition 3.18. *Let μ be a D-ergodic probability measure on* $\mathbf{T}$ *and let L be a D-invariant L-space of $M(\mathbf{T})$. Then either $\mu \in L$ or $\mu \in L^{\perp}$.*

Proof. $*$ We first assume that μ is D-quasi-invariant. If $\mu = \tau + \sigma$ is the unique decomposition of μ in $L \oplus L^{\perp}$, we claim that τ and σ are still D-quasi-invariant : indeed, if $d \in D$, $\mu * \delta_d = \tau * \delta_d + \sigma * \delta_d$ is a decomposition of $\mu * \delta_d$ in $L \oplus L^{\perp}$, by the D-invariance of L and $L^{\perp}$. Now $\mu * \delta_d \ll \mu$, say $\mu * \delta_d = f\mu$, and, by the L-property of an L-space, $f\tau + f\sigma$ is another decomposition of $\mu * \delta_d$ in $L \oplus L^{\perp}$. It follows that

$$\tau * \delta_d \ll \tau \quad \text{and} \quad \sigma * \delta_d \ll \sigma,$$

which was claimed.

We are able to conclude now : since $\sigma \perp \tau$, there exists a Borel set A such that $\sigma(A) = 0$ and $\tau(A^c) = 0$ and we consider $B = \cup_{d \in D}(A + d)$; by the D-quasi-invariance of σ, $\sigma(B) = 0$ and by the D-ergodicity of μ, $\mu(B) = 0$ or 1 whence the discussion :
If $\mu(B) = 0$, then $\mu(A) = 0$ and τ, supported on A, is the zero measure. In this case, $\mu \in L$. Otherwise $\mu(B^c) = 0$ and $\sigma(B^c) = 0$ so that σ is the zero measure and $\mu \in L^{\perp}$.

$*$ If μ is not D-quasi-invariant, we replace μ by λ according to (3.12), and we obtain either $\lambda \in L$ or $\lambda \in L^{\perp}$; now $\mu \ll \lambda$ gives the same conclusion. □

As a consequence, we get a first purity law for D-ergodic probability measures.

Corollary 3.5. *Let μ be a D-quasi-invariant and D-ergodic probability measure on* $\mathbf{T}$*; then μ is either discrete, or continuous singular, or equivalent to the Lebesgue measure on* $\mathbf{T}$.

Proof. Taking $L = M_d(\mathbf{T})$ in the foregoing proposition gives the dichotomy : μ is either discrete or continuous. Now, in the same way, the decomposition of the continuous measures $M_c(\mathbf{T}) = M_{cs}(\mathbf{T}) \oplus L^1(\mathbf{T})$ into L-subspaces (1.20) leads to the following : if μ is continuous, μ is either continuous singular or absolutely continuous with respect to the Lebesgue measure on $\mathbf{T}$.

We have to discuss on D to get the desired conclusion : If D is an infinite subgroup of $\mathbf{T}$, D is dense in $\mathbf{T}$ and the Lebesgue measure is D-ergodic; applying proposition 3.17, we deduce that μ must be equivalent to the Lebesgue measure if it is not singular. If now D is a finite subgroup of $\mathbf{T}$, any D-quasi-invariant and D-ergodic probability measure is necessarily discrete. The proof is complete. □

Remark 3.9. Brown and Moran [49] proved in fact a strengthened result that we will admit.

Theorem 3.11 (Brown & Moran). *Let μ be a D-ergodic probability measure on* $\mathbf{T}$*. Then :*
either $\mu \in M_d(\mathbf{T})$,
or $\mu^n \in L^1(\mathbf{T})$ for some integer $n \geq 1$,
or μ is continuous singular, satisfying : $\delta_x * \mu^n \perp \mu^m, \quad \forall x \in \mathbf{T}, \ \forall n \neq m.$

In [125] Host and Parreau obtained a simpler proof of this generalized purity law by proving the following equivalent statement

Theorem 3.12 (Host & Parreau). *Let μ be a D-ergodic probability measure on* $\mathbf{T}$ *such that μ and $\mu * \mu$ are not mutually singular. Then μ is either discrete or absolutely continuous.*

3.5.3 Applications of Purity Laws

1. We rediscover from proposition 3.16 that the auto-similar Riesz product

$$\rho = \prod_{n \geq 0} (1 + r \cos q^n t)$$

with $0 < |r| \leq 1$ and $q \geq 2$, is singular.
2. Let us return now to the so-called generalized Riesz products which we introduced in section 1.3 :

$$\rho = \prod_{n \geq 0} P(q^n t)$$

where $P(t) = \frac{1}{q} \left| \sum_{k=0}^{q-1} z_k e^{ikt} \right|^2$, $q \geq 2$, $|z_k| = 1$ and $z_0 = 1$.

Proposition 3.19. *Denote by D the countable subgroup of* $\mathbf{T}$ *generated by* $(\frac{2\pi}{q^n},\ n \geq 0)$. *Then ρ is D-ergodic, thus either discrete or continuous singular; in this latter case ρ is D-quasi-invariant.*

Proof. For each $N \geq 0$, we consider D_N the subgroup generated by $\frac{2\pi}{q^N}$, and ω_N the normalized Haar measure of D_N; recall that $\widehat{\omega}_N(k) = 0$ if $q^N \not| \, k$ and $= 1$ otherwise. Hence, for each $N \geq 0$,

$$\rho = \prod_{n=0}^{N-1} P(q^n t) \cdot \prod_{n \geq N} P(q^n t) =: P_N \cdot \rho_N$$

and ρ satisfies the following identity

$$\rho = P_N(\rho * \omega_N). \tag{3.13}$$

Let now A be a D-invariant set in $\mathbf{T}$; we shall prove that $\mathbf{1}_A$ is constant ρ-a.e. For every $d \in D$, and $t \in \mathbf{T}$,

$$\mathbf{1}_A(t) = \mathbf{1}_{A-d}(t) = \mathbf{1}_A(t+d),$$

so that

$$\mathbf{1}_A(\rho * \delta_d) = (\mathbf{1}_A \, \rho) * \delta_d$$

and, ω_N being a finite combination of point masses δ_d,

$$\mathbf{1}_A(\rho * \omega_N) = (\mathbf{1}_A\, \rho) * \omega_N. \tag{3.14}$$

Combining (3.13) with (3.14), we obtain

$$\mathbf{1}_A\rho = P_N\big((\mathbf{1}_A\, \rho) * \omega_N\big), \quad \forall N \geq 0.$$

We reproduce the computations performed in section 1.3, with now $\rho_A := \mathbf{1}_A\rho$ instead of ρ, to get, for $k \leq q^N$,

$$\widehat{\rho}_A(k) = \widehat{P}_N(k)\widehat{\rho}_A(0) + \widehat{P}_N(k - q^N)\widehat{\rho}_A(q^N);$$

letting N go to infinity, $\widehat{P}_N(k)$ tends to $\hat{\rho}(k)$ and $\widehat{P}_N(k-q^N)$ to zero so that, finally,

$$\widehat{\rho}_A(k) = \hat{\rho}(k)\widehat{\rho}_A(0) \quad \text{for all } k \geq 0. \tag{3.15}$$

Observe that $\widehat{\rho}_A(0) = \rho(A)$ and (3.15) means that $\mathbf{1}_A\rho = C\rho$ where the constant $C = \rho(A)$, whence $\mathbf{1}_A = C$, ρ-a.e. This implies $C = 0$ or 1 and ρ is D-ergodic. For proving the dichotomy, we just have to remark that ρ can never be equivalent to the Lebesgue measure : since both measures are q-invariant, corollary 3.5 gives the conclusion.

Finally, if ρ is continuous, so is ρ_N for every $N \geq 0$, so that $P_N(t) \neq 0$ ρ_N-a.e. This implies $\rho_N \ll \rho$, and, if $d \in D_N$,

$$\delta_d * \rho \ll \delta_d * \rho_N = \rho_N \ll \rho$$

which completes the proof. □

3. Let us turn back to the operator W defined on $L^2(X,\mu)$ by

$$Wf(x) = \varphi(x)f(Tx)$$

T being an ergodic rotation on the metric compact abelian group X with its Haar measure μ, and $\varphi : X \to \mathbf{U}$ a measurable function.

Proposition 3.20. *The maximal spectral type of W is Λ-quasi-invariant and Λ-ergodic, where Λ denotes the subgroup of eigenvalues of T.*

Proof. Without loss of generality we assume $T := R_\alpha$, an irrational rotation on $\mathbf{T}$. The maximal spectral type of W is well-known (subsection 2.5.3) :

$$\sigma_{\max} \sim \sum_{k \in \mathbf{Z}} 2^{-|k|}\, \sigma * \delta_{k\alpha},$$

where σ is the spectral measure of the function 1, whose Fourier coefficients are

$$\widehat{\sigma}(n) = \int_{\mathbf{T}} \varphi^{(n)}(x) \, \frac{dx}{2\pi}.$$

Recall the fundamental relation satisfied by the spectral family

$$\sigma_{fe^{ix}} = \sigma_f * \delta_\alpha. \tag{3.16}$$

We just need to prove that σ is T-ergodic.

Let A be a T-invariant Borel set in $\mathbf{T} : A + \alpha = A$; we aim to get finally $\sigma(A) = 0$ or 1. If $P := \mathbf{1}_A(U)$ denotes the orthogonal projection in $L^2(\mathbf{T})$, we have

$$< \mathbf{1}_A(U)g, g >= \sigma_g(A)$$

and, by (3.16),

$$< \mathbf{1}_A(U)(e^{ix}g), e^{ix}g >= \sigma_{e^{ix}g}(A) = (\sigma_g * \delta_\alpha)(A) = \sigma_g(A + \alpha).$$

A being T-invariant, we get

$$< e^{-ix}P(e^{ix}g), g >= \sigma_g(A) =< Pg, g > .$$

We deduce that P commutes with the operator of multiplication by e^{ix}. From proposition 2.6, P is itself an operator of multiplication, namely, by $\mathbf{1}_E$, for some Borel set E in $\mathbf{T}$.

We can precise the link between E and A. On the one hand,

$$m(E) =< \mathbf{1}_E, 1 >=< P(1), 1 >$$

while

$$< P(1), 1 >=< \mathbf{1}_A(U)1, 1 >= \sigma(A),$$

on the other hand.

Therefore we get the proposition if we prove that E in turn is T-invariant. To that effect, we use the fact that P and U are commuting; thus,

$$PU(1) = P(\varphi) = \mathbf{1}_E\varphi$$

is also equal to

$$UP(1) = U(\mathbf{1}_E) = \varphi \cdot \mathbf{1}_E \circ T.$$

It follows that $\mathbf{1}_E = \mathbf{1}_E \circ T$ and, by ergodicity of T, $m(E) = 0$ or 1, which remained to be proved. □

Corollary 3.6. *The maximal spectral type of W (as well as σ) is either discrete or continuous singular or absolutely continuous.*

This purity law has already been observed by H. Helson [113].

3.6 Group Extensions

Let $(X, \mathscr{B}, \mu, T)$ be a dynamical system, T being invertible, and let G be a metric compact abelian group equipped with its Haar measure m_G.

Definition 3.17. We call an *additive cocycle* any measurable map $\phi : X \times \mathbf{Z} \to G$ satisfying, for every $x \in X$ and $n, m \in \mathbf{Z}$, $\phi(x, n+m) = \phi(x, n) + \phi(T^n x, m)$.

This definition can easily be extended to more general group actions. In case of a $\mathbf{Z}$-action, note that $\phi(\cdot, n)$ is nothing but a formalization of a Birkhoff's sum; indeed, the cocycle is determined by the knowledge of

$$\varphi(x) = \phi(x, 1)$$

and the identities for $n > 0$:

$$\varphi^{(n)}(x) := \phi(x, n) = \varphi(x) + \varphi(Tx) + \cdots + \varphi(T^{n-1}x), \quad \varphi^{(-n)}(x) = -\varphi^{(n)}(T^{-n}x);$$

and also $\varphi^{(0)}(x) = 0$. We keep in mind the *cocycle relation*

$$\varphi^{(n+m)}(x) = \varphi^{(n)}(x) + \varphi^{(m)}(T^n x), \quad x \in X,\ n \in \mathbf{Z}. \tag{3.17}$$

A cocycle φ is a *coboundary* if there exists a measurable function $f : X \to G$ such that

$$\varphi = f \circ T - f.$$

Those definitions are easily transposed in the multiplicative context (see subsection 2.5.3).

We have already introduced the transformation T_φ on the product space $(X \times G, \mathscr{B} \otimes \mathscr{B}_G, \mu \otimes m_G)$ defined by

$$T_\varphi(x, z) = (Tx, z + \varphi(x))$$

that we called a *G-extension* or a *skew-product* (subsection 3.2.1). Those systems have been intensively studied in view to exhibit dynamical systems with various prescribed spectral properties. We focus afterwards on particular cases.

Remark 3.10. At the periphery of these problems we quote an interesting problem by Anosov : we consider the irrational rotation R_α on $\mathbf{T}$ and the coboundary equation

$$f(x+\alpha) - f(x) = \varphi(x) \quad \text{with } \varphi \in L^1(\mathbf{T}). \tag{3.18}$$

The questions raised are the following :

1) Does there exist a measurable f solving this equation?
2) If such a solution happens to exist, does f inherit any of the properties of g?

Actually, equation (3.18) may fail to have measurable solutions, even when $\varphi \in L^2(\mathbf{T})$, but if f is a solution in $L^1(\mathbf{T})$, or simply measurable, then φ must have zero mean. If one supposes that $f \in L^1(\mathbf{T})$, we obtain by identification of Fourier coefficients, that

$$\hat{f}(k) = \frac{\hat{\varphi}(k)}{1 - e^{ik\alpha}}, \quad k \in \mathbf{Z}.$$

Anosov has constructed an analytic function φ and α with $\alpha/\pi \notin \mathbf{Q}$ for which (3.18) has a measurable though non-integrable solution.

M. Herman proved that this cannot happen when φ is a lacunary Fourier series : a measurable solution necessarily belongs to $L^2(\mathbf{T})$.

3.6.1 Two-Points Extensions of an Irrational Rotation

In this subsection, we restrict ourselves to the irrational rotation : $(\mathbf{T}, \mathscr{B}_{\mathbf{T}}, m, R_\alpha)$ with $\alpha/\pi \notin \mathbf{Q}$, and we assume that φ is a non-constant, ± 1-valued function. In other words, G is the multiplicative group $\{-1, 1\}$ equipped with the probability measure $\mu = \frac{1}{2}(\delta_1 + \delta_{-1})$, and the skew-product T_φ is defined on $\mathbf{T} \times \{-1, 1\}$ by

$$T_\varphi(x, y) = (x + \alpha \; (\operatorname{mod} 2\pi), \; \varphi(x) \cdot y).$$

Note that T_φ is ergodic if and only if there is no measurable ± 1-valued solution F to the equation : $F(x + \alpha) = \varphi(x) F(x)$ (Anzai's criterion).

We wish to achieve the spectral analysis of the operator associated to the skew-product T_φ.

Let

$$H_1 = \{f \in L^2(\mathbf{T} \times G, m \times \mu), \; f(x, -y) = f(x, y)\}$$

and

$$H_2 = \{f \in L^2(\mathbf{T} \times G, m \times \mu), \; f(x, -y) = -f(x, y)\}, \tag{3.19}$$

so that $H = H_1 \oplus^\perp H_2$; we are thus reduced to study $U := U_{T_\varphi}$ in restriction to the U-invariant subspaces H_1 and H_2.

But, if $f \in H_1$, f is independent of y and U is conjugate to the irrational rotation operator; thus $U_{|H_1}$ has a discrete spectrum. If $f \in H_2$, then $f(x, y) = y f(x, 1)$, and $U f(x, y) = \varphi(x) \cdot y f(x + \alpha, 1)$; we define $\Phi : H_2 \to L^2(\mathbf{T})$ by $\Phi f = F$ where

$$F(x) = y f(x, y);$$

it follows that, under Φ, U is conjugate on H_2 to the well-known operator W defined on $L^2(\mathbf{T})$ by

$$Wf(x) = \varphi(x)f(x+\alpha).$$

We deduce that the maximal spectral type of U on H_2

$$\sigma_{\max} \sim \sum_{k\in\mathbf{Z}} 2^{-|k|} \, \sigma * \delta_{k\alpha},$$

where σ is the spectral measure of the function 1, with Fourier coefficients :

$$\widehat{\sigma}(n) = \int_{\mathbf{T}} \varphi^{(n)}(x) \, \frac{dx}{2\pi}.$$

Thus, the maximal spectral type of U is the sum of the discrete spectral type of $T = R_\alpha$ and the pure spectral type of W (corollary 3.6), and the multiplicity function is the sum of both multiplicity functions.

The first elementary results on the spectral invariants of W (or U) on H_2 are gathered in the following propositions :

Proposition 3.21. *The multiplicity function of U is constant.*

Proof. If U has a purely discrete spectrum, U has simple spectrum if T_φ is ergodic (3.4.1), and has multiplicity 2 if not. If W has a continuous (reduced) spectrum, then U on H_2 (3.19) and W have the same multiplicity function which is a spectral invariant, so that we just have to examine the multiplicity function M of W. Instead of identity (3.16) we prefer the equivalent operator one :

$$WV = e^{i\alpha}VW, \tag{3.20}$$

where V is as usual the operator of multiplication by e^{ix}. Since W and $e^{i\alpha}W$ are conjugate under V, so are their spectral projectors.

Let now m be an integer such that $B = \{M(t) = m\}$ has a positive $\sigma_{\max}$-measure, and E_B the corresponding spectral projector. From (3.20) we derive

$$E_B V = V E_{B-\alpha}$$

and $M(t) = m$ when $t \in B - \alpha$ also. It follows that B is a T-invariant set and, σ as well as $\sigma_{\max}$ being ergodic, $\sigma_{\max}(B) = 1$. □

Proposition 3.22. *If σ is discrete, σ is supported by $\mathbf{Z}\alpha$ or $\alpha/2 + \mathbf{Z}\alpha$. If σ is continuous singular, $\sigma * \sigma \perp \sigma$. The same holds for the maximal spectral type $\sigma_{\max}$ of W.*

Proof. If there exists $\lambda \in \mathbf{T}$ such that $\sigma\{\lambda\} \neq 0$, then $\sigma(\lambda + \mathbf{Z}\alpha) = 1$ by ergodicity of σ, and σ is supported by $\lambda + \mathbf{Z}\alpha$; now, the Fourier transform $\widehat{\sigma}$ is real-valued just like φ, and σ is symmetric : $\sigma(-B) = \sigma(B)$. It follows that $2\lambda \in \mathbf{Z}\alpha$, then $\lambda = 0$ or $\alpha/2$ modulo π.

Otherwise, by purity law, σ is absolutely continuous or continuous singular; in this last case, $\sigma * \sigma \perp \sigma$ (theorem 3.12). □

From now on, we specify the cocycle φ to be the step function

$$\varphi(x) = \begin{cases} -1 & \text{if} \quad 0 \leq x < \beta \\ 1 & \text{if} \quad \beta \leq x < 2\pi \end{cases}$$

β being a fixed number in $[0, 2\pi)$. These transformations $T_\varphi =: T_\beta$ have been introduced by Katok and Stepin [140] in view to exhibit, by exploiting the parameters, counter-examples to the famous "group law conjecture" [111]: *does a maximal spectral type τ of an ergodic dynamical system always obey the relation : $\tau * \tau \ll \tau$?*

General results have been obtained previously. It remains to decide the nature of the spectrum according to arithmetical properties of α and β. When is T_β ergodic? When is the spectrum discrete or continuous singular? Veech described in [236] a set of β for which T_β has a discrete spectrum (see also [185]). One of his results is the following : Suppose that α has bounded partial quotients (see Appendix B); then T_β has a discrete spectrum if and only if T_β is not ergodic.

After Katok and Stepin, many contributions have flourished, that make use of the cyclic approximation method [100, 212, 213], but they were the first to exhibit a large class of pairs (α, β) for which σ is a continuous Dirichlet, and thus singular, measure. In those cases, T_β has a singular spectrum.

Theorem 3.13 (Katok & Stepin). *Suppose that there exist infinitely many irreducible rational numbers (p_n/q_n), (r_n/q_n) such that the r_n are odd integers, and*

$$(1) \qquad |\alpha - 2\pi \frac{p_n}{q_n}| = o(\frac{1}{q_n^2}),$$

$$(2) \qquad |\beta - 2\pi \frac{r_n}{q_n}| = o(\frac{1}{q_n}).$$

Then $\widehat{\sigma}(q_n) \to -1$ and σ is a continuous Dirichlet measure. As a consequence, the spectrum of T_β is singular.

Proof. Recall that σ is the spectral measure of 1 of the operator W acting on $L^2(\mathbf{T})$; σ being a probability measure, the property $\widehat{\sigma}(q_n) \to -1$ is equivalent to the apparently stronger one

$$||\gamma_{q_n} + 1||_{L^2(\sigma)} \to 0,$$

which means that $||W^{q_n}1 + 1||_2$ tends to zero.

We make use of the rational approximations (1) and (2) to get an approximation of

$$\widehat{\sigma}(q_n) = \int_{\mathbf{T}} \varphi(x) \cdots \varphi(x + (q_n - 1)\alpha)\, dm(x).$$

We consider

$$\theta_n(x) = \prod_{j=0}^{q_n - 1} \varphi(x + 2j\pi \frac{p_n}{q_n})$$

where the q_n points $x+2j\pi\frac{p_n}{q_n}$, $j=0,\ldots,q_n-1$, are distinct; we have

$$\int_{\mathbf{T}} |\varphi(x+2j\pi\frac{p_n}{q_n}) - \varphi(x+j\alpha)|\, dm(x) \leq 2j|\alpha - 2\pi\frac{p_n}{q_n}|$$

so that

$$\int_{\mathbf{T}} |\varphi^{(q_n)} - \theta_n|\, dm \leq \sum_{j=0}^{q_n-1} \int_{\mathbf{T}} |\varphi(x+2j\pi\frac{p_n}{q_n}) - \varphi(x+j\alpha)|\, dm(x) \leq q_n^2|\alpha - 2\pi\frac{p_n}{q_n}|,$$

and this last term tends to zero by (1).

Now consider

$$\omega(x) = \begin{cases} -1 & \text{if } 0 \leq x < 2\pi\frac{r_n}{q_n} \\ 1 & \text{if } 2\pi\frac{r_n}{q_n} \leq x < 2\pi \end{cases}$$

and

$$\omega_n(x) = \prod_{j=0}^{q_n-1} \omega(x+2j\pi\frac{p_n}{q_n}).$$

In the same way,

$$\int_{\mathbf{T}} |\omega_n - \theta_n|\, dm \leq 2q_n|\beta - 2\pi\frac{r_n}{q_n}|,$$

and this tends to zero by (2).

But, ω_n is easily seen to be constantly equal to $(-1)^{r_n} = -1$, since one assumed r_n to be odd. Therefore,

$$\int_{\mathbf{T}} |\varphi^{(q_n)} + 1|\, dm = \int_{\mathbf{T}} |\varphi^{(q_n)} - \omega_n|\, dm \to 0$$

which was to be proved. Obviously σ is supported neither by $\mathbf{Z}\alpha$ nor by $\alpha/2+\mathbf{Z}\alpha$ and σ is a continuous Dirichlet measure. Remembering that the maximal spectral type of U on H_2, $\sigma_{\max}$, is equivalent to the probability measure $\sum_{k\in\mathbf{Z}} c_k\ \sigma * \delta_{k\alpha}$, we see that $\widehat{\sigma}_{\max}(q_n) = \sum_{k\in\mathbf{Z}} 2^{-|k|} e^{ik\alpha q_n}\widehat{\sigma}(q_n)$ in turn tends to -1 and $\sigma_{\max}$ is a continuous Dirichlet measure. In particular T_β is mildly mixing for almost all β. □

Moreover, the authors proved that the multiplicity of T_β is equal to one or two, and is one if and only if T_β is ergodic.

The singularity of the spectrum in the general context has been proved by M. Guénais [106]:

Theorem 3.14. *Let $T = R_\alpha$ be an irrational rotation, and $\beta \in [0,2\pi)$. Then T_β has a singular spectrum as soon as it is ergodic.*

Conditions on α and β for ergodicity of the skew product have been found very recently and we state the beautiful result obtained by M. Guénais and F. Parreau :

Theorem 3.15. *Let $T = R_\alpha$ be an irrational rotation, and $\beta \in [0, 2\pi)$. Let $(q_n) := (q_n(\alpha))$ be the sequence of the denominators of α. Define*

$$H_1(\alpha) = \{x \in \mathbf{T},\ \sum_{n\geq 1} ||q_n x|| < \infty\}$$

where $||\cdot||$ is the distance to the nearest integer. Then T_β has discrete spectrum if and only if

$$\beta \in 2H_1(\alpha) + \mathbf{Z} + \mathbf{Z}\alpha;$$

T_β is ergodic if and only if

$$\beta \notin 2H_1(\alpha).$$

3.6.2 Two-Points Extensions of an Odometer

Consider the set $\{0,1\}^{\mathbf{N}} = \{(\omega_j)_{j\geq 0},\ \omega_j = 0,1\}$ with addition modulo 2 on each component and carry over; endowed with the product of the discrete topologies on each factor, this is a compact group sometimes denoted by $\mathbf{Z}_2$, the 2-adic integers. The Borel σ-algebra is generated by the cylinders sets :

$$[\alpha_0\alpha_1\cdots\alpha_k] := \{\omega,\ \omega_j = \alpha_j,\ 0 \leq j \leq k\}$$

where $\alpha_j = 0,1$ are fixed and $k \geq 0$. We consider now the transformation τ defined on $\{0,1\}^{\mathbf{N}_0}$ by

$$\tau\omega = \omega + 1, \quad \text{where } 1 = (1,0,0,\ldots).$$

The probability measure $\mu = \otimes_j((\delta_0 + \delta_1)/2)$ is preserved by τ.

Definition 3.18. The dynamical system $(\mathbf{Z}_2, \mathscr{B}, \mu, \tau)$ is called the *two-odometer* or the *two-adding machine*.

Observe that $\tau(1,1,\ldots) = 0$ so that -1 can be identified to $(1,1,\ldots)$, then $-2 = (0,1,1,\ldots)$ and so on. We deduce that $\mathbf{Z}_2$ contains $\mathbf{Z}$ as a dense subgroup and thus, it is a monothetic compact group. The two-odometer has discrete spectrum and the group of its eigenvalues is

$$\Lambda = \{e^{2i\pi p 2^{-k}},\ k \geq 1,\ 1 \leq p \leq 2^k\}.$$

If $q \geq 2$, the q-odometer can be defined in the same way (see subsection 9.1.1)

Some dynamical systems arising from arithmetic sequences will appear as extensions over an odometer and we just mention in this subsection a pioneer result of Ledrappier that we revisit in a next chapter [158]. We keep the notations of the section : if $\varphi : \mathbf{Z}_2 \to \{\pm 1\}$ is measurable, τ_φ denotes the skew-product defined on $\mathbf{Z}_2 \times \{\pm 1\}$ by

$$\tau_\varphi(x, \varepsilon) = (\tau x, \varepsilon\varphi(x)),$$

and W, the associated operator defined on $L^2(\mathbf{Z}_2)$ by

$$W(f) = \varphi \cdot f \circ \tau.$$

Theorem 3.16 (Ledrappier). *Let us consider φ the step function defined by*

$$\varphi = (-1)^{m+1} \quad \textit{on the cylinder set } [\underbrace{1\cdots 1}_{m}0].$$

Then the maximal spectral type of the associated operator W is the generalized Riesz product

$$\rho = \prod_{n=0}^{\infty}(1 - \cos(2^n t)).$$

Chapter 4
Dynamical Systems Associated with Sequences

In this chapter, we recall elementary facts on topological dynamical systems, and we focus later on a particular case of such systems naturally associated with a bounded sequence. When the sequence takes its values in a finite set, we give an explicit characterization of the dynamical properties of the system. The last section is devoted to the spectral study of sequences with the aid of correlation measures, and to a quick survey on uniform distribution and related sets of integers.

4.1 Topological Dynamical Systems

In this section, X is a compact metric space, and T a continuous map from X to itself; we say in short that T is a *transformation* of X. The pair (X,T) is usually called a *topological dynamical system.* If $x \in X$, $O(x)$ denotes the *orbit* of x under T, which is the set of all $T^n x$, with n a non-negative integer. These sets are invariant in the following sense :

$$\text{If } E := O(x), \quad T(E) \subset E.$$

We shall use a different terminology and say that such a set E is *topologically T-invariant*, that E is *strongly T-invariant* if $T(E) = E$, and just T-invariant if $T^{-1}E = E$, to be consistent with the measurable case. Since W. Gottschalk's results [104], topological dynamics developed as an independent discipline, but we shall rather be interested in topological systems from a measure-theoretical point of view.

4.1.1 Minimality and Topological Transitivity

We now introduce topological analogues of ergodic properties in measurable dynamics.

Definition 4.1. (X,T) is said to be *minimal* (or just T is minimal) if the only closed topologically T-invariant sets in X are X and $\emptyset$.

M. Queffélec, *Substitution Dynamical Systems – Spectral Analysis: Second Edition*, Lecture Notes in Mathematics 1294, DOI 10.1007/978-3-642-11212-6_4,
© Springer-Verlag Berlin Heidelberg 2010

It is quite easy to establish the following characterization :

Proposition 4.1. *(X,T) is minimal if and only if $O(x)$ is dense in X for every $x \in X$.*

Remark 4.1. 1. If X is a finite set, "(X,T) is minimal" means that X consists of a single orbit. If X is an infinite set and (X,T) is minimal, then T has no periodic points, since the orbit of any periodic point is a finite set and is by no means dense in X.

2. Suppose that (X,T) is minimal and $f \in C(X)$, the space of all complex-valued continuous functions; if f is T-invariant on X, then f is constant on the dense subsets $O(x)$, $x \in X$, thus constant on X by continuity.

Notice that the second remark remains valid as soon as there exists *one* $x \in X$ with a dense orbit. Thus the less restrictive definition :

Definition 4.2. T is said to be *topologically transitive* on X if $O(x)$ is dense in X for some $x \in X$.

In this case, it can be shown that, in fact, the set of $x \in X$ with a non-dense orbit is a set of first category in the Baire sense [241].

Theorem 4.1 (Gottschalk & Hedlund). *Let T be a minimal transformation of the compact metric set X, and $g \in C(X)$. The following assertions are equivalent :*
(a) $g = f \circ T - f$, for some $f \in C(X)$.
(b) There exists $x_0 \in X$ for which

$$\sup_n \Big| \sum_{j=0}^{n-1} g \circ T^j(x_0) \Big| < \infty.$$

Proof. Assuming (b), we have to construct $f \in C(X)$ realizing (a). We consider the continuous function $F : X \times \mathbf{R} \to X \times \mathbf{R}$ defined by

$$F(x,t) = (Tx,\ t + g(x)).$$

$F^n(x,t) = (T^n x,\ t + \sum_{j<n} g(T^j x))$ and, under our assumption, the F-orbit of (x_0,t) is totally bounded for every real t. In particular, $K = \overline{\{F^n(x_0,t), n \geq 0\}}$ is a topologically F-invariant compact subset of $X \times \mathbf{R}$; it must then contain a topologically F-invariant compact set M such that the system (M,F) is minimal.

We claim that M is a graph : If $\pi : X \times \mathbf{R} \to X$ is the projection onto X, we have to prove that $\pi_{|M} : M \to X$ is a bijection. Since $F(M) \subset M$, $T(\pi(M)) \subset \pi(M)$ in turn; but (X,T) is minimal and $\pi(M) = X$, i.e. $\pi_{|M}$ is onto. For every $a \in \mathbf{R}$, the translation $\tau_a : (x,t) \mapsto (x, t+a)$ commutes with F so that the system $(\tau_a M, F)$ is minimal; suppose now that $\pi_{|M}$ fails to be one-to-one; then, there exists $a \in \mathbf{R}$, $a \neq 0$, such that $M \cap \tau_a M \neq \emptyset$; but $M \cap \tau_a M$ is still an F-invariant compact subset of both M and $\tau_a M$, thus $M = \tau_a M$; to get a contradiction, we observe that $M = \tau_a^k M = \tau_{ka} M$ for all $k \geq 1$ and M cannot be bounded.

Consequently, M is a graph, more precisely, the graph of a continuous function f by the closed graph theorem. Since $F(M) \subset M$, for every $x \in X$ one can find $y \in X$ such that

$$F(x, f(x)) = (y, f(y));$$

remembering that

$$F(x, f(x)) = (Tx, g(x) + f(x)),$$

we must have

$$y = Tx, \quad \text{and} \quad g(x) + f(x) = f(Tx)$$

which was to be proved. □

Turning back to the coboundary equation for the irrational rotation, we deduce the following [114].

Corollary 4.1 (M. Herman). *Let α be such that $\alpha/\pi \notin \mathbf{Q}$ and $\varphi \in C(\mathbf{T})$. Then, any L^∞-solution of the equation*

$$f(x+\alpha) - f(x) = \varphi(x), \ \textit{for almost all } x \in \mathbf{T}$$

must be a.e. equal to a continuous function.

Proof. Consider the functions $f_k : x \mapsto f(x+k\alpha) - f(x)$, $k \geq 0$; since $f_k = \varphi^{(k)} := \sum_{j<k} \varphi \circ R_\alpha^j$ where R_α is the irrational rotation, each f_k is continuous and the sequence (f_k) is bounded in $C(\mathbf{T})$; by the previous theorem, there exists $g \in C(\mathbf{T})$ such that

$$g(x+k\alpha) - g(x) = f_k(x) \text{ for all } k \geq 0,$$

in particular $g(x+\alpha) - g(x) = f(x+\alpha) - f(x)$; this means that $g - f$ is R_α-invariant thus constant and the corollary follows. □

Definition 4.3. The complex number $a \in \mathbf{C}^*$ is a *(continuous) eigenvalue* of (X, T) (or just of T), if there exists $f \in C(X)$, $f \neq 0$, such that

$$f(Tx) = af(x), \quad \text{for every } x \in X.$$

Proposition 4.2. *Suppose that T is a topologically transitive transformation of the compact metric set X. Then*

(a) The eigenvalues of (X, T) have modulus one and the eigenfunctions have constant modulus.

(b) The eigenvalues are simple and form a countable subgroup of $\mathbf{U}$.

Proof. (a) If $f(Tx) = af(x)$ for all $x \in X$, then $\sup_{x \in X} |f(Tx)| = |a| \sup_{x \in X} |f(x)|$; but T needs to be surjective since it is topologically transitive and $|a| = 1$; it follows that $|f|$ is T-invariant thus constant. Note that f never vanishes.

(b) If f and g are two eigenfunctions associated with a, $(f/g)(Tx) = (f/g)(x)$ and f/g is a T-invariant continuous function thus constant. Now denote by f_a the unique eigenfunction of modulus one associated to the eigenvalue a. If $a \neq 1$, we can find $n \geq 1$ and $x \in X$ such that $|a^n f_a(x) - 1| \geq 1/4$; therefore $||f_a - 1||_\infty \geq |f_a(T^n x) - 1| = |a^n f_a(x) - 1| \geq 1/4$. If $a \neq b$, this implies that $||f_a - f_b||_\infty \geq 1/4$ and we deduce from the separability of $C(X)$ that the set of eigenvalues is at most countable. □

4.1.2 Invariant Measures and Unique Ergodicity

A priori, no T-invariant measure is specified on a given topological system (X,T). In order to get statistical results, we shall construct a lot of such probability measures on X, the interesting systems being, of course, the uniquely ergodic ones, when there exists only one T-invariant probability measure on X.

Theorem 4.2 (Krylov & Bogolioubov). *If T is a transformation of the compact metric set X, there exist Borel probability measures on X preserved by T.*

Proof. Fix x an arbitrary point of X, and consider μ a weak-star cluster point in $M(X)$ of the sequence of probability measures (μ_N) :

$$\mu_N = \frac{1}{N} \sum_{n<N} \delta_{T^n x}.$$

Clearly, μ, in turn, is a probability measure, satisfying

$$\int_X f \circ T \, d\mu = \int_X f \, d\mu$$

if $f \in C(X)$. □

Each T-invariant probability measure μ gives rise to a dynamical system (X,μ,T).

Definition 4.4. A topological system (X,T) (or just T) is said to be *uniquely ergodic* if there exists a unique T-invariant probability measure μ on X.

We recall, using the notations of section 3.3, that the convex set $\mathcal{M}_T$ being reduced to a singleton, this probability measure must necessarily be ergodic with respect to T, whence the terminology. The canonical example of such topological systems consists of the irrational rotation on $\mathbf{T}$, more generally, the ergodic rotation on a monothetic compact group G.

Remark 4.2. If T is uniquely ergodic, the closed support of the unique invariant probability measure is the unique minimal compact subset of X : if K is a compact subset of X such that $T(K) \subset K$, T is thus a transformation of K and, by the Krylov-Bogolioubov theorem applied with $T_{|K}$, there exists a T-invariant probability measure ν_K supported in K. If now T is uniquely ergodic, for every topologically invariant K, $\nu_K = \nu$ the unique T-invariant probability measure on X; that means that ν is supported in any such K.

The unique ergodicity property leads to a strengthened version of the Birkhoff ergodic theorem, related to Hermann Weyl's theorem on uniform distribution, which will play an important role in the study of substitution dynamical systems.

Theorem 4.3 (Oxtoby). *The following statements are equivalent*

(a) The topological system (X,T) is uniquely ergodic with μ as its unique T-invariant probability measure.

(b) For every $f \in C(X)$, the Birkhoff's averages $(M_N f)$ with

$$M_N f(x) = \frac{1}{N} S_N f(x) := \frac{1}{N} \sum_{n<N} f(T^n x)$$

converge uniformly on X to a constant.

(c) For every $f \in C(X)$, the Birkhoff's averages $(M_N f)$ converge pointwise on X to $\int_X f\, d\mu$.

Proof. ▷ Suppose that (a) holds but (b) does not hold : we can find $\varepsilon > 0$, $g \in C(X)$ and a sequence (x_{n_j}) of points of X such that $n_j \to \infty$ and

$$\left| (M_{n_j} g)(x_{n_j}) - \int_X g\, d\mu \right| > \varepsilon.$$

Let then ν be a weak-star closure point of the sequence of probability measures $(\frac{1}{n_j} \sum_{k<n_j} \delta_{T^k x_{n_j}})_j$; refining (n_j) if necessary, we have

$$\lim_{j\to\infty} (M_{n_j} f)(x_{n_j}) = \int_X f\, d\nu$$

for every $f \in C(X)$. Hence

$$\left| \int_X f\, d\mu - \int_X f\, d\nu \right| > \varepsilon$$

and the measures μ and ν are distinct. We have got a contradiction and $(a) \Longrightarrow (b)$.

▷ $(b) \Longrightarrow (c)$ is obvious since the constant limit in (b) must be $\int_X f\, d\mu$.

▷ If $f \in C(X)$ and $(M_N f)$ converge pointwise on X to $\int_X f\, d\mu$, for any $\nu \in \mathcal{M}_T$ we have

$$\lim_{N\to\infty} \int_X M_N f\, d\nu = \int_X f\, d\nu = \int_X f\, d\mu$$

by the dominated convergence theorem; this implies $\mu = \nu$ and $(c) \Longrightarrow (a)$.

Remark 4.3.

1. It is sufficient to suppose that the convergence in (b) and (c) holds on a dense part of $C(X)$.
2. These conditions are equivalent to the following one : *The closure of the range of the operator defined on $C(X)$ by $f \mapsto f \circ T - f$ has codimension one.* Actually,

this means that there exists a unique T-invariant linear form on $C(X)$, normalized by $\varphi(1) = 1$, which is continuous, while (a) means that there exists a unique positive normalized linear form φ on $C(X)$ which is T-invariant.

Definition 4.5. Let (X,T) be a topological system, and μ be a T-invariant probability measure on X. We say that $a \in X$ is a *T-generic point for μ* if

$$\frac{1}{N}\sum_{n<N} f(T^n a) \to \int_X f\, d\mu$$

for every $f \in C(X)$.

Example 4.1. The generic points of the q-shift $(\mathbf{T}, S_q, m)$, $q \geq 2$, are precisely the numbers which are normal to base q.

Birkhoff's theorem can be re-formulated as follows :

For an ergodic system (X,μ,T), μ-almost all points of X are T-generic for μ.
And the theorem of Oxtoby says exactly :

The system (X,μ,T) is uniquely ergodic if and only if every point of X is T-generic for μ.

We shall now investigate the links between ergodic properties and topological dynamical properties.

Proposition 4.3. *Let (X,T) be a topological system, and μ be a T-invariant probability measure on X. Assume that μ gives positive mass to every non-empty open set of X. If T is ergodic with respect to μ, then*

$$\mu\{x \in X,\ \overline{O(x)} = X\} = 1.$$

In particular, T is topologically transitive.

Proof. Let (ω_n) be a countable basis of the topology, consisting in non-empty open sets of X; we may write

$$\{x \in X,\ \overline{O(x)} \neq X\} = \cup_{n=1}^{\infty} \cap_{k=0}^{\infty} T^{-k}(X \setminus \omega_n) =: \cup_{n=1}^{\infty} A_n$$

where $T^{-1}A_n$ contains A_n, so that $\mu(T^{-1}A_n \Delta A_n) = 0$. This property, associated with the ergodicity, implies that $\mu(A_n) = 0$ or 1. But $\mu(A_n^c) \geq \mu(\omega_n) > 0$ from our assumption and $\mu(A_n) = 0$ for every n. It follows that $\mu\{x \in X,\ \overline{O(x)} \neq X\} = 0$ and the proposition is proved. □

Thus, topological transitivity is related to ergodicity while minimality is related to unique ergodicity as we shall see.

Proposition 4.4. *Let (X,μ,T) be a uniquely ergodic system. This system is minimal if and only if μ gives positive mass to every non-empty open set of X.*

Proof. ▷ Suppose that (X,T) is minimal. If ω is a non-empty open set of X, then $X = \cup_{n=0}^{\infty} T^{-n}\omega$ and $\mu(\omega) > 0$, otherwise μ would be zero.

◁ Conversely suppose that there exists a non-empty, topologically T-invariant, closed set E in X, distinct from X. Applying theorem 4.2 to the system $(E, T_{|E})$, we get a T-invariant probability measure ν_E supported in E; now ν_E provides a Borel probability measure ν on X by putting

$$\nu(B) = \nu_E(B \cap E),$$

which is still preserved by T. But ν and μ are distinct since $\nu(X \backslash E) = 0$ while $\mu(X \backslash E) > 0$ from our assumption. This violates the unique ergodicity and the proposition is proved. □

Note this ultimate proposition [207]

Proposition 4.5. *Let (X,T) be a minimal topological system, and assume that there exists, for every $x \in X$, a probability measure μ_x for which x is T-generic. Then (X,T) is uniquely ergodic.*

Proof. By assumption, for every $x \in X$ and every $f \in C(X)$,

$$\frac{1}{N} \sum_{n<N} f(T^n x) \to \int_X f \, d\mu_x = C(f,x).$$

We have to show that $C(f,x)$ does not depend on x. For this, we fix f in $C(X)$ and we prove that $x \mapsto C(f,x) =: C_f(x)$ is a continuous and T-invariant function. Suppose that C_f fails to be continuous at $x_0 \in X$. Clearly, $T^n x_0$, for every n, is a discontinuity point for the function C_f; but the system is minimal and the orbit of x_0 is dense in X. It follows that C_f is nowhere continuous. However, C_f is a pointwise limit of a sequence of continuous functions, and Baire's theorem asserts that for such a function, the set of discontinuity points is a set of first category. This leads to a contradiction and C_f must be continuous. Since it is T-invariant, we conclude, using once more the minimality, that C_f is a constant function. □

Definition 4.6. A minimal and uniquely ergodic topological system is said to be *strictly ergodic*.

4.1.3 Examples and Application to Asymptotic Distribution

1. The main example of uniquely ergodic (in fact strictly ergodic) system has been discovered by H. Weyl and consists of the irrational rotation on **T**. Recall the definition :

   Definition 4.7. A sequence of real numbers $(x_n)_{n \geq 0}$ is said to be *uniformly distributed* (mod 2π) if, for every subinterval $I \subset \mathbf{T}$,

   $$\lim_{N \to \infty} \frac{1}{N} \operatorname{card}\{n < N, \, x_n \in I \ (\operatorname{mod} 2\pi)\} = m(I),$$

   m denoting the Lebesgue measure on **T**.

H. Weyl proved that the sequence $(n\alpha)$ is uniformly distributed (mod 2π) if and only if $\alpha/\pi \notin \mathbf{Q}$ by establishing in fact the following very useful equivalence [244].

Theorem 4.4 (Weyl's criterion). *The sequence of real numbers $(x_n)_{n\geq 0}$ is uniformly distributed (mod 2π) if and only if one of the following equivalent properties holds :*

(a) $\frac{1}{N}\sum_{n<N} f(x_n) \to \int_{\mathbf{T}} f\, dm$, *for every continuous function f on* $\mathbf{T}$.

(b) $\frac{1}{N}\sum_{n<N} e^{ikx_n} \to 0$, *for every integer $k \neq 0$.*

Taking $X = \mathbf{T}$, $T = R_\alpha := x \mapsto x+\alpha$ (mod 2π) and $x_n = T^n 0$ in theorem 4.3, we see that condition (a) means exactly that the system (X,T) is uniquely ergodic, with unique probability measure m (what one already knew). Actually, for every $f \in C(\mathbf{T})$,

$$\frac{1}{N}\sum_{n<N} f(x+n\alpha) \to \int_{\mathbf{T}} f\, dm$$

uniformly on $\mathbf{T}$.

2. Following R. Lyons [171], we consider more generally :

Definition 4.8. A sequence of real numbers $(x_n)_{n\geq 0}$ is said to be *Weyl-distributed* (mod 2π) if there exists a probability measure $\nu \neq m$ on $\mathbf{T}$ such that,

$$\lim_{N\to\infty} \frac{1}{N} \operatorname{card}\{n < N,\ x_n \in I \ (\operatorname{mod} 2\pi)\} = \nu(I),$$

for every subinterval $I \subset \mathbf{T}$ whose end points are not mass-points of ν.
We shall write in this case : $(x_n) \sim \nu$.

A similar version of Weyl's criterion for asymptotic distribution (mod 2π) can be proved in the same way [153, 171]

Theorem 4.5. *Let $(x_n)_{n\geq 0}$ be a real sequence. The following assertions are equivalent :*

(a) $(x_n) \sim \nu$.

(b) $\frac{1}{N}\sum_{n<N} f(x_n) \to \int_{\mathbf{T}} f\, d\nu$, *for every* $f \in C(\mathbf{T})$.

(c) $\frac{1}{N}\sum_{n<N} f(x_n) \to \int_{\mathbf{T}} f\, d\nu$, *for every bounded function f on* $\mathbf{T}$, *Riemann-integrable with respect to ν.*

(d) $\frac{1}{N}\sum_{n<N} e^{ikx_n} \to \hat{\nu}(k)$, *for every integer $k \neq 0$.*

This leads to a new formulation of genericity : with our previous notations, "x is T-generic for μ" means "$(T^n x) \sim \mu$". If now μ is any T-invariant Borel probability measure on the compact metric set X, the Birkhoff theorem ensures that for μ-a.e. $x \in X$, $(T^n x)$ admits an asymptotic distribution, say μ_x, which is T-invariant. By integration, we get for $f \in C(X)$

$$\int_X (\lim_{N\to\infty} \frac{1}{N} \sum_{n<N} f(T^n x))\, d\mu(x) = \int_X (\int_X f\, d\mu_x)\, d\mu(x);$$

besides, by exchanging the limit and the integral, we also have

$$\int_X (\lim_{N\to\infty} \frac{1}{N} \sum_{n<N} f(T^n x))\, d\mu(x) = \int_X f\, d\mu.$$

Therefore

$$\mu = \int_X \mu_x\, d\mu(x) \quad \text{in a weak sense.}$$

3. An important result about uniform distribution can be deduced from the unique ergodicity of some well-chosen system. This has been found by Furstenberg and we reproduce his proof [90].

Theorem 4.6. *Let P be a polynomial with real coefficients, one of the non-constant coefficients being rationally independent of π. Then, the sequence $(P(n))$ is uniformly distributed (mod 2π).*

Proof. We write $P(X) = 2\pi(a_0 + a_1 X + \cdots + a_p X^p)$ and we suppose a_j to be irrational while $a_{j+1}, \ldots, a_p$ are rational numbers. We claim that we may suppose $j = p$: if $0 < j < p$, let us choose s integer such that $sa_{j+1}, \ldots, sa_p$ are integers. Now we split the sequence $(P(n))_{n\geq 0}$ into s subsequences $(P(sq+r))q \geq 0$, with $r = 0, 1, \ldots, s-1$. For any fixed r,

$$\begin{aligned}
\tfrac{1}{2\pi} P(sq+r) &\equiv a_0 + \cdots + a_j (sq+r)^j + a_{j+1}(r)^{j+1} + \cdots + a_p (r)^p \pmod 1 \\
&\equiv b_0 + \cdots + b_j q^j \pmod 1
\end{aligned}$$

where $b_j = a_j s^j$ is irrational. Thus, each of these s polynomials $P(sX+r)$ has a π-independent dominant coefficient, whence the reduction.

From now on we assume a_p to be an irrational number and we consider the transformation S on the p-dimensional torus $\mathbf{T}^p$, defined by

$$\begin{aligned}
S(x_1, \ldots, x_p) = (\ & x_1 + \alpha \pmod{2\pi}, \\
& x_2 + 2\pi r_{21} x_1 \pmod{2\pi}, \\
& x_3 + 2\pi r_{31} x_1 + 2\pi r_{32} x_2 \pmod{2\pi}, \\
& \cdot \\
& \cdot \\
& \cdot \\
& x_p + 2\pi r_{p1} x_1 + \cdots + 2\pi r_{pp-1} x_{p-1} \pmod{2\pi})
\end{aligned}$$

where the integers (r_{ij}) have to be specified.

Then, given the polynomial P, we can choose the numbers α, (r_{ij}), and (x_j) in such a way to have $\alpha/\pi \notin \mathbf{Q}$ together with the last component of $S^n(x_1,\ldots,x_p) \equiv P(n) \pmod{2\pi}$. We omit the details (see [61]).

If S happened to be uniquely ergodic, we would obtain

$$\lim_{N\to\infty} \frac{1}{N} \sum_{n<N} f(S^n x)) = \frac{1}{(2\pi)^p} \int_{\mathbf{T}^p} f(y)\, dy$$

for every continuous function on the p-torus, and for every $x \in \mathbf{T}^p$. In particular, taking $f : (y_1,\ldots,y_p) \mapsto e^{iky_p}$ with $k \neq 0$, and $(x_1,\ldots,x_p)$ as above, we would get

$$\lim_{N\to\infty} \frac{1}{N} \sum_{n<N} e^{ikP(n)} = 0$$

which is our assertion according the Weyl theorem. We are just led to prove by induction on p that S is uniquely ergodic and this will result from the following key lemma.

Lemma 4.1. *Let (X,T) be a uniquely ergodic system with μ as unique T-invariant probability measure and let $\phi : X \to \mathbf{T}$ be a continuous function. Consider the transformation S of $X \times \mathbf{T}$ defined by*

$$S(x, e^{it}) = (Tx, \phi(x).e^{it}).$$

Then, S is uniquely ergodic if and only if S is ergodic with respect to $\mu \times m$.

Proof. We rapidly sketch the proof. If ν is any S-invariant probability measure, it is sufficient to prove that ν coincides with $\mu \times m$ on the following complete set in $C(X \times \mathbf{T})$ of functions

$$(x,t) \to f(x)e^{int}, \quad n \in \mathbf{Z}, \quad f \in C(X).$$

If $n = 0$ this is simply due to the unique ergodicity of (X,μ,T).

If now $n \neq 0$, we have to show that

$$J(f) = \int\int_{X\times\mathbf{T}} f(x)e^{int}\, d\nu(x,t) = 0.$$

Observe that J is a continuous linear functional on $L^2(X \times \mathbf{T})$, and, if the operator U acts on $C(X)$ or $L^2(X,\mu)$ by the formula: $Uf(x) = \phi(x) \cdot f(Tx)$, we have

$$J(Uf) = \int\int_{X\times\mathbf{T}} \phi(x)f(Tx)e^{int}\, d\nu(x,t) = J(f)$$

since ν is S-invariant. Therefore,

$$J(f) = \int_X f\overline{g}\, d\mu$$

for some U-invariant function $g \in L^2(X,\mu)$. This entails that the function $(x,e^{it}) \mapsto g(x)e^{it}$ is S-invariant, thus must be constant since S is ν-ergodic. This imposes $g = 0$ and the lemma is proved. □

In the case $p = 2$, note that S is defined on $\mathbf{T}^2$ by

$$S(x,y) = (x+\alpha \pmod{2\pi},\ x+y \pmod{2\pi})$$

and the ergodicity with respect to the Haar product measure is a consequence of the Anzai theorem (3.1). The theorem follows by induction. □

4. Note that the q-adic transformation on $\mathbf{T}$ admits a lot of invariant probability measures and is far from being uniquely ergodic. Thus nonnormal numbers do exist (evident with the symbolic definition).

4.2 Dynamical Systems Associated with Finitely Valued Sequences

In this section we bring a very important combinatorial aspect of dynamics into play. Consider a dynamical system (X,T) and suppose that $X = \cup_{i=1}^{g} X_i$ where (X_i) is a finite partition of X. To every $x \in X$, we can then associate its itinerary, that is a finitely valued sequence u_x, by putting

$$u_x(n) = i \ \text{if} \ T^n x \in X_i;$$

we thus define, by coding, a new dynamical system which, at least in good cases, accurately reflects the properties of the initial system [152].

We begin with very classical notations in symbolic dynamics.

4.2.1 Subshifts

1. In this section, A is a finite set, called an *alphabet*, of $s \geq 2$ symbols called *letters*, i.e. s =Card A is the cardinality of A.
2. A *word* on A is a finite (non-empty) sequence of letters; $A^* = \cup_{k\geq 0} A^k$ consists of all words on the alphabet A and the empty word . For $\omega = \omega_1 \cdots \omega_n \in A^*$, $|\omega| = n$ is the *length* of the word ω. The word ω is a *prefix* of $W \in A^*$ if $W = \omega\omega'$ for some word ω', which in turn, is a *suffix* of W.
3. We consider $A^{\mathbf{N}}$, the set of all sequences $x = (x_n)_{n\geq 0}$ whose components belong to A and such a sequence would sometimes be called an *infinite word* on A.
4. $A^{\mathbf{N}}$ is endowed with the product of the discrete topologies on each factor, which coincides with the topology of pointwise convergence. This topology in fact is metrizable, and is for example defined by the ultrametric

$$d(x,y) = \begin{cases} 2^{-\inf\{k\geq 0,\, x_k \neq y_k\}} & \text{if } x \neq y \\ 0 & \text{if } x = y \end{cases}$$

Consequently $A^{\mathbf{N}}$ is a compact metric set, actually a Cantor set. Two elements x, y of $A^{\mathbf{N}}$ are close to each other if there exists some integer $n \geq 1$ such that $x_j = y_j$ for $0 \leq j \leq n-1$; so, given $\alpha_0, \ldots, \alpha_{n-1}$ some fixed letters in A, the set

$$\{x \in A^{\mathbf{N}}, x_j = \alpha_j,\ 0 \leq j \leq n-1\} =: [\alpha_0 \cdots \alpha_{n-1}],$$

is called a *cylinder set* of $A^{\mathbf{N}}$. Cylinder sets form an open and closed (clopen) basis for the topology of $A^{\mathbf{N}}$.

5. We define on $A^{\mathbf{N}}$ the *(one-sided) shift* transformation T by

$$(Tx)_k = x_{k+1}, \quad \forall k \geq 0.$$

The T-orbit of the sequence $u \in A^{\mathbf{N}}$ is denoted by $O(u) = \{T^n u,\ n \geq 0\}$.

T is clearly continuous and onto but it is not an homeomorphism of $A^{\mathbf{N}}$. If X is a non-empty, T-invariant closed subset of $A^{\mathbf{N}}$, the system $(X, T_{|X})$ (or simply (X,T)) is called a *subshift* on A. Note that the system $(A^{\mathbf{Z}}, T)$, where T now is the bilateral shift map, is usually called the *full shift*.

Definition 4.9. Let u be a fixed sequence in $A^{\mathbf{N}}$. The topological system $(\overline{O(u)}, T)$ associated to u is called the *subshift spanned by u*.

Definition 4.10. An element $x \in A^{\mathbf{N}}$ is said to be *recurrent* for the shift map T (or T-recurrent) if there exists a sequence of integers (n_j) going to infinity, such that

$$x = \lim_{j\to\infty} T^{n_j} x,$$

or, equivalently, every word in u occurs at least twice.

For a T-recurrent point x, observe that T maps $\overline{O(x)}$ onto $\overline{O(x)}$; indeed, if $x = \lim_{j\to\infty} T^{n_j} x$, x is also the limit of $T(T^{n_j - 1}x)$ so that x can be written $x = Ty$ for some cluster point $y \in \overline{O(x)}$. This proves that x belongs to $T(\overline{O(x)})$ and the surjectivity of T follows.

4.2.2 Minimality and Unique Ergodicity

From now on, we restrict our attention to the topological system (X,T) associated with some fixed sequence u in $A^{\mathbf{N}}$. We need some additional notations in this framework.

The non-empty words $u_m u_{m+1} \cdots u_n =: u_{[m,n]}, m \leq n$, occurring in u are also called *factors* of u and the set of all factors of u, called the *language* of u, is denoted by $\mathscr{L}(u)$. If X is a subshift on A, $\mathscr{L}(X)$, the *language* of X, is the collection of all words occurring in elements of X. The following simple remark is useful :

Proposition 4.6. *If* $X = \overline{O(u)}$ *where* $u \in A^{\mathbf{N}}$, *then*
(i) $\mathscr{L}(X) = \mathscr{L}(u)$.
(ii) $X = \{x \in A^{\mathbf{N}},\ \mathscr{L}(x) \subset \mathscr{L}(u)\}$.
In this case, the language of u characterizes X.

Proof. If $x \in X$, $x = \lim_{i\to\infty} T^{n_i}u$, and, according to the topology on X, for every j, there exists $n(j)$ such that

$$x_{[0,j]} := x_0 \cdots x_j = u_{n(j)} \cdots u_{n(j)+j} =: u_{[n(j),\, n(j)+j]} \in \mathscr{L}(u).$$

Thus $x_{[m,n]} \in \mathscr{L}(u)$ for every $m \leq n$. Conversely, this property characterizes the closed orbit of u. □

If $\omega \in \mathscr{L}(u)$

$$[\omega] = \{x \in X,\ x_{[0,\,|\omega|-1]} = \omega\}$$

is the cylinder set of the sequences in X whose initial factor is ω. These are the elementary cylinder sets of X, and the clopen sets

$$\{T^{-j}[\omega],\ \omega \in \mathscr{L}(u),\ j \geq 0\}$$

span the topology of X.

We are now ready to describe minimality and unique ergodicity of such a subshift, in terms of the language of u.

Definition 4.11. A sequence $x \in A^{\mathbf{N}}$ is said to be *uniformly recurrent* or *almost periodic* if for every neighbourhood V of x, the set $\{n \geq 0,\ T^n x \in V\}$ is relatively dense (or "syndetic") .

This means that *every factor of x occurs in x with bounded gaps,* the bound depending on the factor. Considering, as neighbourhood V of x, the cylinder sets $[x_0 \cdots x_n]$, $n \geq 0$, leads directly to this equivalence.

Proposition 4.7. *Let* $X = \overline{O(u)}$ *where* $u \in A^{\mathbf{N}}$*; the system* (X,T) *is minimal if and only if u is uniformly recurrent.*

Proof. ◁ Suppose that u is uniformly recurrent, without the system being minimal. There must exist $x \in X$ such that $\overline{O(x)} \neq X$ and, for that reason, u must lie outside $\overline{O(x)}$. So, according to the previous proposition, one can find j such that $u_{[0,j]} \notin \mathscr{L}(x)$. Now, $x = \lim_{k\to\infty} T^{n_k}u$ and u being uniformly recurrent,

$$u_{[n_k+s,\, n_k+s+j]} = u_{[0,j]}$$

for some s and infinitely many n_k. By continuity of T, $T^{n_k+s}u$ tends to $T^s x$ so that

$$x_{[s,\, s+j]} = u_{[n_k+s,\, n_k+s+j]}$$

for k large enough; it follows that $u_{[0,\, j]} = x_{[s,\, s+j]}$ and a contradiction.

▷ Conversely, we suppose that (X,T) is minimal and we fix $y \in X$; then $X = \overline{O(y)}$ and for every neighbourhood V of u, $V \cap O(y) \neq \emptyset$; this means that $T^k y \in V$ for some k and $X = \cup_{k \geq 0} T^{-k} V$. By the Borel-Lebesgue property,

$$X = T^{-k_1}V \cup T^{-k_2}V \cup \cdots \cup T^{-k_n}V$$

so that, setting $K = \max_{1 \leq i \leq n} k_i$, every y enters V after at most K steps under T. In particular, taking $y = T^j u$, we deduce that one of the points $T^j u, \ldots, T^{j+K} u$ must lie in V; in other words, if we choose $V = [u_0 \cdots u_\ell] =: [\omega]$, for every $j > 0$, ω is one of the factors $u_{[j,\, j+\ell]}, \ldots, u_{[j+K,\, j+K+\ell]}$ and u is uniformly recurrent. □

We already proved that there exist Borel probability measures preserved by T. The next proposition is just a re-formulation of theorem 4.3.

Proposition 4.8. *The system (X,T) is uniquely ergodic if and only if, for every $f \in C(X)$,*

$$\lim_{N \to \infty} \frac{1}{N} \sum_{n<N} f(T^{n+j}u)$$

exists uniformly in $j \geq 0$.

Proof. We show that the condition is sufficient. If $f \in C(X)$ and $x \in X$, we wish to prove that the limit of

$$\frac{1}{N} \sum_{n<N} f(T^n x)$$

exists and does not depend on x. Let (n_k) be such that $x = \lim_{k \to \infty} T^{n_k} u$. By our uniformity assumption, the Birkhoff averages

$$\frac{1}{N} \sum_{n<N} f(T^{n+n_k} u)$$

converge, as N goes to infinity, to a limit a independent of k; we thus obtain

$$\begin{aligned}
\lim_{N \to \infty} \tfrac{1}{N} \textstyle\sum_{n<N} f(T^n x) &= \lim_{N \to \infty} \lim_{k \to \infty} \tfrac{1}{N} \textstyle\sum_{n<N} f(T^{n+n_k} u) \\
&= \lim_{k \to \infty} \lim_{N \to \infty} \tfrac{1}{N} \textstyle\sum_{n<N} f(T^{n+n_k} u) \\
&= a
\end{aligned}$$

□

Note that this result still holds when A is an arbitrary compact metric set.

Corollary 4.2. *(a) The system (X,T) is uniquely ergodic if and only if each factor of u occurs in u with a uniform frequency.*

(b) The system (X,T) is strictly ergodic if and only if each factor of u occurs in u with a uniform positive frequency.

Proof. (a) Referring to theorem 4.3, we have just to test in the previous proposition the convergence of the Birkhoff averages for a dense set of functions in $C(X)$. So is the linear span of the functions $\{\mathbf{1}_{[B]} \circ T^k,\ k \geq 0,\ B \in A^*\}$. Thus we get : (X,T) is uniquely ergodic if and only if $\frac{1}{N}\sum_{n<N} \mathbf{1}_{[B]}(T^{n+j}u)$ converges to a constant, uniformly in $j \geq 0$, for every word $B \in A^*$. This means that

$$\frac{1}{N}\ \mathrm{card}\{n \leq N,\ u_{[n+j,\ldots,n+j+|B|-1]} = B\}$$

converges to a constant, uniformly in j, or, equivalently, the frequency of occurrence of the word B in $T^j u$ exists uniformly in j.

(b) If μ is the unique T-invariant probability measure on X, for every word B,

$$\mu([B]) = \lim_{N\to\infty} \frac{1}{N}\ \mathrm{card}\{n \leq N,\ u_{[n+j,\ldots,n+j+|B|-1]} = B\}; \tag{4.1}$$

if, in addition, (X,T) is minimal,

$$\mu([B]) > 0 \quad \text{for every} \quad B \in \mathscr{L}(u). \tag{4.2}$$

Conversely, properties (4.1) and (4.2) together imply the strict ergodicity by (a) and proposition 4.4. □

4.2.3 Examples

Consider the two well-known sequences, $m \in \{0,1\}^{\mathbf{N}}$ and $r \in \{-1,1\}^{\mathbf{N}}$ defined inductively by

$$\begin{cases} m_0 = 0 \\ m_{2n} = m_n, \quad m_{2n+1} = 1 - m_n \end{cases}$$

and

$$\begin{cases} r_0 = 1 \\ r_{2n} = r_n, \quad r_{2n+1} = (-1)^n r_n \end{cases}$$

The first one is called the *Thue-Morse sequence* on the alphabet $\{0,1\}$, and the second one, the *Rudin-Shapiro sequence*. Both sequences enjoy extremal properties and play important roles in many different areas (see the bibliography).

The Thue-Morse sequence and the Rudin-Shapiro sequence admit an arithmetical definition involving the 2-adic representation of integers. If we denote by $S_q(n)$ the sum of digits in the q-adic representation of the natural integer n, $(q \geq 2)$, then, one easily verifies that

$$m_n \equiv S_2(n) \pmod 2.$$

On the other hand

$$r_n = (-1)^{f(n)}$$

where $f(n)$ is the occurrence number of "11" in the 2-adic representation of $n \in \mathbf{N}$.

We shall prove in chapter 5 that the system arising from each of them is strictly ergodic. In fact both appear as an automaton-generated sequence, and we shall establish a general result for such sequences, when the automaton is primitive.

4.3 Spectral Properties of Bounded Sequences

In this section, we suppose, more generally, that A is a compact subset of $\mathbf{C}$, and that $u = (u_n)_{n\geq 0}$ is a sequence taking its values in A. T still designs the one-sided shift on $A^{\mathbf{N}}$ equipped with the product topology. We are interested in the spectral study of the dynamical system (X,μ,T), where $X = \overline{O(u)}$ is the orbit closure of $u \in A^{\mathbf{N}}$, and μ is any T-invariant probability measure on X. Recall that the operator $U := U_T$, when unitary, is spectrally determined by the spectral family of measures (σ_f) (chapter 2); in the case of a one-sided sequence, U may fail to be unitary, and we are sometimes led to extend the elements of X towards the "left" in a natural way to make T a homeomorphism. Nevertheless, in the general case, U is an isometry and we could still define σ_f by putting, for all $k \geq 0$,

$$\hat{\sigma}_f(k) = < U^k f, f >$$

and

$$\hat{\sigma}_f(-k) = \overline{\sigma_f(k)}.$$

Moreover, according to the von Neumann theorem (2.5), for every $f \in L^2(X,\mu)$ and $\lambda \in \mathbf{T}$,

$$\sigma_f\{\lambda\} = ||P_\lambda f||^2$$

where P_λ is the orthogonal projection onto the eigen-subspace corresponding to the eigenvalue $e^{i\lambda}$.

4.3.1 Correlation Measures

Definition 4.12. We call a *correlation function* of $u \in A^{\mathbf{N}}$, any cluster point of the sequence of the arithmetical functions (γ_N) defined on $\mathbf{N}$ by

$$\gamma_N(k) = \frac{1}{N} \sum_{n<N} u_{n+k}\overline{u_n}.$$

If C is a correlation function of u, by using a diagonal process we can find a sequence of integers (N_j) such that

$$C(k) = \lim_{j\to\infty} \frac{1}{N_j} \sum_{n<N_j} u_{n+k}\overline{u_n}$$

for every $k \in \mathbf{N}$. As we already did with spectral measures, we extend C to the negative numbers by the formula

$$C(-k) = \overline{C(k)},\ k \in \mathbf{N}.$$

So defined, C is a positive-definite sequence and thus, by Bochner's theorem (section 1.1), C is the Fourier transform of a positive measure on $\mathbf{T}$, of total mass $\lim_{j\to\infty} \frac{1}{N_j} \sum_{n<N_j} |u_n|^2$, that we call a *correlation measure.*

• Consider now $M(u)$, the set of all T-invariant probability measures on X, which are weak-star cluster points of the following sequence

$$\left(\frac{1}{N} \sum_{n<N} \delta_{T^n u}\right)_{N\geq 1}.$$

Thanks to the separability of $C(X)$, if $\mu \in M(u)$ one can find a sequence of integers (N_j) such that

$$\lim_{j\to\infty} \frac{1}{N_j} \sum_{n<N_j} f(T^n u) = \int_X f\, d\mu$$

for every $f \in C(X)$; therefore, if $k \geq 0$,

$$\widehat{\sigma}_f(k) := \int_X f \circ T^k \cdot \bar{f}\, d\mu$$

is also

$$= \lim_{j\to\infty} \frac{1}{N_j} \sum_{n<N_j} f(T^{k+n} u)\overline{f(T^n u)}.$$

We denote by π_m, $m \geq 0$, the projection $x \mapsto x_m$ onto the mth-component of the sequence x; then, $\sigma_{\pi_m} = \sigma_{\pi_0}$ for every $m \geq 0$, and

$$\widehat{\sigma}_{\pi_0}(k) = \lim_{j\to\infty} \frac{1}{N_j} \sum_{n<N_j} u_{n+k}\overline{u_n}.$$

Thus, the spectral measure σ_{π_0} on the associated system (X,μ,T) is a correlation measure.

• If $M(u) = \{\mu\}$ (in particular if the system (X,T) is uniquely ergodic), u admits a unique correlation measure, and

$$\lim_{N\to\infty} \frac{1}{N} \sum_{n<N} u_{n+k}\overline{u_n} \quad \text{exists for all } k \geq 0. \tag{4.3}$$

N. Wiener has first introduced those sequences and we usually say, under assumption (4.3), that u belongs to the *Wiener space* $\mathscr{S}$ [245]. We shall use in a repetitive way the following useful description of the unique correlation measure :

Proposition 4.9. *If σ is the unique correlation measure of the sequence u, σ is the weak-star limit point of the sequence of absolutely continuous measures $R_N \cdot m$, where*

$$R_N(t) = \frac{1}{N}\Big|\sum_{n<N} u_n e^{int}\Big|^2.$$

This is nothing but a re-formulation of the definition since, for $k \geq 0$,

$$\widehat{R}_N(k) = \frac{1}{N} \sum_{n+k<N} u_{n+k}\overline{u_n}$$

and

$$\widehat{R}_N(-k) = \overline{\widehat{R}_N(k)}.$$

As a consequence of this remark combined with proposition 1.1 we get the following [59]:

Corollary 4.3. *Let a and b be two sequences in the Wiener space $\mathscr{S}$ with correlation measures σ_a and σ_b respectively; then,*

$$\limsup_{N\to\infty} \frac{1}{N}\Big|\sum_{n=0}^{N-1} a_n\overline{b_n}\Big| \leq \rho(\sigma_a,\sigma_b),$$

where $\rho(\sigma_a,\sigma_b)$ is the affinity between both measures.

The affinity between two measures has been defined in section 1.1.

Proof. By the above proposition, σ_a and σ_b are the weak-star limit of the sequences $(\sigma_a^{(N)})$ and $(\sigma_b^{(N)})$ respectively, with $\sigma_a^{(N)} = \frac{1}{N}\Big|\sum_{n<N} a_n e^{int}\Big|^2 \cdot m$ and $\sigma_b^{(N)} =$

$\frac{1}{N}\left|\sum_{n<N} b_n e^{int}\right|^2 \cdot m$; it follows from proposition 1.1 that

$$\begin{aligned}\rho(\sigma_a,\sigma_b) &\geq \limsup_N \rho(\sigma_a^{(N)},\sigma_b^{(N)}) \\ &= \limsup_N \frac{1}{N}\int_{\mathbf{T}}\left|\sum_{n<N} a_n e^{int}\right|\left|\sum_{n<N}\overline{b_n}e^{-int}\right| dm(t) \\ &\geq \limsup_N \frac{1}{N}\left|\int_{\mathbf{T}}\sum_{n<N,m<N} a_n\overline{b_m}e^{i(n-m)t}\right| dm(t) \\ &= \limsup_N \frac{1}{N}\left|\sum_{n<N} a_n\overline{b_n}\right|.\end{aligned}$$

□

The correlation measure in turn, gives informations on the sequence u, more specifically on the so-called *Fourier-Bohr spectrum* of this sequence, which is

$$\{\lambda \in \mathbf{T}, \quad \limsup_{N\to\infty}\frac{1}{N}\left|\sum_{n<N} u_n e^{-in\lambda}\right| \neq 0\}.$$

A sequence with a unique correlation measure may fail to have an average value, but the following estimate can easily be derived from the previous corollary [27,59]

Proposition 4.10. *Assume that σ is the unique correlation measure of the sequence $u \in \mathscr{S}$. Then*

$$\limsup_{N\to\infty}\frac{1}{N}\left|\sum_{n<N} u_n e^{-in\lambda}\right| \leq \sigma\{\lambda\}^{1/2}$$

for every $\lambda \in \mathbf{T}$.

Proof. We give a direct proof. We may consider only the case $\lambda = 0$, since the sequence v with components $v_n = u_n e^{-in\lambda}$ for $n \geq 0$, admits too a unique correlation measure : $\tau = \sigma * \delta_\lambda$; we get readily the general result by noticing that $\tau\{0\} = \sigma\{\lambda\}$.

Let now (N_j) be a sequence such that $\frac{1}{N_j}\sum_{n<N_j} u_n$ tends to $\limsup_{N\to\infty}\frac{1}{N}\sum_{n<N} u_n$, and let $\mu \in M(u)$ be a weak-star limit point of the sequence of probability measures $(\frac{1}{N_j}\sum_{n<N_j}\delta_{T^n u})$. The Fourier coefficients of the corresponding σ_{π_0} are equal to

$$\begin{aligned}\widehat{\sigma}_{\pi_0}(k) &= \int_X \pi_0\circ T^k\cdot\overline{\pi_0}\,d\mu \\ &= \lim_{j\to\infty}\frac{1}{N_j}\sum_{n<N_j} u_{n+k}\overline{u_n} \\ &= \widehat{\sigma}(k)\end{aligned}$$

and $\sigma_{\pi_0} = \sigma$ for every limit point $\mu \in M(u)$. Consequently, if P is the orthogonal projection onto the T-invariant functions in $L^2(X,\mu)$,

$$\sigma\{0\} = \sigma_{\pi_0}\{0\} = ||P\pi_0||^2$$

while

$$\int_X \pi_0 \, d\mu = \lim_{j\to\infty} \frac{1}{N_j} \sum_{n<N_j} u_n.$$

Since $\pi_0 - P\pi_0$ is orthogonal to constants, we finish the proof by using the inequality

$$\left|\int_X \pi_0 \, d\mu\right|^2 \le \left|\int_X P\pi_0 \, d\mu\right|^2 \le ||P\pi_0||^2, \tag{4.4}$$

with equality if and only if $P\pi_0$ is constant. □

With a slight modification of this proof, we deduce easily the following [166]

Corollary 4.4. *Suppose that* $\limsup_{N\to\infty} \frac{1}{N}\left|\sum_{n<N} u_n e^{-in\lambda}\right| > 0$ *for some* $\lambda \in \mathbf{T}$. *Then there exists* $\mu \in M(u)$ *such that* $e^{i\lambda}$ *is an eigenvalue of the dynamical system* (X,μ,T).

Remark 4.4. In order to have simple formulations of these results, we often assume that $M(u) = \{\mu\}$, which means

$$\limsup_{N\to\infty} \frac{1}{N} \sum_{n<N} f(T^n u) = \int_X f \, d\mu, \quad \forall f \in C(X);$$

but, generally, this does not imply the unique ergodicity of (X,T) (which requires the uniform convergence on X), not even the ergodicity of the system (X,μ,T).

• We are looking for conditions on the system (or, still better, on the sequence u) ensuring the equality in the previous proposition.

Proposition 4.11. *Assume that* $M(u) = \{\mu\}$ *and let* σ *be the unique correlation measure of the sequence* u.

(a) If the system (X,μ,T) *is ergodic,*

$$\sigma\{0\}^{1/2} = \lim_{N\to\infty} \frac{1}{N}\left|\sum_{n<N} u_n\right|.$$

(b) If the system (X,μ,T) *is uniquely ergodic, for every continuous eigenvalue* $e^{i\lambda}$ *of* T,

$$\sigma\{\lambda\}^{1/2} = \lim_{N\to\infty} \frac{1}{N}\left|\sum_{n<N} u_n e^{-in\lambda}\right|. \tag{4.5}$$

Proof. (a) We keep the notations of the above proposition; if the system (X,μ,T) is ergodic, $P\pi_0$ is now constant and we get equality in (4.4) as already observed.

(b) Let f_λ be the continuous eigenfunction corresponding to the eigenvalue $e^{i\lambda}$, normalized by $f_\lambda(u) = 1$. The eigen-subspaces of $C(X)$ are one-dimensional, and

$$\lim_{N\to\infty} \frac{1}{N} \sum_{n<N} \pi_0(T^n u) \cdot \overline{f_\lambda(T^n u)} = \int_X \pi_0 \cdot \overline{f_\lambda}\, d\mu.$$

Now

$$\begin{aligned}
\sigma_{\pi_0}\{\lambda\} &= \lim_{K\to\infty} \tfrac{1}{K} \textstyle\sum_{k<K} e^{-ik\lambda} \widehat{\sigma}_{\pi_0}(k) \\
&= \lim_{K\to\infty} \tfrac{1}{K} \textstyle\sum_{k<K} e^{-ik\lambda} \left(\int_X \pi_0 \circ T^k \cdot \overline{\pi_0}\, d\mu \right) \\
&= \lim_{K\to\infty} \tfrac{1}{K} \textstyle\sum_{k<K} \left(\lim_{N\to\infty} \tfrac{1}{N} \sum_{n<N} (\pi_0 \overline{f_\lambda})(T^{k+n}u) \cdot \overline{(\pi_0 \overline{f_\lambda})(T^n u)} \right).
\end{aligned}$$

Since the system has been supposed to be uniquely ergodic, the averages $\frac{1}{N}\sum_{n<N} f(T^{n+k}u)$ converge to $\int_X f\, d\mu$ uniformly in k, for every $f \in C(X)$. Thereby, exchanging the limits, we get

$$\begin{aligned}
\sigma_{\pi_0}\{\lambda\} &= \lim_{N\to\infty} \tfrac{1}{N} \textstyle\sum_{n<N} \overline{(\pi_0 \overline{f_\lambda})(T^n u)} \left(\lim_{K\to\infty} \tfrac{1}{K} \sum_{k<K} (\pi_0 \overline{f_\lambda})(T^{k+n}u) \right) \\
&= \left| \textstyle\int_X \pi_0 \overline{f_\lambda}\, d\mu \right|^2 \\
&= \lim_{N\to\infty} \tfrac{1}{N} \left| \textstyle\sum_{n<N} u_n e^{-in\lambda} \right|^2.
\end{aligned}$$

□

Remark 4.5. 1. When the operator U is unitary, the ergodicity of the system (X,μ,T) suffices to imply identity (4.5) for every $\lambda \in \mathbf{T}$; in this case indeed, each eigen-subspace is one-dimensional and thus, $P_\lambda \pi_0 = f_\lambda < \pi_0, f_\lambda >$, where f_λ is the normalized eigenfunction in $L^2(X,\mu,T)$ corresponding to the eigenvalue $e^{i\lambda}$. We compute then directly

$$\sigma\{\lambda\} = \sigma_{\pi_0}\{\lambda\} = ||P_\lambda \pi_0||^2 = \left| \int_X \pi_0 \overline{f_\lambda}\, d\mu \right|^2$$

2. If U fails to be unitary or if the system (X,μ,T) fails to be ergodic, identity (4.5) can be derived from the stronger assumption

$$\lim_{N\to\infty} \frac{1}{N} \sum_{n<N} u_{n+k} \quad \text{exists uniformly in } k.$$

3. The two previous propositions admit generalized formulations : instead of σ_{π_0}, we can consider σ_f where f is any function in $L^2(X,\mu)$ and get similar results; in addition, we may remove the somewhat restrictive hypothesis $M(u) = \{\mu\}$, and thus obtain for any $(u_n) \in A^{\mathbf{N}}$,

$$\limsup_{N\to\infty} \frac{1}{N} \Big| \sum_{n<N} u_n e^{-in\lambda} \Big| \le \sup\{\sigma\{\lambda\}^{1/2},\ \sigma \text{ correlation measure of } (u_n)\} \quad (4.6)$$

4. Improvements of these results lead to Wiener-Wintner's theorem, in a forthcoming item.

4.3.2 Applications of Correlation Measures

4.3.2.1 Uniform Distribution Modulo 2π

As a first application of correlation measures, we shall give a proof of theorem 4.6, in fact the initial proof of this result, whose dynamical context has been exploited by Furstenberg as we said before. This proof makes use of the van der Corput lemma, already stated in the Hilbert framework (subsection 3.3.3), and whose initial formulation goes as follows.

Lemma 4.2. *If the real sequences $(u_{n+h} - u_n)_{n\ge 0}$ are uniformly distributed modulo 2π for every $h \in A$, where $A \subset \mathbf{N}^*$ has density one, then the sequence $(u_n)_{n\ge 0}$ itself is uniformly distributed modulo 2π.*

Proof. Using the Weyl criterion, the assumption means

$$\lim_{N\to\infty} \frac{1}{N} \sum_{n<N} e^{ik(u_{n+h}-u_n)} = 0, \quad \text{for every } k \in \mathbf{Z}^* \text{ and } h \in A. \quad (4.7)$$

For $k \neq 0$, let σ_k be any correlation measure of the sequence α_k whose components are $\alpha_k(n) = e^{iku_n}$, $n \ge 0$. From (4.7),

$$\widehat{\sigma_k}(h) = 0 \quad \text{when} \quad h \in A.$$

Now,

$$\frac{1}{N} \sum_{n<N} |\widehat{\sigma_k}(n)|^2 \le \frac{1}{N} \sum_{\substack{n<N \\ n\notin A}} 1$$

which tends to zero since A^c has density zero. It follows from Wiener's lemma that any correlation measure σ_k of the sequence α_k must be continuous. In particular, $\sigma_k\{0\} = 0$, and, according to (4.6),

$$\lim_{N\to\infty}\frac{1}{N}\sum_{n<N}e^{iku_n}=0,$$

which was to be proved. □

We deduce from this a new proof of theorem 4.6 : *Let P be a polynomial with real coefficients, one of the non-constant coefficients being rationally independent of π. Then, the sequence $(P(n))$ is uniformly distributed (mod 2π).*

Proof. Without loss of generality, we may assume that

$$P(n) = 2\pi(a_0 + a_1 n + \cdots + a_p n^p)$$

where a_p is an irrational number, as already observed. Let P_1 be the polynomial defined by

$$P_1(X) = P(X+h) - P(X), \quad h \in \mathbf{N}^*.$$

Then P_1 has degree $p-1$ and the coefficient in the term of highest degree of $\frac{1}{2\pi}P_1$, pha_p, is still irrational for every integer $h \neq 0$. Using inductively this argument and the van der Corput lemma, we see that the theorem follows from the uniform distribution (mod 2π) of the sequence $(\alpha n + \beta)$, where α/π is any irrational number. □

4.3.2.2 Van der Corput Sets and Recurrence Sets

The link between dynamical systems and distribution of sequences has been already mentioned at the end of chapter 2, with weakly ergodic and ergodic sequences. We introduce now sets of integers at the interplay between recurrence properties of dynamical systems and uniform distribution (mod 2π).

Following Kamae, Mendès France [138] and Rusza [219], we define the following sets :

Definition 4.13. The set $\Lambda \subset \mathbf{N}^*$ is called a *van der Corput set* (in short, *vdC set*) if it "satisfies" the van der Corput lemma : for every sequence $(u_n)_{n\geq 0}$ of real numbers, if the sequences $(u_{n+h} - u_n)_{n\geq 0}$ are uniformly distributed modulo 2π for every $h \in \Lambda$, then the sequence $(u_n)_{n\geq 0}$ is uniformly distributed modulo 2π.

(Such a set is also called a *correlative set.*)

The vdC-property can be formulated, in a more suitable way, in terms of positive measures.

Theorem 4.7. *$\Lambda \subset \mathbf{N}^*$ is a* vdC set *if and only if it enjoys the following property : If $\sigma \in M(\mathbf{T})$ is a positive measure,*

$$\hat{\sigma}(h) = 0 \textit{ for every } h \in \Lambda \quad \text{implies} \quad \sigma\{0\} = 0; \tag{4.8}$$

or, equivalently,

$$\hat{\sigma}(h) = 0 \text{ for every } h \in \Lambda \quad \text{implies} \quad \sigma \text{ continuous} \tag{4.9}$$

Proof. The equivalence between (4.8) and (4.9) is just a trick : suppose that (4.8) holds and that $\widehat{\sigma}(h) = 0$ for all $h \in \Lambda$; if $\tau = \sigma * \check{\sigma}$ where $\check{\sigma}(E) = \sigma(-E)$, then $\widehat{\tau}(h) = |\widehat{\sigma}(h)|^2 = 0$ on Λ and $0 = \tau\{0\} = \sum_{x \in \mathbf{T}} |\sigma\{x\}|^2$ so that σ is continuous, and the implication follows.
⋆ As in the proof of the van der Corput lemma, we easily show that a set Λ satisfying condition (4.8) must be a vdC set.
⋆ Conversely, we shall prove that $c_+ \leq b$ where

$$c_+ = \sup\{\sigma\{0\},\ \sigma \text{ probability measure},\ \hat{\sigma}_{|\Lambda} = 0\}$$

and

$$b = \sup\{\limsup_{N \to \infty} \frac{1}{N} \big|\sum_{n<N} y_n\big|,\ |y_n| \leq 1,\ \lim_{N \to \infty} \frac{1}{N} \sum_{n<N} y_{n+k}\overline{y_n} = 0 \text{ if } k \in \Lambda\}.$$

Note that the proof would be easy if every probability measure could be realized as a spectral measure σ_f, associated to some **ergodic** dynamical system $(X, \mathscr{B}, \mu, T)$; indeed, taking $y_n = f(T^n x_0)$ for a suitable x_0, the result would follow from proposition 4.11 (a).

Replacing the ergodic theorem by the strong law of large numbers, we are led to consider a sample (X_n) of X with σ as probability law. Thus, for every $k \in \mathbf{Z}$,

$$\lim_{N \to \infty} \frac{1}{N} \sum_{j<N} e(kX_j) = \hat{\sigma}(k), \tag{4.10}$$

where $e(x) = e^{ix}$. Now, we define the sequence (Y_n) by blocks :

$$Y_n = e(rX_m) \quad \text{if } n = m^2 + r,\ 0 \leq r \leq 2m.$$

By considering

$$\frac{1}{N^2} \sum_{n<N^2} Y_{n+k}\overline{Y_n} = \frac{1}{N^2} \sum_{j<N} (2j+1) e(kX_j),$$

one deduces from (4.10) that $\lim_{N \to \infty} \frac{1}{N} \sum_{n<N} Y_{n+k}\overline{Y_n} = \widehat{\sigma}(k)$ (for a Cesaro convergent bounded sequence (u_n), $\lim_N \frac{1}{N^2} \sum_{n<N} (2n+1) u_n$ is equal to the Cesaro mean). We are left with the determination of $\limsup_{N \to \infty} \frac{1}{N} \sum_{n<N} Y_n$. We write $S_j = \sum_{k \leq 2j+1} e(kX_j)$ and we decompose

$$\begin{aligned} \tfrac{1}{N^2} \textstyle\sum_{n<N^2} Y_n &= \tfrac{1}{N^2} \textstyle\sum_{j<N} S_j \\ &= \tfrac{1}{N^2} (\textstyle\sum_{j \in J_1} S_j + \sum_{j \in J_2} S_j + \sum_{j \in J_3} S_j) \end{aligned}$$

where

$$J_1 = \{j < N,\ X_j = 0\},\ \ J_2 = \{j < N,\ X_j \neq 0,\ |X_j| < 1/\sqrt{j+1}\}$$

and

$$J_3 = \{j < N,\ \ |X_j| \geq 1/\sqrt{j+1}\}.$$

Note that $|S_j| \leq 2j+1$ if $j \in J_1 \cup J_2$ while $|S_j| \leq 2/|e(X_j)-1| \leq 2\sqrt{j+1}$ if $j \in J_3$. Combining those estimates, we get

$$\Big|\sum_{n<N^2} Y_n\Big| \leq \sum_{j\in J_1\cup J_2} (2j+1) + \sum_{j\in J_3} 2\sqrt{j+1}.$$

By the strong law of large numbers,

$$\lim_N \frac{1}{N}\sum_{j<N} \mathbf{1}_{J_1} = P(X=0) = \sigma\{0\};$$

in the same way

$$\lim_N \frac{1}{N}\sum_{j<N} \mathbf{1}_{J_2}(j) = \frac{1}{N}\sum_{j<N} P(0 < |X_j| < 1/\sqrt{j+1}) = \sigma(]0, 1/\sqrt{j+1}[),$$

while the second sum is $O(m^{3/2})$. From the remark on Cesaro means, we deduce that $\lim_{N\to\infty} \frac{1}{N}\sum_{n<N} Y_n = \sigma\{0\}$ and, finally, $c_+ \leq b$. The theorem follows. □

By Wiener's lemma 1.1, the Fourier transform of a continuous measure must tend to 0 on a sequence of integers of density one; this naturally leads to the following objects.

Definition 4.14. $\Lambda \subset \mathbf{Z}$ is a FC^+ *set* if it "forces" the continuity of positive measures : If $\sigma \in M(\mathbf{T})$ is a positive measure,

$$\hat{\sigma}(h) \to 0 \text{ as } |h| \to \infty \text{ and } h \in \Lambda, \quad \text{implies} \quad \sigma \text{ continuous.} \tag{4.11}$$

This definition is equivalent, as above, to the one with the weaker conclusion : $\sigma\{0\} = 0$. By extension, $\Lambda \subset \mathbf{N}$ is a FC^+ *set* in $\mathbf{N}$ if $\Lambda \cup (-\Lambda)$ is a FC^+ *set* in $\mathbf{Z}$. Clearly, from the previous theorem, a FC^+ set in $\mathbf{N}$ is a vdC set.

Classical Examples

1. *The set* $\Lambda = \{n^2,\ n \in \mathbf{Z}\}$ *of squares is* FC^+.

   Proof. Let σ be a positive measure on $\mathbf{T}$ such that

$$\lim_{n\to\infty} \widehat{\sigma}(n^2) = 0.$$

We shall prove that $\sigma\{0\} \leq \varepsilon$ for any given $\varepsilon > 0$. We consider, for $m \geq 1$, $E_m = \{\frac{2j\pi}{m},\ j = 0, 1, \ldots m-1\}$ and we decompose

$$\frac{1}{N}\sum_{n=1}^{N} \widehat{\sigma}(n^2m^2) = \int_{\mathbf{T}} \frac{1}{N}\sum_{n=1}^{N} e^{in^2m^2t}\, d\sigma(t)$$
$$= \int_{t \notin \mathbf{Q}\pi} + \int_{t \in (\mathbf{Q}\pi)\setminus E_m} + \int_{E_m} =: A_N + B_N + C_N.$$

We choose m large enough to ensure that $\sigma(\mathbf{Q}\pi \setminus E_m) \leq \varepsilon$ so that $|B_N| \leq \varepsilon$ for every N; clearly

$$C_N = \sum_{t \in E_m} \sigma\{t\}$$

and, by the uniform distribution (mod 2π) of (n^2t) when $t/\pi \notin \mathbf{Q}$,

$$\lim_{N \to \infty} A_N = 0.$$

Finally, taking the limit on N, we deduce from the assumption on σ that

$$\sum_{t \in E_m} \sigma\{t\} \leq \varepsilon, \quad \text{for every } \varepsilon > 0$$

and $\sigma\{0\} = 0$. □

2. *If F is an infinite subset of $\mathbf{Z}$, then $\Lambda = F - F$ is FC^+.*

Proof. We proceed in the same way. Let σ be a positive measure on $\mathbf{T}$ such that

$$\lim_{\substack{n \to \infty \\ n \in F - F}} \widehat{\sigma} = 0.$$

If $F = \{\lambda_1 < \lambda_2 < \cdots < \lambda_N < \cdots\}$, we have

$$\sigma\{0\} \leq \int_{\mathbf{T}} \left| \frac{1}{N}\sum_{j=1}^{N} e^{i\lambda_j t} \right|^2 d\sigma(t) = \frac{1}{N^2} \sum_{1 \leq i,j \leq N} \widehat{\sigma}(\lambda_i - \lambda_j),$$

hence

$$\sigma\{0\} \leq \frac{1}{N^2} \sum_{n=1}^{N} \sum_{\substack{|i-j|=n \\ i \leq N, j \leq N}} |\widehat{\sigma}(\lambda_i - \lambda_j)|.$$

There exist at most $2N$ pairs (i, j) with $|i - j| = n$, $i \leq N, j \leq N$, so that, given $\varepsilon > 0$, one can find n_0 such that

$$|\widehat{\sigma}(\lambda_i - \lambda_j)| \leq \varepsilon \quad \text{if} \quad |i - j| \geq n_0.$$

Thus,

$$\sigma\{0\} \leq \frac{1}{N^2} \sum_{n=1}^{n_0} \sum_{\substack{|i-j|=n \\ i \leq N, j \leq N}} |\widehat{\sigma}(\lambda_i - \lambda_j)| + \frac{1}{N^2} \sum_{n=n_0+1}^{N} 2N\varepsilon$$

and

$$\sigma\{0\} \leq 2\varepsilon.$$

The result follows. □

A shorter proof involves technics of generalized characters (section 1.2).

3. By making use of a very useful criterion, Kamae and Mendès France exhibit in [138] a lot of vdC sets; the following example is rather surprizing : *if $\mathscr{P}$ is the set of prime numbers, $\mathscr{P} - 1$ and $\mathscr{P} + 1$ are vdC sets but this is false for any other translate of $\mathscr{P}$.*
4. We identify a sequence of integers with the set of its values. As a consequence of the dominated convergence theorem, *any ergodic sequence is FC^+* (see the end of chapter 2).

We now introduce another class of sets of integers in a dynamical context. Let $(X, \mathscr{B}, \mu, T)$ be a dynamical system and $A \in \mathscr{B}$ with $\mu(A) > 0$.

Definition 4.15. We say that $h \geq 1$ is a *return time* into A if $\mu(A \cap T^{-h}A) > 0$ and we denote by $\mathscr{R}_A$ this set of integers.

If we consider the spectral measure $\sigma_{\mathbf{1}_A}$, defined by

$$\widehat{\sigma}_{\mathbf{1}_A}(k) = \mu(A \cap T^{-k}A) \quad \text{for} \quad k \geq 0$$

and

$$\widehat{\sigma}_{\mathbf{1}_A}(-k) = \overline{\widehat{\sigma}_{\mathbf{1}_A}(k)},$$

then, by (4.6),

$$\mu(A) = \widehat{\sigma}_{\mathbf{1}_A}(0) \leq \sigma_{\mathbf{1}_A}\{0\}^{1/2}.$$

Let now Λ be a vdC set; since $\sigma_{\mathbf{1}_A}$ is not continuous, there must be some $h \in \Lambda$ such that

$$\widehat{\sigma}_{\mathbf{1}_A}(h) = \mu(A \cap T^{-h}A) > 0$$

and $h \in \mathscr{R}_A$. This motivates the definition below due to Furstenberg.

Definition 4.16. The set $\Lambda \subset \mathbf{N}^*$ is called a *recurrence set* (also called *Poincaré set*) if the following holds : for every dynamical system $(X, \mathscr{B}, \mu, T)$ and $A \in \mathscr{B}$ with $\mu(A) > 0$, there exists $h \in \Lambda$ such that $\mu(A \cap T^{-h}A) > 0$; or, equivalently, for every dynamical system $(X, \mathscr{B}, \mu, T)$ and $A \in \mathscr{B}$,

$$\widehat{\sigma}_{\mathbf{1}_A} = 0 \quad \text{on } \Lambda \Longrightarrow \sigma_{\mathbf{1}_A}\{0\} = 0.$$

We have just noticed that a vdC set is a recurrence set. It was also observed by Kamae and Mendès France [138] that vdC sets are *intersective* sets, the definition of which follows.

Definition 4.17. $\Lambda \subset \mathbf{N}^*$ is an *intersective set* if :
for every set $B \subset \mathbf{N}^*$ with upper density $\bar{d}(B) > 0$, then $\Lambda \cap (B - B) \neq \emptyset$.

(Recall that $\bar{d}(B) = \limsup_{N\to\infty} \frac{1}{N}\text{Card}(B \cap [1,N])$.)

Following [25, 219], we shall prove that these two definitions, in fact, are equivalent.

Proposition 4.12. *Λ is a recurrence set if and only if Λ is an intersective set.*

Proof. ▷ Suppose that Λ is a recurrence set. If $B \subset \mathbf{N}^*$ with $\bar{d}(B) > 0$, we consider $u \in \{0,1\}^{\mathbf{N}^*}$ defined by $u_n = \mathbf{1}_B(n)$, and the associated dynamical system $(X, \mathscr{B}, \mu, T)$, where $X = \overline{O(u)}$ and μ any probability measure in $M(u)$. We may assume that μ is the weak-star limit of

$$\frac{1}{N_j} \sum_{n<N_j} \delta_{T^n u}$$

with, in addition,

$$\lim_{j\to\infty} \frac{1}{N_j} \sum_{n<N_j} u_n = \bar{d}(B) > 0.$$

But $\frac{1}{N_j}\sum_{n<N_j} u_n = \frac{1}{N_j}\sum_{n<N_j} \pi_0(T^n u)$ converges to $\int_X \pi_0 \, d\mu$ since π_0 is continuous on the compact set X. It follows that

$$\bar{d}(B) = \mu\{x \in X,\ \pi_0(x) = 1\} =: \mu(A).$$

Applying the recurrence assumption with $(X, \mathscr{B}, \mu, T)$ and the set A, we deduce that $\mu(A \cap T^{-h}A) > 0$ for some $h \in \Lambda$. Now

$$\begin{aligned}\mu(A \cap T^{-h}A) &= \int_X \pi_0(T^h x)\pi_0(x) \, d\mu(x) = \lim_{j\to\infty} \frac{1}{N_j} \sum_{n<N_j} \mathbf{1}_B(n+h)\mathbf{1}_B(n) \\ &= \lim_{j\to\infty} \frac{1}{N_j} \sum_{n<N_j} \mathbf{1}_{(B-h)\cap B}(n) > 0;\end{aligned}$$

a fortiori $\Lambda \cap (B - B) \neq \emptyset$ and Λ is intersective.

◁ Conversely, Λ is now an intersective set. We consider any dynamical system $(X, \mathscr{B}, \mu, T)$ and $A \in \mathscr{B}$ with $\mu(A) > 0$. Since

$$\mu(A) = \int_X \frac{1}{N} \sum_{n<N} \mathbf{1}_A(T^n x) \, d\mu(x),$$

the set

$$E = \{x \in X,\ \lim_{N\to\infty} \frac{1}{N} \sum_{n<N} \mathbf{1}_A(T^n x) > 0\}$$

need to have positive μ-measure; otherwise, by using Birkhoff's theorem and the dominated convergence theorem, $\mu(A) = \int_X E^{\mathscr{I}}(\mathbf{1}_A)\, d\mu$ would be equal to zero.

Besides, for every $x \in E$, there exists $B_x =: B \subset \mathbf{N}^*$ with positive density, such that $T^n x \in A$ whenever $n \in B$. We fix $x \in X$ and we use the assumption on Λ : since $\Lambda \cap (B - B) \neq \emptyset$, there exist $\ell, m \in B$ such that $T^\ell x \in A$, $T^m x \in A$, and $h = \ell - m \in \Lambda$. Thus $T^m x$ belongs to $A \cap T^{-h}A$.

Finally, recall that $\mu(E) > 0$. There must exist $m \in \mathbf{N}^*$ and $h \in \Lambda$ such that

$$\mu\{x \in E,\ T^m x \in A \cap T^{-h}A\} > 0;$$

in turn, $\mu(A \cap T^{-h}A) > 0$, and Λ is a recurrence set. □

Corollary 4.5. *If $B \subset \mathbf{N}^*$ has an upper density $\bar{d}(B) > 0$, then $B - B$ must contain some square.*

Remark 4.6. 1. It is natural to ask whether recurrence sets are vdC sets ?

This would be true if, for every positive measure σ with $\sigma\{0\} > 0$, the set

$$E_\varepsilon = \{k,\ \widehat{\sigma}(k) > \varepsilon\}$$

would contain, for $\varepsilon > 0$ small enough, a difference set $B - B$ where $\bar{d}(B) > 0$ (and not only for measures of the form $\sigma_{\mathbf{1}_A}$, $\mu(A) > 0$). In this case, Λ being intersective would intersect E_ε and thus $\widehat{\sigma}(h) > 0$ for some $h \in \Lambda$. Λ would be vdC.

Observe now, by proposition 2.17, that the set E_ε has bounded gaps for every $\varepsilon < \sigma\{0\}$ and the same is true for any difference set $B - B$ with $\bar{d}(B) > 0$:

Lemma 4.3. *Let $B \subset \mathbf{N}^*$ with $\bar{d}(B) > 0$; then $B - B$ has bounded gaps.*

Proof. We consider as above a correlation measure σ of the sequence $u = (\mathbf{1}_B(n))$. Since $\mathbf{1}_B(n+k)\mathbf{1}_B(n) = 0$ if $k \notin B - B$, the spectrum of σ is included in $B - B$. If now $B - B$ has arbitrarily large gaps, we proved in the first chapter (1.3) that σ must be continuous; but, from (4.6),

$$\sigma\{0\}^{1/2} \geq \limsup_{N\to\infty} \Big| \frac{1}{N} \sum_{n<N} u_n \Big| \geq \bar{d}(B) > 0,$$

and we get a contradiction. □

2. However, J. Bourgain constructed in [35] a recurrence set which is not a vdC set.

Theorem 4.8. *Define the infinite convolution*

$$\nu = *_{j=1}^{\infty}\left(\frac{1}{2}(\delta_{-\frac{2\pi}{j!}} + \delta_{\frac{2\pi}{j!}})\right)$$

and let $\mu = \delta_0 + \nu$. *Let now* $\alpha : \mathbf{N} \to \mathbf{R}^+$ *satisfy* $\lim_{n\to\infty} \alpha(n) = \infty$. *Then the set*

$$\Lambda = \bigcup_N \{1 \le n \le N,\ |\hat{\mu}(n)| < \frac{\alpha(N)}{N}\}$$

is a recurrence set, but is not FC^+ *provided that* $\lim_{n\to\infty} \frac{\alpha(n)}{n} = 0$.

3. Nevertheless problems subsist in the classification of these sets of integers, also in connection with the classical thin sets in Harmonic Analysis (Sidon sets, Bohr-dense sets, and so on.) See [23] for a recent overview on the subject.

4.3.2.3 Wiener-Wintner's Theorem

A simple use of Fubini's theorem leads to the following weighted ergodic theorem : *Let* $(X, \mathscr{B}, \mu, T)$ *be an ergodic dynamical system and let* $f \in L^1(X, \mu)$. *Then* $\lim_{n\to\infty} \frac{1}{n}\sum_{j=0}^{n-1} z^j f(T^j x)$ *exists for almost every* $x \in X$ *and for* **almost** *every* $z \in \mathbf{U}$. Actually, this result has been improved by N. Wiener and H. Wintner, and more recently, different versions emphasize the link with the concept of *disjoint systems* [91, 163, 215]. In this subsection, we give a proof of the initial Wiener-Wintner theorem as a beautiful application of the correlation measures [22, 246], and we quote some possible extensions.

Theorem 4.9 (Wiener & Wintner). *Let* $(X, \mathscr{B}, \mu, T)$ *be an ergodic dynamical system and let* $f \in L^1(X, \mu)$. *Then there exists* $X_f \subset X$ *of measure one such that*

$$\lim_{n\to\infty} \frac{1}{n}\sum_{j=0}^{n-1} z^j f(T^j x) \quad \text{exists} \tag{4.12}$$

for every $x \in X_f$ *and* **every** $z \in \mathbf{U}$.

Proof. $\star$ Consider first the case where $f \in L^2(X, \mu)$ and, as usual, denote by σ_f the spectral measure of f. By Birkhoff's theorem, for each $k \ge 0$

$$\lim_{n\to\infty} \frac{1}{n}\sum_{j=0}^{n-1} f(T^{j+k}x)\overline{f}(T^j x) = \widehat{\sigma}_f(k) \ \ a.e. \tag{4.13}$$

Putting X_f as the set of $x \in X$ for which the limit in (4.13) exists for all the k's, we have $\mu(X_f) = 1$ and we claim that the theorem is satisfied with this choice.

Recall that $L^2(X, \mu)$ can be decomposed into a sum $L^2(X, \mu) = H_d \oplus H_c$ where H_d is generated by the eigenfunctions of T, and H_c consists of the functions with continuous spectral measure. If f is an eigenfunction of T, (4.12) is obviously satisfied, thus the claim is already proved for $f \in H_d$. Suppose now that $f \in H_c$; in this case, $\lim_{n\to\infty} \frac{1}{n}\sum_{j=0}^{n-1} z^j f(T^j x) = 0$ for every $x \in X_f$ as a consequence of the following lemma 4.4, applied with $a_j = z^j$ and $b_j = f(T^j x)$, $j \geq 0, x \in X_f$.

Lemma 4.4. *Let a and b be two sequences in the Wiener space $\mathscr{S}$ with mutually singular correlation measures : $\sigma_a \perp \sigma_b$. Then,*

$$\lim_{n\to\infty} \frac{1}{n} \sum_{j=0}^{n-1} a_j \overline{b_j} = 0.$$

Note that the lemma is nothing but a particular case of corollary 4.3.

$\star$ If $f \in L^1(X,\mu)$ only, let (g_k) be a sequence of functions in $L^2(X,\mu)$ approximating f in $L^1(X,\mu)$ and for each k, denote by X_k the full measure set associated with g_k by the first step and such that, in addition,

$$\lim_{n\to\infty} \frac{1}{n} \sum_{j=0}^{n-1} \left| f(T^j x) - g_k(T^j x) \right| = \int_X |f - g_k| \, d\mu \quad \text{if } x \in X_k;$$

this is possible thanks to the pointwise ergodic theorem. We now show that $X_f = \cap_k X_k$ does the job; indeed fix $\varepsilon > 0$ and $k := k_\varepsilon$ such that $\int_X |f - g_k| \, d\mu \leq \varepsilon$. If $x \in X_f$, we have both

$$\limsup_{n\to\infty} \frac{1}{n} \left| \sum_{j=0}^{n-1} z^j f(T^j x) - \sum_{j=0}^{n-1} z^j g_k(T^j x) \right| \leq \varepsilon$$

and

$$\lim_{n\to\infty} \frac{1}{n} \sum_{j=0}^{n-1} z^j g_k(T^j x) \quad \text{exists.}$$

It follows that $\frac{1}{n}\sum_{j=0}^{n-1} z^j f(T^j x)$ in turn admits a limit, by Cauchy's criterion for example. □

The following improvement for uniquely ergodic systems has been brought out by E. Arthur Robinson [215]:

Theorem 4.10. *Let (X,T) be a topological dynamical system with unique T-invariant probability measure μ. Then, for every continuous eigenvalue $\lambda \in \mathbf{U}$ of T and $f \in C(X)$,*

$$\lim_{N\to\infty} \frac{1}{N} \sum_{n<N} f(T^n x) \lambda^{-n} = P_\lambda f \in C(X), \quad \textit{uniformly in } x \in X$$

where P_λ is the projection to the eigenspace corresponding to λ; the limit is zero whenever λ is not even a L^2-eigenvalue.

The proof combines the previous theorem and the Oxtoby theorem on uniform convergence of ergodic means.

Extension of Wiener-Wintner theorem to more general weights has been established by A. Bellow and V. Losert [22].

Definition 4.18. A sequence (a_n) is called a *good universal weight* if

$$\lim_{N\to\infty} \frac{1}{N} \sum_{n=0}^{N-1} a_n f(T^n x) \quad \text{exists a.e.}$$

for any ergodic dynamical system and every integrable f.

Sufficient conditions for a bounded sequence a to be a good universal weight are given in [22]. Note the following ones : *a admits a discrete correlation measure and* $\lim_{N\to\infty} \frac{1}{N} \sum_{n<N} a_n e^{int}$ *exists for every* $t \in \mathbf{T}$.

The next theorem is interesting since it provides (rather constraining) quantitative spectral conditions for the pointwise convergence of weighted Birkhoff's sums ([31] and [22] for an improvement).

Theorem 4.11 (Blum & Reich). *Let* (a_n) *be a sequence of* ± 1. *Assume that, for every* $t \in \mathbf{T}$, *one can find* $C(t) > 0$ *and* $\varepsilon(t) > 0$ *such that, for each* N,

$$\left| \sum_{n<N} a_n e^{int} \right| \leq C(t) N^{1-\varepsilon(t)}; \tag{4.14}$$

then, for any dynamical system $(X, \mathscr{B}, \mu, T)$ *and any* $f \in L^1(X, \mu)$,

$$\frac{1}{N} \sum_{n<N} a_n\, f \circ T^n \to 0 \ \ \mu - a.e. \tag{4.15}$$

Proof. Fix $(X, \mathscr{B}, \mu, T)$ and, for $\varepsilon > 0$ and $C > 0$, put

$$E_{\varepsilon,C} = \{t \in \mathbf{T},\ \ C(t) < C,\ \varepsilon(t) < \varepsilon\}.$$

Finally, consider

$$H_{\varepsilon,C} = \{f \in L^2(X, \mu),\ \ \sigma_f(E^c_{\varepsilon,C}) = 0\}.$$

$\star$ We claim that (4.15) holds for every f in some $H_{\varepsilon,C}$. Indeed, by definition of σ_f,

$$\begin{aligned}
\left\| \frac{1}{N} \sum_{n<N} a_n\, f \circ T^n \right\|_2^2 &= \int_{\mathbf{T}} \left| \frac{1}{N} \sum_{n<N} a_n e^{int} \right|^2 d\sigma_f(t) \\
&= \int_{E_{\varepsilon,C}} \left| \frac{1}{N} \sum_{n<N} a_n e^{int} \right|^2 d\sigma_f(t) \\
&\leq \int_{E_{\varepsilon,C}} C^2 N^{-2\varepsilon}\, d\sigma_f(t) \\
&\leq C^2 ||f||_2^2 N^{-2\varepsilon}.
\end{aligned}$$

Choosing $N_k = [k^{1/\varepsilon}]$, clearly,

$$\sum_k \left\| \frac{1}{N_k} \sum_{n<N_k} a_n \, f \circ T^n \right\|_2^2 < \infty$$

and $\frac{1}{N_k} \sum_{n<N_k} a_n \, f \circ T^n$ tends to 0 μ-a.e. We conclude with a classical interpolation argument.

$\star$ As a consequence of proposition 3.4, note that the set of $f \in L^1(X,\mu)$ for which (4.15) holds must be closed in L^1. The proof will thus be finished if we show that the set $\bigcup_{\varepsilon,C} H_{\varepsilon,C}$ is complete in L^2, thus in L^1.
Observe first, with notations of chapter 2 and $U = U_T$, that

$$\mathbf{1}_{E_{\varepsilon,C}}(U)(L^2(X,\mu)) \subset H_{\varepsilon,C}$$

since

$$\sigma_{\mathbf{1}_{E_{\varepsilon,C}}(U)f} = \mathbf{1}_{E_{\varepsilon,C}} \sigma_f \ \text{ if } f \in L^2(X,\mu).$$

Let now $g \in L^2(X,\mu)$ be orthogonal to the $H_{\varepsilon,C}$ for every $\varepsilon > 0$ and $C > 0$. In particular,

$$g \perp \mathbf{1}_{E_{\varepsilon,C}}(U)U^k g \ \text{ for any } \ k \in \mathbf{N}$$

which implies

$$\sigma_{g,\mathbf{1}_{E_{\varepsilon,C}}(U)g} = \mathbf{1}_{E_{\varepsilon,C}} \sigma_g = 0,$$

for every $\varepsilon > 0, C > 0$.
But, by assumption on the sequence (a_n), $\bigcup_{\varepsilon,C} E_{\varepsilon,C} = \mathbf{T}$. It follows that $\sigma_g = 0$ and g itself is zero. This proves that (4.15) holds for every $f \in L^1(X,\mu)$. □

This allows us to turn back to ergodic and weakly ergodic sequences in connection with the ergodic theorem along subsequences (section 2.6 and subsection 3.2.2). Analogously we introduce :

Definition 4.19. A sequence of positive integers $\{n_1 < n_2 < \ldots\}$ is said to be a *good universal subsequence* if the pointwise ergodic theorem holds along this subsequence, in other terms, if

$$\lim_{K\to\infty} \frac{1}{K} \sum_{k=1}^{K} f(T^{n_k}x) \quad \text{exists a.e.}$$

for any ergodic dynamical system and every integrable f.

Thanks to uniform integrability, such a sequence needs be weakly ergodic, since the L^2-convergence of ergodic means holds along a weakly ergodic sequence (with the expected limit if exactly ergodic).

Along subsequences having positive density, the pointwise ergodic theorem may be reduced to a weighted pointwise ergodic one by putting $a = \mathbf{1}_{(n_k)}$.

Remark 4.7. 1. Good universal subsequences with zero density do exist : such are the squares and the prime numbers [35, 36].

2. A general process to exhibit good universal subsequences led to the so-called *return time sequences* [38]:

   Theorem 4.12 (Bourgain, Furstenberg, Katznelson, Ornstein). *Let $(Y,\mathscr{C},\nu,S)$ be an ergodic dynamical system and $A\in\mathscr{C}$ with positive measure : $\nu(A)>0$. For $y\in Y$ we denote by Λ_y the return time sequence to A :*

   $$\Lambda_y=\{n\in\mathbf{N},\ T^n y\in A\}.$$

   Then, for $\nu-$ almost all y, Λ_y is a good universal subsequence.

3. A complete survey on these Wiener-Wintner theorems can be found in [18].

4.3.3 Examples of Correlation Measures

1. We turn back to the two-points extension over an irrational rotation (subsection 3.6.1). The skew product $T=T_\phi$ was defined on $\mathbf{T}\times\{-1,1\}$ by

   $$T_\phi(x,y)=(x+\alpha\ (\operatorname{mod}2\pi),\ \phi(x)\cdot y),$$

   with a ±1-valued step function ϕ. The spectral study of T has been derived directly from the spectral properties of σ, the spectral measure of the function $\mathbf{1}$, characterized by

   $$\widehat{\sigma}(n)=\int_{\mathbf{T}}\phi^{(n)}dm,$$

   with, as usual, $\phi^{(n)}(x)=\phi(x)\phi(x+\alpha)\cdots\phi(x+(n-1)\alpha)$.
   Then σ *is the common correlation measure of all the sequences* $(\phi^{(n)}(x))$, $x\in\mathbf{T}$. Actually, this is a consequence of the cocycle identity below, together with the unique ergodicity of R_α. From

   $$\phi^{(n+k)}(x)=\phi^{(n)}(x)\cdot\phi^{(k)}(x+n\alpha)$$

   we get

   $$\frac{1}{N}\sum_{n<N}\phi^{(k)}(x+n\alpha)=\frac{1}{N}\sum_{n<N}\phi^{(n+k)}(x)\cdot\phi^{(n)}(x)$$

   since $\phi(x)=\pm1$. The function ϕ being Riemann-integrable, we obtain by the Weyl theorem (theorem 4.5),

   $$\widehat{\sigma}(k):=\int_{\mathbf{T}}\phi^{(k)}dm=\lim_{N\to\infty}\frac{1}{N}\sum_{n<N}\phi^{(n+k)}(x)\cdot\phi^{(n)}(x)$$

   for every $x\in\mathbf{T}$.

2. Let q be an integer ≥ 2 and $u = (u_n)_{n \geq 0}$ be a sequence of unimodular complex numbers, satisfying

$$u_0 = 1, \quad u_{aq^n+b} = u_a \cdot u_b$$

for every $n \geq 0$, $a \geq 0$ and $0 \leq b < q^n$. Such a sequence is said to be *strongly q-multiplicative* [59, 99]. We shall prove that such a sequence u admits a unique correlation measure, more precisely (see section 1.3):

Proposition 4.13. *Let u be a strongly q-multiplicative sequence; then u admits a unique correlation measure which is the generalized Riesz product*

$$\sigma = w^* - \lim_{N \to \infty} \prod_{n<N} P(q^n t),$$

where

$$P(t) = \frac{1}{q} \left| 1 + u_1 e^{it} + \cdots + u_{q-1} e^{i(q-1)t} \right|^2.$$

Proof. We write $u(n)$ instead of u_n for sake of clearness. For the unicity of the correlation measure, we observe, following [28], that, for every fixed $k \geq 0$, there exists a partition of $\mathbf{N} = \cup_m \mathscr{P}_m$ into arithmetical progressions such that

$$u(n+k)\overline{u(n)} = e^{i\lambda_m} := constant$$

when n runs over $\mathscr{P}_m = \{a_m + q^{r_m}\mathbf{N}\}$, $a_m < q^{r_m}$, (we omit the details). Now, if $N \geq 1$, $\mathscr{P}_m \cap [0, N-1] \neq \emptyset$ implies $a_m \leq N-1$ necessarily, and thus,

$$\text{Card}\,\{m, \mathscr{P}_m \cap [0, N-1] \neq \emptyset\} = O(\ln N).$$

From this remark, we deduce that

$$\begin{aligned}
\tfrac{1}{N} \textstyle\sum_{n<N} u(n+k)\overline{u(n)} &= \tfrac{1}{N} \textstyle\sum_{m;\ \mathscr{P}_m \cap [0,N-1] \neq \emptyset} e^{i\lambda_m}\, \text{Card}\,(\mathscr{P}_m \cap [0, N-1]) \\
&= \tfrac{1}{N} \textstyle\sum_m e^{i\lambda_m} \left(\frac{N-1}{q^{r_m}}\right) + O\!\left(\frac{\ln N}{N}\right).
\end{aligned}$$

The existence of the limit $\lim_{N \to \infty} \frac{1}{N} \sum_{n<N} u(n+k)\overline{u(n)}$ for every k follows now from the convergence of the series $\sum_m \frac{e^{i\lambda_m}}{q^{r_m}}$.

It remains to identify this unique measure. From proposition 4.9, σ is the weak-star limit point of the sequence $(R_n \cdot m)$, with

$$R_N(t) = \frac{1}{q^N} \Big| \sum_{n<q^N} u(n) e^{int} \Big|^2.$$

Letting then, as in section 1.3,

$$P(t) = \frac{1}{q} \Big| \sum_{j<q} u(j) e^{ijt} \Big|^2$$

and

$$P_m(t) = \prod_{j=0}^{m-1} P(q^j t)$$

we shall prove by induction on $m \geq 1$ that

$$P_m = R_m \quad \text{for all } m \geq 1. \tag{4.16}$$

Identity (4.16) is clear for $m = 1$; now suppose that (4.16) is checked for some $m \geq 1$. We readily have

$$\begin{aligned} P_{m+1}(t) &= P(t)P_m(qt) \\ &= \tfrac{1}{q}\tfrac{1}{q^m}\left|\sum_{j<q}\sum_{k<q^m} u(j)u(k)e^{ijt}\, e^{ikqt}\right|^2 \\ &= \tfrac{1}{q^{m+1}}\left|\sum_{j+qk<q^{m+1}} u(j)u(k)e^{i(j+kq)t}\right|^2 \\ &= \tfrac{1}{q^{m+1}}\left|\sum_{\ell<q^{m+1}} u(\ell)e^{i\ell t}\right|^2 \end{aligned}$$

We used the description $\{\ell < q^{m+1}\} = \{j + qk, j < q, k < q^m\}$ and the q-multiplicativity property of the sequence.

It follows that σ is the weak-star limit point of the sequence $(\prod_{j<N} P(q^j t)\cdot m)_{N\geq 1}$ and, thereby, $\sigma = \rho$, the generalized Riesz product constructed on $u_0 = 1, u_1, \ldots, u_{q-1}$. □

3. An example of such a q-multiplicative sequence is given by

$$u_n = e^{2i\pi c S_q(n)}, \quad n \in \mathbf{N},$$

where $S_q(n)$ is the sum of the digits of n in the q-adic representation, and c is a real constant. In this case,

$$P(t) = \frac{1}{q}\left|\sum_{j<q} e^{2i\pi jc} e^{ijt}\right|^2$$

and σ is discrete if and only if $|\widehat{\sigma}(1)| = 1$; a straightforward computation leads to $\widehat{\sigma}(1) = \dfrac{(q-1)e^{2i\pi qc}}{qe^{2\pi i(q-1)c} - 1}$; thus σ is discrete if and only if $c(q-1) \in \mathbf{Z}$. Otherwise σ is continuous by D-ergodicity (see [59, 202]). In this case, the topological system associated to the sequence is uniquely ergodic [134, 166], and, in any case we have

$$\sigma\{\lambda\}^{1/2} = \limsup_{N\to\infty} \frac{1}{N}\left|\sum_{n<N} u_n e^{-in\lambda}\right|.$$

If we particularize $c = -1$ and $q = 2$, the sequence becomes

$$u_n = (-1)^{S_2(n)} =: t_n$$

and we get the Thue-Morse sequence on the alphabet $\{-1, 1\}$, defined by $t_0 = 1$, $t_{2n} = t_n$, $t_{2n+1} = -t_n$, as easily seen. The unique correlation measure of this sequence is thus the generalized product (see also [143])

$$\rho = \prod_{n \geq 0} (1 - \cos 2^n t).$$

This provides the last result of chapter 3.

4. In the next chapter we focus on fixed points of primitive substitutions and prove for these sequences, according to proposition 4.11 b)

$$\sigma\{\lambda\}^{1/2} = \lim_{N \to \infty} \frac{1}{N} \left| \sum_{n<N} u_n e^{-in\lambda} \right|$$

where σ is the unique correlation measure of u.

The Thue-Morse sequence and the Rudin-Shapiro sequence both arise from a primitive substitution and we rediscover in this way both correlation measures.

4.3.4 Back to Finite-Valued Sequences

As in subsection 4.3.1, $(X, \mathscr{B}, \mu, T)$ is the dynamical system associated to the sequence $u \in A^{\mathbf{N}}$, where A is a compact subset of $\mathbf{C}$. For sake of simplicity, we suppose that $M(u) = \{\mu\}$ and we denote by $\sigma_{\max}$ the maximal spectral type of the system. If σ denotes the correlation measure of the sequence u, that is $\sigma := \sigma_{\pi_0}$ with our previous notations, σ inherits properties from $\sigma_{\max}$ (by absolute continuity). When A is finite, we can prove in the opposite direction the following result.

Proposition 4.14. *Assume Card $A < +\infty$. If the correlation measure σ is discrete, the system $(X, \mathscr{B}, \mu, T)$ has discrete spectrum.*

Proof. Since $\sigma = \sigma_{\pi_0}$, σ is discrete if and only if π_0 belongs to the closed span of the eigenfunctions (f_λ). Let B denote the algebra generated by these eigenfunctions; we shall prove that in fact $\overline{B} = L^2(X, \mu)$.

We may assume that the (f_λ) are bounded by 1. Let now $\mathscr{A}$ be the σ-algebra : $\sigma\{(f_\lambda)\}$; if $f \in B$, f is $\mathscr{A}$-measurable and $\overline{B} \subset L^2(\mathscr{A})$. If, conversely, $f \in L^2(\mathscr{A})$, we can approach f by some function $\psi(f_{\lambda_1}, \dots, f_{\lambda_r})$ where $\psi \in L^2(D^r)$ and D is the closed unit disc. Then, approaching ψ by a polynomial in $L^2(D^r)$, we deduce that $f \in \overline{B}$. We have thus proved that $\overline{B} = L^2(\mathscr{A})$.

From our assumption, the projections π_n, $n \geq 0$, are $\mathscr{A}$-measurable, so that, every cylinder set of the form

$$C = [\alpha_0 \cdots \alpha_n] := \{x \in X,\ x_j = \alpha_j,\ 0 \leq j \leq n\} = (\pi_0, ..., \pi_n)^{-1}(\alpha_0, ..., \alpha_n)$$

with $\alpha_0, ..., \alpha_n \in A$, belongs to $\mathscr{A}$. It follows that every simple function $f = \sum_{finite} c_i \mathbf{1}_{C_i}$, finite-valued and constant on cylinder sets, is $\mathscr{A}$-measurable. But, the alphabet A being finite, the cylinder sets are generating the Borel σ-algebra $\mathscr{B}$ on X and the class of simple functions is dense in $L^2(X,\mu)$. This proves that $L^2(X,\mu) = L^2(\mathscr{A})$.

Finally $L^2(X,\mu) = \overline{B}$ and the system $(X, \mathscr{B}, \mu, T)$ has discrete spectrum. □

Chapter 5
Dynamical Systems Arising from Substitutions

In this chapter, and in the two forthcoming chapters, we are dealing with particular sequences, generated by a substitution on a finite alphabet (substitutive and automatic sequences). There exist a lot of contributions around these sequences, many of which are concerned with their combinatorial and arithmetical properties, others with the properties of the associated topological system and more recently, geometric aspects have been intensively studied. J.P. Allouche and J. Shallit have devoted a substantial book to Automatic Sequences [14] while all recent results on substitutions have been gathered in an impressive collective work [88].

We present, in this chapter, the basic notions and properties of dynamical systems, associated to a wide class of substitutions, in order to achieve later their spectral analysis. Most of the results of this chapter are already contained in [68] and [188] where the authors deal with two-sided subshifts, but we begin with one-sided sequences which lead to simpler formulations.

5.1 Definitions and Notations

Throughout this chapter, we refer to the notations of section 4.2 : A is a finite set with Card $A = s \geq 2$, which may be identified with $A = \{0, 1, \ldots, s-1\}$. $A^* = \cup_{k \geq 0} A^k$ consists in all words on the alphabet A and the empty word . For $\omega = \omega_1 \cdots \omega_n \in A^*$, $|\omega| = n$ is the *length* of the word ω, 0 for the empty word. Henceforth, unless explicit mention, a word will be non-empty.

Definition 5.1. A *substitution* ζ on A (or, on s symbols) is a map from A to A^+, the non-empty words on A, which associates to the letter $\alpha \in A$, the word $\zeta(\alpha)$, with length $|\zeta(\alpha)| =: \ell_\alpha$.

Thus $\ell_\alpha \geq 1$ for every α and it is assumed to be ≥ 2 for at least one α.

The substitution ζ induces a *morphism of the monoid* A^* endowed with concatenation, by putting

$$\zeta(B) = \zeta(b_0)\zeta(b_1)\cdots\zeta(b_n) \quad \text{if} \quad B = b_0 \cdots b_n \in A^+$$

M. Queffélec, *Substitution Dynamical Systems – Spectral Analysis: Second Edition*, Lecture Notes in Mathematics 1294, DOI 10.1007/978-3-642-11212-6_5,
© Springer-Verlag Berlin Heidelberg 2010

and $\zeta(\emptyset) = \emptyset$. It follows that we can iterate the substitution ζ on A and we denote by $\zeta^k = \zeta \circ \zeta^{k-1}$ the k-times iterated map. At this stage, we are able to define the *language* of ζ :

$$\mathscr{L}_\zeta = \{\text{factors of } \ \zeta^n(\alpha), \ \text{ for some } n \geq 0, \ \alpha \in A\}$$

and the *subshift* (X_ζ, T) where

$$X_\zeta = \{x \in A^{\mathbf{N}}, \ \mathscr{L}(x) \subset \mathscr{L}_\zeta\}.$$

Likewise, we define a map from $A^{\mathbf{N}}$ to $A^{\mathbf{N}}$, the set of all infinite words on A, that we also denote by ζ and still call "substitution", by the formula

$$\zeta(x) = \zeta(x_0)\zeta(x_1)\cdots \quad \text{if} \quad x = (x_n)_{n\geq 0} \in A^{\mathbf{N}}.$$

Notice that this map $\zeta : A^{\mathbf{N}} \to A^{\mathbf{N}}$ is continuous but not onto. We are interested in the possible fixed points or periodic points of ζ, that we call *substitutive sequences* (associated to ζ). Such points do exist (periodic points of ζ must not be confused with periodic sequences which preferably will be called *shift-periodic* if necessary).

Proposition 5.1. *Let ζ be a substitution on the alphabet A, such that,*

$$\lim_{n\to\infty} |\zeta^n(\alpha)| = +\infty, \quad \forall \alpha \in A.$$

Then, ζ admits periodic points : there exist $u \in A^{\mathbf{N}}$ and $k \geq 1$ such that $u = \zeta^k(u)$.

Proof. Since A is a finite set, we can find $k \geq 1$ and $\alpha \in A$ such that the word $\zeta^k(\alpha)$ begins with α. For an arbitrary sequence $x \in A^{\mathbf{N}}$ with $x_0 = \alpha$, the infinite word $\zeta^{kn}(x)$ begins with $\zeta^{kn}(\alpha)$ so that $\zeta^{kn}(x)$ and $\zeta^{kn+kp}(x)$ begin with the same word $\zeta^{kn}(\alpha)$, whose length grows to infinity as $n \to \infty$. It follows that $(\zeta^{kn}(x))_{n\geq 1}$ is a Cauchy sequence in the compact $A^{\mathbf{N}}$ and converges to a sequence u satisfying $u = \zeta^k(u)$ and $u_0 = \alpha$. The sequence u can be obtained by iterating the substitution ζ^k on the initial letter α. □

Inspired by recent contributions, we shall call a *morphic sequence*, any letter-to-letter projection of a substitutive one.

From now on, we make the following assumptions on ζ (replacing ζ by an iterate if necessary) :

(i) $\lim_{n\to\infty} |\zeta^n(\beta)| = +\infty$ for every $\beta \in A$.

(ii) There exists a letter $\alpha_0 \in A$ such that $\zeta(\alpha_0)$ begins with α_0.

According to the previous proposition, ζ admits a fixed point that we denote by u. To avoid artificial and useless intricacy, we only keep in the alphabet, the letters which really appear in u. Thus we define now the language of $u = \zeta^\infty(\alpha_0)$:

$$\mathscr{L}(u) = \{\text{factors of } \ \zeta^n(\alpha_0), \ \text{ for some } n \geq 0\}$$

Definition 5.2. A substitution is said to be *of constant length* (or the associated morphism is *uniform*) if there exists $q \geq 2$ such that

$$\ell_i := |\zeta(i)| = q, \quad \text{for every } 0 \leq i \leq s-1.$$

A *q-automaton* consists in q *instructions* on a finite alphabet A, namely q maps : $\phi_j : A \to A$, $0 \leq j \leq q-1$, together with an output function $f : A \to B$; the automaton is acting on A as an adding machine and, starting from an initial element (or state) $a \in A$, generates the sequence $v = (v_n)_{n \geq 0}$ where

$$v_n = \phi_{n_0} \circ \phi_{n_1} \circ \cdots \circ \phi_{n_{k-1}}(a),$$

if

$$n = n_0 + n_1 q + \cdots + n_{k-1} q^{k-1}, \; n_{k-1} \neq 0.$$

The sequence $u = f(v)$ with values in B is called *q-automatic*.

If now ζ is a substitution of length q and if $m, j, r \in \mathbf{N}$ are such that $mq^j + r < q^k$, note that

$$\zeta^k(\alpha) = \zeta^j(\zeta^{k-j}(\alpha)_m)_r, \tag{5.1}$$

where $\zeta(\beta)_j$ refers to as the jth letter of the word $\zeta(\beta)$. A fixed point of some substitution ζ of length q thus comes from the q-automaton whose instructions are

$$\phi_j(\alpha) = \zeta(\alpha)_j, \quad \alpha \in A.$$

In fact, for uniform morphisms, the morphic sequences are exactly the automatic sequences according to the seminal result [55, 57]:

Theorem 5.1 (Cobham). *Let $u = (u_n)_{n \geq 0}$ be a sequence with values in a finite alphabet. Then u is q-automatic if and only if u is the image by a letter-to-letter projection of a fixed point of a substitution of constant length q.*

This theorem could provide a definition of an automatic sequence from the "substitution" point of view. There is an interesting characterization of automatic sequences [80]:

Theorem 5.2 (Eilenberg). *$u = (u_n)_{n \geq 0}$ is q-automatic if and only if the set of subsequences*

$$\{(u_{nq^k+m})_{n \geq 0},\; k \geq 0,\; 0 \leq m < q^k - 1\}$$

is a finite set.

Basic Examples

1. ζ defined on $A = \{0, 1\}$ by $\zeta(0) = 010$ and $\zeta(1) = 101$ admits two fixed points $\zeta^\infty(0)$ and $\zeta^\infty(1)$ which are shift-periodic sequences.

2. ζ defined on $A = \{0,1\}$ by $\zeta(0) = 01$ and $\zeta(1) = 10$ also admits two fixed points

$$u = \zeta^{\infty}(0) = 0110100110010110\cdots$$

and $v = \zeta^{\infty}(1)$ obtained by exchanging in u the letters 0 and 1; but we shall see that these sequences are not shift-periodic. The sequence u is the *Thue-Morse sequence* on the alphabet $\{0,1\}$; indeed, since $u = \zeta(u)$,

$$u = \zeta(u_0)\zeta(u_1)\cdots\zeta(u_k)\cdots = \underbrace{u_0u_1}\underbrace{u_2u_3}\cdots\underbrace{u_{2k}u_{2k+1}}\cdots$$

so that $u_{2k} = u_k$ and $u_{2k+1} = 1 - u_k$ for every $k \geq 0$.
3. ζ defined on $A = \{0,1\}$ by $\zeta(0) = 01$ and $\zeta(1) = 0$ admits a unique fixed point $u = \zeta^{\infty}(0) = 0100101001001\cdots$ called the *Fibonacci sequence* (because of the length of the basic words $\zeta^n(0)$ satisfying the inductive relation of the Fibonacci numbers) [208].
4. The Baum-Sweet sequence, defined by

$$u_n = \begin{cases} 0 & \text{if the dyadic expansion of n contains at least one block of 0's} \\ & \text{of odd length,} \\ 1 & \text{if not} \end{cases}$$

is a 2-automatic sequence. Consider, indeed, ζ, the substitution on the alphabet $\{a,b,c,d\}$, defined by

$$a \to ab,\ b \to cb,\ c \to bd,\ d \to dd.$$

We get two fixed points by starting with $\alpha = a$ or $\alpha = d$, while b and c lead to periodic points (on a three-letters alphabet). We obtain the Baum-Sweet sequence by letting $a = b = 1$ and $c = d = 0$ in $\zeta^{\infty}(a)$ [88].

5.2 Minimality of Primitive Substitutions

We consider ζ fulfilling the conditions (i) and (ii) with $\alpha_0 = 0$, and we associate to the infinite word $u = \zeta^{\infty}(0)$ the topological system (subshift) (X,T), where $X = \overline{O(u)}$ is the orbit closure of u in $A^{\mathbf{N}}$ and T the one-sided shift on X.

Proposition 5.2. *The system (X,T) is minimal if and only if, for every $\alpha \in A$, there exists $k \geq 0$ such that 0 occurs in $\zeta^k(\alpha)$.*

Proof. ▷ Suppose that the system (X,T) is minimal : every factor of u occurs in u with bounded gaps (proposition 4.2). Every letter α occurs in $u = \zeta(u)$ so that $\zeta^k(\alpha)$ is a factor of u for every $\alpha \in A$ and every $k \geq 1$. Now, $\lim_{k\to\infty} |\zeta^k(\alpha)| = +\infty$ by our assumption, and for k large enough, $\zeta^k(\alpha)$ must contain the letter 0 whatever α.

◁ Observe first that the system is minimal as soon as 0 occurs in u with bounded gaps. If it is the case indeed, every word $\zeta^n(0)$ with $n \geq 1$ will occur in u with bounded gaps; now, if B is any factor of u, the same will hold for B since $B \subset \zeta^n(0)$ for n large enough.

Suppose that 0 occurs in $\zeta^k(\alpha)$ for $\alpha \in A$ and some $k \geq 0$. Then, $\zeta^{k+1}(\alpha)$ contains $\zeta(0)$, therefore contains 0. Consider now

$$K = \sup_{\alpha \in A} \inf_{k \geq 1} \{k, \ \zeta^k(\alpha) \text{ contains } 0\}; \tag{5.2}$$

by definition of K, $\zeta^K(\alpha)$ contains 0 for every $\alpha \in A$. Since $u = \zeta^K(u)$, it follows that u is the juxtaposition of words of the form $\zeta^K(\alpha)$, all of them containing 0. This proves that 0 occurs in u with bounded gaps and the minimality is established. □

If $u = \zeta(u)$ without any additional assumption on ζ, the condition "$u_0 = 0$ occurs in u with bounded gaps" is sufficient to ensure the minimality of the system (X,T).

Example 5.1. Consider the following substitutions on $\{0,1,2\}$

$$\begin{aligned}
\zeta_1 &: 0 \to 01, \quad 1 \to 20, \quad 2 \to 11\\
\zeta_2 &: 0 \to 01, \quad 1 \to 22, \quad 2 \to 11\\
\zeta_3 &: 0 \to 010, \quad 1 \to 02, \quad 2 \to 1.
\end{aligned}$$

The systems arising from ζ_1 and ζ_3 are minimal, but minimality is failing for ζ_2.

Remark 5.1. A characterization of minimal morphic sequences has been worked out by F. Durand, by coding minimal sequences with help of the *return words* : for a given factor W of an infinite word u, a return word over W is a word beginning with an occurrence of W and ending exactly before the next occurrence of W in u. This can be seen as a symbolic version of the first return map in a dynamical system. If $W \in \mathscr{L}(u)$, let $\Omega := \Omega_W$ be the set of all distinct return words over W. By minimality, Ω is a finite set $\{\omega_i, \ 1 \leq i \leq |\Omega|\}$, and we consider the sequence of successive return words over W occurring in u. The new sequence obtained by identifying ω_i with the letter i is called the *derivated sequence* of u in W. Let D be the set of all derivated sequences of u. The following has been proved in [76].

Theorem 5.3 (F. Durand). *A minimal sequence is a morphic sequence if and only if D is a finite set.*

This result has to be compared with the above Eilenberg's theorem.

Given a substitution with a fixed point u, we have at our disposal two possible systems : the substitution subshift (X_ζ, T) and the system (X,T) where $X = \overline{O(u)}$. Of course, $\overline{O(u)} \subset X_\zeta$ and we shall consider a wide class of substitutions enjoying the minimality property, for which both systems are identical. The following definitions come from the theory of nonnegative matrices and Markov chains.

Definition 5.3. 1. The substitution ζ is said to be *irreducible* on the alphabet A, if, for every pair α, β of letters in A, one can find $k := k(\alpha, \beta)$ such that β occurs in $\zeta^k(\alpha)$.

2. The substitution ζ is said to be *primitive* if there exists k such that, for every α and β in A, β occurs $\zeta^k(\alpha)$ (that is, k can be chosen independent of α and β).

Observe that the initial condition (i) is automatically satisfied by a primitive substitution : if $\alpha \in A$, all the letters of A appear in the word $\zeta^k(\alpha)$ by definition of k; it follows that

$$|\zeta^{km}(\alpha)| \geq (\text{Card}\, A)^m$$

and $|\zeta^n(\alpha)|$, as well as $|\zeta^{km}(\alpha)|$, tends to infinity.

Proposition 5.3. *If u is any fixed point of a primitive substitution ζ, then the systems (X_ζ, T) and $(\overline{O(u)}, T)$ are identical. In particular the system does not depend on the chosen fixed point.*

Proof. If $x \in X_\zeta$, every factor of x occurs in some word $\zeta^n(\alpha)$ with $n \geq 0$ and $\alpha \in A$. But α appears in u and it is the same for $\zeta^n(\alpha)$ since $u = \zeta^n(u)$. It follows that $X_\zeta \subset \overline{O(u)}$. □

From now on, we denote by (X_ζ, T) the substitution subshift for a primitive ζ.

Proposition 5.4. *If ζ is a primitive substitution, then $X_{\zeta^n} = X_\zeta$ for every $n \geq 1$.*

Proof. This follows from the fact that ζ and ζ^n are generating the same language $\mathcal{L}_\zeta$. It is clear from the definition that $\mathcal{L}_{\zeta^{n+1}} \subset \mathcal{L}_{\zeta^n}$ for every $n \geq 1$. Let now $B \in \mathcal{L}_\zeta$; by primitivity, B is a factor of $\zeta^n(\alpha)$ for every $\alpha \in A$ and for every n large enough. It follows that $\mathcal{L}_\zeta \subset \mathcal{L}_{\zeta^n}$ for n large enough, and finally $\mathcal{L}_\zeta = \mathcal{L}_{\zeta^n}$ for every n. □

Thereby, without loss of generality, we may suppose that ζ admits a fixed point. As a direct consequence of proposition 5.2, we have the following.

Proposition 5.5. *The system (X_ζ, T) is minimal if and only if ζ is a primitive substitution.*

Proof. We suppose that (X_ζ, T) is minimal and ζ fulfills the conditions (i) and (ii) with $\alpha_0 = 0$. For every $\alpha \in A$, suppose that there exists $k \geq 0$ such that 0 occurs in $\zeta^k(\alpha)$ and let K be defined by (5.2). Since $\zeta^K(\alpha)$ contains 0 and $\zeta^N(0)$ contains all the letters of A for N large enough, it follows that $\zeta^{N+K}(\alpha)$ contains $\zeta^N(0)$ as a factor for every α, whence the primitivity of ζ. □

This means that *under conditions (i) and (ii), primitivity and irreducibility are equivalent properties.*

5.3 Nonnegative Matrices and ζ-Matrix

Let ζ be a substitution defined on the alphabet A.
Notation : If B and C are two words in A^*, we denote by $L_C(B)$ the *occurrence number* of C in B. In particular, if $\alpha \in A$, $L_\alpha(B)$ is the number of α occurring in B (another standard notation is $|B|_\alpha$ for this number).

Definition 5.4. We call ζ*-matrix* and we denote by $M = M(\zeta)$, the matrix whose entries are $\ell_{\alpha\beta} = L_\alpha(\zeta(\beta))$, for $\alpha, \beta \in A$.

Note that M is a *nonnegative* $s \times s$-matrix (i.e. with nonnegative entries, not all being equal to 0), whose entries are nonnegative integers. For every $\beta \in A$,

$$\sum_{\alpha \in A} L_\alpha(\zeta(\beta)) = |\zeta(\beta)|, \tag{5.3}$$

and

$$M(\zeta^n) = (M(\zeta))^n, \quad \text{for every } n \geq 1. \tag{5.4}$$

If ζ is of constant length q, $\sum_{\alpha \in A} \ell_{\alpha\beta} = q$ for every $\beta \in A$, and $S = M/q$ is a column-stochastic matrix.

If $B \in \mathscr{L}_\zeta$, $L(B)$ denotes the vector in $\mathbf{R}^s$ whose components are the $L_\alpha(B)$, $\alpha \in A$. It is clear that

$$L(\zeta(B)) = M \cdot L(B);$$

in particular, $L(\zeta(\beta)) = (\ell_{\alpha\beta})_{\alpha \in A}$. The map $L : \mathscr{L}_\zeta \to \mathbf{R}^s$ will be called the *composition function*, and, sometimes, M, the *composition matrix*. Its transpose is the *incidence matrix* associated with the graph of the substitution.

The composition matrix does not contain all the informations on the system (X_ζ, T) for the reason that the occurrence order of the letters is ignored (the reduction to the linear system is usually called *abelianization*). However many properties of the system arise from those of the matrix as we shall see (unique ergodicity, discrete part of the spectrum). We are thus led to describe more generally the properties of nonnegative matrices and we shall prove the Perron-Frobenius theorem (see also [224]), whose generalization to nonnegative operators plays an important role in functional analysis of dynamical systems.

Definition 5.5. A nonnegative matrix $M = (m_{ij})_{0 \leq i,j \leq s-1}$ is said to be *irreducible* if, for every i, j, there exists $k \geq 1$ such that $m_{ij}^{(k)} > 0$, denoting by $(m_{ij}^{(k)})$ the entries of M^k. It is said to be *primitive* if M^k is positive for some $k \geq 1$ (i.e. all the $m_{ij}^{(k)} > 0$) and the smaller such k is the *primitive index* of M.

Note that ζ is primitive (resp. irreducible) if and only if $M(\zeta)$ is a primitive (resp. irreducible) matrix by (5.4).

5.3.1 Perron Frobenius' Theorem for Nonnegative Matrices

Theorem 5.4 (Perron-Frobenius). *Let M be a primitive matrix. Then*

a) M admits a positive eigenvalue θ, such that $|\lambda| < \theta$ for any other eigenvalue λ of M.

b) There exists a positive eigenvector corresponding to θ.

c) θ is a simple eigenvalue (also with algebraic multiplicity equal to one).

Proof. a) Let $\tau \in \mathbf{C}$ be an eigenvalue of M whose modulus is maximal :

$$|\lambda| \le |\tau| \text{ for any eigenvalue } \lambda \text{ of } M.$$

If $y \in \mathbf{C}^s$ is an eigenvector associated to τ, then $\tau y_i = \sum_j m_{ij} y_j$ for every i, and $|\tau||y_i| \le \sum_j m_{ij} |y_j|$, so that

$$|\tau| \le \min_{\{i;\ y_i \ne 0\}} \frac{\sum_j m_{ij} |y_j|}{|y_i|}.$$

Consider now the function r, defined on the nonnegative vectors in $\mathbf{R}^s$ by

$$r(x) = \min_{\{i;\ x_i \ne 0\}} \frac{\sum_j m_{ij} x_j}{x_i} \quad \text{if } x \ne 0.$$

This function r is u.s.c. and homogeneous on the set $\{x \in \mathbf{R}^s \backslash \{0\},\ x \ge 0\}$ so that

$$\theta = \sup_{\substack{x \ge 0 \\ x \ne 0}} r(x) = \sup_{\substack{x \ge 0 \\ ||x|| = 1}} r(x)$$

exists and is attained. Since $\min_{\{i;\ x_i \ne 0\}} \frac{\sum_j m_{ij} x_j}{x_i} \le \theta$ for every $x \ge 0$, we have already

$$\theta \ge |\tau| > 0.$$

We claim that θ is an eigenvalue.

Write $\theta = \min_{\{i;\ y_i \ne 0\}} \frac{\sum_j m_{ij} y_j}{y_i}$ for some $y \ge 0$, $||y|| = 1$. If $z = My - \theta y$ were different from zero, we should have

$$M(M^k y) - \theta M^k y = M^k z > 0$$

where $M^k > 0$ by the primitivity property. Thus, setting $x = M^k y$, we should get $\theta x < Mx$ and $\theta x_i < \sum_j m_{ij} x_j$ for every i, which provides a contradiction. Therefore, we have proved that θ is a positive eigenvalue with $|\lambda| \le \theta$ for any other eigenvalue λ of M.

Suppose now that λ is an eigenvalue of M with $|\lambda| = \theta$. If $My = \lambda y$, then $\theta|y| = |My| \le M|y|$ and the above arguments show that $M|y| = \theta|y|$. It follows that

$$M^k|y| = \theta^k|y| = |M^k y|$$

and, for every i,

$$\Big|\sum_j m_{ij}^{(k)} y_j\Big| = \sum_j m_{ij}^{(k)} |y_j|.$$

The components y_j must have identical arguments $e^{i\varphi}$ so that $ye^{-i\varphi}$ is a positive eigenvector corresponding to λ. Necessarily, λ is positive and $\lambda = \theta$.

b) Notice that we have proved in passing the following lemma

Lemma 5.1. *If θ is a positive dominant eigenvalue of the primitive matrix M, and y, a nonnegative vector such that $My \ge \theta y$, then y is a positive eigenvector corresponding to θ;*

which confirms the second assertion by considering $y = |x|$ for any eigenvector x.

c) We start by proving that the geometric multiplicity of θ is one, in other words, that the eigen-subspace associated to θ is one-dimensional. In the opposite case, let x and y be two independent vectors in $\mathrm{Ker}(M-\theta I)$. From the preceding lemma, x and y have non zero components and $z = y_1 x - x_1 y \in \mathrm{Ker}(M-\theta I)$ is such that $z_1 = 0$; a new application of the lemma 5.1 leads to $z = 0$ and to the first claim.

We shall now prove that $\mathrm{Ker}(M-\theta I)$=$\mathrm{Ker}(M-\theta I)^2$ which exactly means that the algebraic multiplicity of θ is one. Suppose that $y \in \mathrm{Ker}(M-\theta I)^2 \setminus \mathrm{Ker}(M-\theta I)$. There must exist $x \in \mathrm{Ker}(M-\theta I)$ such that $My = \theta y + x$, so that, for every $n \ge 1$,

$$M^n y = \theta^n y + n\theta^{n-1} x.$$

By lemma 5.1, $|x|$ is a positive vector and one can find n_0 large enough such that $n_0|x| - \theta|y| \ge \theta|y|$. It follows that

$$M^{n_0}|y| \ge |M^{n_0} y| \ge \theta^{n_0-1}(n_0|x| - \theta|y|) \ge \theta^{n_0}|y|.$$

But M^{n_0} in turn is a primitive matrix with θ^{n_0} as its positive dominant eigenvalue. Using once more the lemma 5.1, we get

$$M^{n_0}|y| = \theta^{n_0}|y|$$

and $M^{n_0} y = \theta^{n_0} y$, which contradicts our assumption. We have thus proved that $\mathrm{Ker}(M-\theta I)$=$\mathrm{Ker}(M-\theta I)^2$ and the theorem follows. □

Corollary 5.1. *Let M be a primitive matrix and θ its positive and maximal eigenvalue. Then there exist a unique right eigenvector $(d_i)_i$ and a unique left eigenvector $(g_j)_j$ such that*

$$\sum_i d_i = 1, \quad \sum_i d_i g_i = 1$$

and, for every i and j,

$$\lim_{n\to\infty}\frac{1}{\theta^n}m_{ij}^{(n)} = d_i g_j,$$

with a geometric speed of convergence.

Proof. We interpret the property of θ to be simple and maximal. We may decompose the operator M into a sum $M = \theta P_\theta + N$, where P is a projection onto the one-dimensional eigen-subspace $\mathrm{Ker}(M - \theta I)$, and N is an operator satisfying : $NP_\theta = P_\theta N = 0$, and, according to assertion a) in the Perron-Frobenius theorem,

$$\max\{|\lambda_i|,\ \lambda_i \text{ eigenvalue of } N\} =: \tau < \theta. \tag{5.5}$$

It follows that $M^n = \theta^n P_\theta + N^n$, for every $n \geq 1$ and $\lim_{n\to\infty}\frac{M^n}{\theta^n} = P_\theta$; more precisely, there exists a constant $c > 0$ such that

$$||M^n - \theta^n P_\theta||_\infty \leq c\tau^n. \tag{5.6}$$

Let $d = (d_i)_i$ be the positive right eigenvector associated with θ and normalized by $\sum_i d_i = 1$; there exists $g = (g_j)_j$ such that $(P_\theta)_{ij} = d_i g_j$ since P is one-dimensional. It follows that the trace of $P_\theta = \sum_i d_i g_i = 1$ and thereby the identity $P_\theta^* g = g$ is easily verified. The rate of convergence is $O((\tau/\theta)^n)$ hence geometric. □

Remark on the irreducible case : A description of irreducible matrices and the version of the Perron-Frobenius theorem for such a matrix can be found in [224]. Actually the statement is almost the same : b) and c) are still valid, only a) must be changed into

a') M admits a positive eigenvalue θ such that $|\lambda| \leq \theta$ for any other eigenvalue λ of M.

Nevertheless, it is possible to precise the form of the eigenvalues $\lambda \neq \theta$ with modulus $|\lambda| = \theta$. We need for that the following definition :

Definition 5.6. An irreducible matrix M is said to be *periodic with period* $d \geq 1$ if, for every i, $d(i) = d$, where $d(i)$ is the g.c.d. of the set $\{k \geq 1,\ m_{ii}^{(k)} > 0\}$.

Proposition 5.6. *An irreducible matrix is primitive if and only if its period is $d = 1$ (=aperiodic matrix).*

Proposition 5.7. *Let M be an irreducible matrix with period $d > 1$. Then M admits exactly d eigenvalues λ satisfying $|\lambda| = \theta$, which are*

$$\theta \cdot e^{2i\pi k/d},\ \ k = 0, 1, \ldots, d-1.$$

5.3.2 Frequency of Letters

We are going to apply the previous results to the composition matrix of some primitive substitution ζ, which is a primitive matrix, as already noticed.

Definition 5.7. The positive and maximal eigenvalue θ of $M(\zeta)$ is called the *Perron-Frobenius eigenvalue* of ζ.

Assume $u = \zeta(u)$ to be a fixed point of ζ. We shall deduce from the Perron-Frobenius theorem that each letter of u occurs in u with a positive frequency, and this is the first step towards the unique ergodicity of the system (X_ζ, T). More precisely :

Proposition 5.8. *Let $\alpha \in A$. Then $\lim_{n\to\infty} \frac{1}{|\zeta^n(\alpha)|} L(\zeta^n(\alpha)) = d$, where d is a positive vector, independent of α, and satisfying $\sum_{\beta\in A} d_\beta = 1$; in addition, the speed of convergence is geometric.*

Proof. It is a re-formulation, in this context, of corollary 5.1. We detail the link between both results. Observe that $e_\alpha := (0,\ldots,0,\underbrace{1}_{\alpha^{th}},0,\ldots,0)$ is also the vector $L(\alpha)$ for $\alpha \in A$, so that, keeping the notations of the corollary,

$$\lim_{n\to\infty} \frac{L(\zeta^n(\alpha))}{\theta^n} = \lim_{n\to\infty} \frac{M^n L(\alpha)}{\theta^n} = P_\theta(L(\alpha)).$$

Each vector $v(\alpha) := P_\theta(L(\alpha))$ is a nonnegative thus positive eigenvector corresponding to θ (theorem 5.4 b)).

Now, denote by $\mathbf{1}$ the unit vector $(1,\ldots,1)$ so that $|\zeta^n(\alpha)| =< L(\zeta^n(\alpha)), \mathbf{1} >= \sum_{\beta\in A} L_\beta(\zeta^n(\alpha))$. Since

$$\frac{|\zeta^n(\alpha)|}{\theta^n} =< \frac{L(\zeta^n(\alpha))}{\theta^n}, \mathbf{1} > \text{ tends to } < v(\alpha), \mathbf{1} > \tag{5.7}$$

as $n \to \infty$, thus

$$\lim_{n\to\infty} \frac{1}{|\zeta^n(\alpha)|} L(\zeta^n(\alpha)) = \frac{v(\alpha)}{< v(\alpha), \mathbf{1} >} =: w(\alpha).$$

But $\sum_\beta w_\beta(\alpha) = 1$ and there is a unique such eigenvector corresponding to θ, whence $w(\alpha) = d$ for every $\alpha \in A$.

According to (5.6), there exists a constant $c > 0$ such that

$$|L_\beta(\zeta^n(\alpha)) - \theta^n v_\beta(\alpha)| \leq c\tau^n \quad \text{for every } \alpha, \beta \in A. \tag{5.8}$$

where

$$\tau = \max\{|\lambda|,\ \lambda \text{ eigenvalue of } M(\zeta),\ \lambda \neq \theta\}. \tag{5.9}$$

It follows that

$$|L_\beta(\zeta^n(\alpha)) - d_\beta|\zeta^n(\alpha)|| \leq c\tau^n \quad \text{for every } \alpha, \beta \in A. \tag{5.10}$$

The proof is complete. □

In particular, considering $u = \zeta^\infty(\alpha_0)$, we deduce that each letter α of u occurs in u with the positive frequency d_α.

When ζ is of constant length q, $|\zeta^n(\alpha)| = q^n$ and $|\zeta^{n+1}(\alpha)| = q|\zeta^n(\alpha)|$, for every $\alpha \in A$. In the general case, we immediately deduce from the limit in (5.7) the analogous :

Corollary 5.2. *For every* $\alpha \in A$, $\lim_{n\to\infty} |\zeta^{n+1}(\alpha)|/|\zeta^n(\alpha)| = \theta$.

More precisely, as a consequence of (5.8), there exists a constant $k > 0$ such that

$$||\zeta^n(\alpha)| - \theta^n \sum_{\beta \in A} v_\beta(\alpha)| \leq k\tau^n \quad \text{for every } \alpha, \beta \in A. \tag{5.11}$$

Remark 5.2. 1. Let us turn back to the irreducible case. Suppose that ζ is an irreducible substitution and let d be the period of its composition matrix. Let us fix an element $\alpha \in A$ and set for $0 \leq k < d$,

$$A_k = \{\beta \in A, \text{ there exists } n \geq 1 \text{ with } n \equiv k \pmod d$$
$$\text{such that } \zeta^n(\alpha) \text{ contains } \beta\}.$$

It is easily seen that (A_k) is a partition of A and the restriction of ζ^d to each A_k defines a primitive substitution on A_k.

2. In an arbitrary morphic sequence, a given letter need not have a natural frequency. In his paper [57], A. Cobham showed, among other interesting results, that the logarithmic frequency of a letter in an automatic sequence u always exists; that is, for any recurrent letter α in u,

$$\lim_{n\to\infty} \frac{1}{\ln n} \sum_{j<n,\ u_j=\alpha} \frac{1}{j} \quad \text{exists.}$$

5.3.3 *Perron Numbers, Pisot Numbers*

A class of interesting algebraic numbers has thus been exhibited and studied, which contains the Pisot-Vijayaraghavan numbers. Recall the following definition [24]:

Definition 5.8. A *Pisot-Vijayaraghavan number* (in short, PV-numbers) is an algebraic integer $\theta > 1$ whose remaining Galois conjugates lie inside the open unit disc.

Let $\theta_1,\dots,\theta_{d-1}$ be the Galois conjugates of the PV-number θ. If $||x||$ denotes the distance of the real number to the nearest integer, then $||\theta^n|| = O(\lambda^n)$, where $\max_{1\leq j\leq d-1}|\theta_j| =: \lambda < 1$. The sequence of powers (θ^n) enjoys extremal approximation properties and this explains the important role those numbers play in harmonic analysis [187, 221].

Definition 5.9. A substitution, the Perron-Frobenius eigenvalue of which being a PV-number, will be called a *Pisot substitution*.

The Fibonacci substitution is a famous example of Pisot substitution and generalizations have been explored by G. Rauzy in [208]. We reconsider this class of substitutions at the end of chapter 6.

Definition 5.10. A *Perron number* is an algebraic integer $\theta \geq 1$ whose remaining Galois conjugates have strictly smaller absolute value.

They have been characterized by D. Lind in [168]. As expected,

Theorem 5.5 (D. Lind). *A Perron number is the spectral radius of some primitive nonnegative integral matrix.*

A symbolic proof of this appeared in [26] and complements can be found in [41].

5.4 Unique Ergodicity

Let $u = \zeta(u)$ be a fixed point of the primitive substitution ζ on A. We shall prove that the system (X_ζ, T) is uniquely ergodic, hence strictly ergodic [188].

5.4.1 Frequency of Factors

We have just before established that any letter of A occurs in u with a positive frequency. More generally, by considering a substitution on the factors of u, we prove the analogue for every word in $\mathscr{L}_\zeta = \mathscr{L}(u)$.

Proposition 5.9. *Let ζ be a primitive substitution on A. Then, for every $\alpha \in A$ and every word $B \in \mathscr{L}_\zeta$, the sequence of nonnegative numbers*

$$\frac{L_B(\zeta^n(\alpha))}{|\zeta^n(\alpha)|}$$

admits a limit as $n \to \infty$, which is independent of α and positive, that we shall denote by d_B; in addition, the speed of convergence is geometric.

Proof. Let $B \in \mathscr{L}_\zeta$ of length $|B| =: \ell \geq 1$. If $\ell = 1$ the result has been proved in proposition 5.8. We suppose $\ell \geq 2$ and we denote by Ω_ℓ the set of all words of length ℓ in $\mathscr{L}_\zeta$ (or factors of u of length ℓ). We consider Ω_ℓ as an alphabet and we define the substitution ζ_ℓ on Ω_ℓ in the following way :

If

$$\omega = \omega_0 \cdots \omega_{\ell-1} \in \Omega_\ell$$

and

$$\zeta(\omega) =: y = y_0 \cdots y_{|\zeta(\omega_0)|-1} y_{|\zeta(\omega_0)|} \cdots y_{|\zeta(\omega)|-1},$$

we set

$$\zeta_\ell(\omega) = (y_0 \cdots y_{\ell-1})(y_1 \cdots y_\ell) \cdots (y_{|\zeta(\omega_0)|-1} \cdots y_{|\zeta(\omega_0)|+\ell-2}). \tag{5.12}$$

In other words, $\zeta_\ell(\omega)$ consists in the ordered list of the first $|\zeta(\omega_0)|$ factors of length ℓ of the word $\zeta(\omega)$. In particular, $|\zeta_\ell(\omega)| = |\zeta(\omega_0)|$, and we extend ζ_ℓ to Ω_ℓ^* and $\Omega_\ell^{\mathbf{N}}$ by juxtaposition.

Lemma 5.2. *If $u = u_0 u_1 \cdots$ is a fixed point of ζ, the sequence $U_\ell \in \Omega_\ell^{\mathbf{N}}$ is a fixed point of ζ_ℓ, where*

$$U_\ell = (u_0 \cdots u_{\ell-1})(u_1 \cdots u_\ell)(u_2 \cdots u_{\ell+1}) \cdots$$

Proof. We set $\omega = u_0 u_1 \cdots u_{\ell-1}$, the first factor in u of length ℓ, and we iterate ζ_ℓ on ω. Since $u = \zeta(u)$, $\zeta(\omega) = u_0 u_1 \cdots u_{|\zeta(\omega)|-1}$ and we get from (5.12)

$$\zeta_\ell(\omega) = (u_0 \cdots u_{\ell-1})(u_1 \cdots u_\ell) \cdots (u_{|\zeta(u_0)|-1} \cdots u_{|\zeta(u_0)|+\ell-2}).$$

$\zeta_\ell(\omega)$ begins with ω and this condition is sufficient to prove the existence of a fixed point for ζ_ℓ, $\zeta_\ell^\infty(\omega)$, obtained by iterating ζ_ℓ on ω. It remains to compare $\zeta_\ell^\infty(\omega)$ and U_ℓ : it is easy to check that $(\zeta^n)_\ell = (\zeta_\ell)^n$ for every $n \geq 1$, i.e.

$$\zeta_\ell^n(\omega) = (u_0 \cdots u_{\ell-1})(u_1 \cdots u_\ell) \cdots (u_{|\zeta^n(u_0)|-1} \cdots u_{|\zeta^n(u_0)|+\ell-2}),$$

and $\zeta_\ell^\infty(\omega) = U_\ell$.

Lemma 5.3. *The substitution ζ_ℓ is primitive if ζ is primitive.*

Proof. As previously observed, it is sufficient to prove that ζ_ℓ is irreducible : given B and ω in Ω_ℓ, we have to exhibit N large enough, such that B occurs in $\zeta_\ell^N(\omega)$. Since $B \in \mathscr{L}_\zeta$, there exist $\alpha \in A$ and $p \geq 1$ such that B occurs in $\zeta^p(\alpha)$. Now, $\zeta^m(\omega_0)$ contains α for $m \geq m_0$ because ζ is primitive, hence B occurs in $\zeta^{m+p}(\omega_0)$. Writing

$$\zeta^n(\omega) = \zeta^n(\omega_0)\zeta^n(\omega_1 \cdots \omega_{\ell-1}) = y_0 y_1 \cdots y_{|\zeta^n(\omega_0)|-1} \alpha_0 \alpha_1 \cdots$$

we get, iterating (5.12),

$$\zeta_\ell^n(\omega) = (y_0 \cdots y_{\ell-1})(y_1 \cdots y_\ell) \cdots (y_{|\zeta^n(\omega_0)|-1} \alpha_0 \cdots \alpha_{\ell-2}).$$

The factors of length ℓ of $\zeta^n(\omega_0)$ occur as "letters" in $\zeta_\ell^n(\omega)$. If we choose $n = m + p$ with $m \geq m_0$, $\zeta_\ell^n(\omega)$ surely contains B, and the second lemma is proved. □

As a consequence of the properties of ζ_ℓ, we get the

Lemma 5.4. *If ζ is primitive, then, for every $\ell \geq 2$, ζ_ℓ has the same Perron-Frobenius eigenvalue as ζ.*

Proof. This can be derived from corollary 5.2 : if θ_ℓ is the dominant eigenvalue of M_ℓ, then

$$\lim_{n\to\infty} \frac{|\zeta_\ell^{n+1}(\omega)|}{|\zeta_\ell^n(\omega)|} = \theta_\ell$$

for every $\omega \in \Omega_\ell$, the set of ℓ-words in $\mathscr{L}_\zeta$. But $|\zeta_\ell^n(\omega)| = |\zeta^n(\omega_0)|$ if $\omega = \omega_0 \cdots \omega_{\ell-1}$ and the lemma follows. □

We now finish the proof of the proposition. Applying the proposition 5.9 to this substitution ζ_ℓ, we readily obtain that

$$\lim_{n\to\infty} \frac{L_B(\zeta_\ell^n(\omega))}{|\zeta_\ell^n(\omega)|}$$

exists, is positive and independent of $\omega \in \Omega_\ell$. We call d_B this limit. By definition of ζ_ℓ, notice that

$$L_B(\zeta_\ell^n(\omega)) \sim L_B(\zeta^n(\omega_0)) \quad \text{as } n \to \infty.$$

Finally, since $|\zeta_\ell^n(\omega)| = |\zeta^n(\omega_0)|$, we deduce that

$$\lim_{n\to\infty} \frac{L_B(\zeta_\ell^n(\omega))}{|\zeta_\ell^n(\omega)|} = \lim_{n\to\infty} \frac{L_B(\zeta^n(\omega_0))}{|\zeta^n(\omega_0)|} = d_B > 0$$

is independent of ω_0.

The rate of convergence can be precised, as in proposition 5.9, thanks to the Perron-Frobenius theorem applied to $M(\zeta_\ell)$ instead of $M(\zeta)$; there exists a constant $C > 0$ such that

$$|L_B(\zeta^n(\alpha)) - d_B|\zeta^n(\alpha)|| \leq C\tau^n \quad \text{for every } \alpha, \beta \in A. \tag{5.13}$$

where $\tau := \tau_\ell < \theta$ is the maximal modulus of the eigenvalues of $M(\zeta_\ell)$, distinct from θ; and the proof is complete. □

Example : Let u be the Thue-Morse sequence on the alphabet $\{0,1\}$. Recall that $u = \zeta(u)$ with $\zeta(0) = 01$, $\zeta(1) = 10$. The eigenvalues of the composition matrix $M := M(\zeta) = \begin{pmatrix} 1 & 1 \\ 1 & 1 \end{pmatrix}$ are $\theta = 2$ (the Perron-Frobenius eigenvalue) and $\lambda = 0$. We shall describe ζ_2 on the two-words alphabet $\Omega_2 = \{(00), (01), (10), (11)\}$.

$$\zeta(00) = 01 \cdot 01 \quad \text{so that} \quad \zeta_2((00)) = (01)(10)$$

and, also,

$$\zeta_2((01)) = (01)(11)$$
$$\zeta_2((10)) = (10)(00)$$
$$\zeta_2((11)) = (10)(01)$$

Thus, the composition matrix M_2 of ζ_2 has the following form

$$\begin{pmatrix} 0 & 0 & 1 & 0 \\ 1 & 1 & 0 & 1 \\ 1 & 0 & 1 & 1 \\ 0 & 1 & 0 & 0 \end{pmatrix}$$

and the eigenvalue of M_2 are $\theta = 2,\ \lambda = 0,\ 1,\ -1$.

The normalized positive eigenvector d corresponding to θ is $(1/6, 1/3, 1/3, 1/6)$, so that, the frequencies of the two-words are respectively

$$d_{(00)} = \frac{1}{6} = d_{(11)}, \quad d_{(01)} = \frac{1}{3} = d_{(10)}.$$

Remark 5.3. We shall describe later the spectrum of the composition matrix $M(\zeta_\ell)$ and we shall derive from this observation, an algorithmic machinery to compute the occurrence frequency of any factor of u.

5.4.2 The Unique Invariant Measure

If $B \in \mathscr{L}_\zeta$ is any factor of u, the cylinder set $[B]$, generated by B, is

$$[B] = \{x \in X,\ x_{[0,\ |B|-1]} = B\}.$$

If, now, $\mu \in M(u)$ (see subsection 4.3.1), there exists, by definition, an increasing sequence (N_j) of integers such that, for every $B \in \mathscr{L}_\zeta$,

$$\mu([B]) = \lim_{j\to\infty} \frac{1}{N_j} \text{ Card } \{n < N_j,\ u_{[n,\ n+|B|-1]} = B\}$$

The last proposition asserts that, for every letter $\alpha \in A$,

$$\mu = \lim_{j\to\infty} \frac{1}{|\zeta^j(\alpha)|} \sum_{n<|\zeta^j(\alpha)|} \delta_{T^n u} \in M(u),$$

and thus (N_j), with $N_j = |\zeta^j(\alpha)|$, is a suitable sequence. In particular, μ is a T-invariant probability measure on X_ζ.

We are now looking forward conditions on ζ ensuring the unicity of the T-invariant measure μ on X_ζ. The following has been proved by P. Michel [188]

Theorem 5.6 (P. Michel). *For a primitive substitution ζ, the system (X_ζ, T) is uniquely ergodic.*

Proof. Since X_ζ is the orbit closure of u, we wish to prove, referring to the criterion in corollary 4.2, that *each factor of u occurs in u with a uniform frequency.*

More precisely, if $B \in \mathscr{L}_\zeta$, we must prove that

$$\lim_{N \to \infty} \frac{L_B(u_{[k,\, k+N]})}{N+1} = d_B$$

uniformly in k.

In order to make use of proposition 5.9, we aim to compare the word $u_{[k,\, k+N]}$ with an elementary word of the form $\zeta^n(\omega)$, for some factor $\omega \in \mathscr{L}_\zeta$ and some $n \geq 0$. The following lemma is interesting by itself.

Lemma 5.5. *Let $V \in \mathscr{L}_\zeta$ and $K = \max_{\alpha \in A} |\zeta(\alpha)|$. There exist $n \geq 0$ and $2n+1$ possibly empty words $v_i \in \mathscr{L}_\zeta$ $(0 \leq i \leq n)$, $w_i \in \mathscr{L}_\zeta$ $(0 \leq i < n)$, but $v_n \neq \emptyset$, such that*

$$V = v_0 \zeta(v_1) \cdots \zeta^{n-1}(v_{n-1}) \zeta^n(v_n) \zeta^{n-1}(w_{n-1}) \cdots \zeta(w_1) w_0, \tag{5.14}$$

$$|v_i| \leq K \text{ for } i = 0, \ldots, n \quad \text{and} \quad |w_i| \leq K \text{ for } i = 0, \ldots, n-1. \tag{5.15}$$

Proof. This decomposition (5.14) can be established step by step. By definition of $\mathscr{L}_\zeta$, there exists $\alpha \in A$ such that V occurs in $\zeta^{n+1}(\alpha) = \zeta(\zeta^n(\alpha))$, n minimal with this property; one can thus find V_1, v_0 and w_0 in $\in \mathscr{L}_\zeta$ such that

$$V = v_0 \zeta(V_1) w_0$$

and V_1 maximal for this property so that

$$|v_0|, |w_0| \leq K.$$

We continue with V_1 as a factor of $\zeta^n(\alpha)$ and we exhibit, in this way, a sequence $(V_j) \in \mathscr{L}_\zeta$ with $|V_{j+1}| < |V_j|$. The process must stop with $V_n =: v_n$ non empty and the decomposition is obtained by substituting $V_j = v_j \zeta(V_{j+1}) w_j$ inside V_{j-1}. □

Applying the lemma with $V = u_{[k,\, k+N]}$, we get (assuming $N > |B|$),

$$L_B(V) = \sum_{j \leq n} L_B(\zeta^j(v_j)) + \sum_{j < n} L_B(\zeta^j(w_j)),$$

and

$$N+1=\sum_{j\leq n}|\zeta^{j}(v_{j})|+\sum_{j<n}|\zeta^{j}(w_{j})|.$$

By the estimate (5.13), there exists a constant $C>0$ such that, for every j and α,

$$\left|L_{B}(\zeta^{j}(\alpha))-d_{B}|\zeta^{j}(\alpha)|\right|\leq C\tau^{j};$$

we derive from (5.15) and the triangular inequality,

$$\left|L_{B}(\zeta^{j}(v_{j}))-d_{B}|\zeta^{j}(v_{j})|\right|\leq KC\tau^{j}$$

for $0\leq j\leq n$ (and the analogue for the words w_j); then

$$\begin{aligned}\left|L_{B}(V)-d_{B}(N+1)\right| &= \left|L_{B}(V)-d_{B}(\Sigma_{j\leq n}|\zeta^{j}(v_{j})|+\Sigma_{j<n}|\zeta^{j}(w_{j})|)\right|\\ &\leq 2KC\Sigma_{j=0}^{n}\tau^{j}\\ &\leq M\tau^{n}\quad (M \text{ if } \tau=0).\end{aligned}$$

If $0\leq\tau\leq 1$ we are done; if $\tau>1$, we finish the proof by making use of the estimate (5.11) : for some constant $D>0$,

$$N+1\geq|\zeta^{n}(v_{n})|\geq D\theta^{n};$$

this implies

$$n\ln\tau=n\ln\theta(\ln\tau/\ln\theta)\ll(\ln\tau/\ln\theta)\ln N$$

and

$$\left|\frac{L_{B}(V)}{N+1}-d_{B}\right|\leq\frac{M}{N+1}\tau^{n}\ll N^{(\ln\tau/\ln\theta)}/N+1. \tag{5.16}$$

The theorem follows, since $\ln\tau/\ln\theta<1$. □

Corollary 5.3. *The unique invariant probability measure μ on X_ζ is characterized by*

$$\mu([B])=\lim_{N\to\infty}\frac{1}{N}\text{ Card }\{n<N,\ u_{[n,\,n+|B|-1]}=B\}$$

for every $B\in\mathscr{L}_\zeta$.

In particular,

Corollary 5.4. *The vector on $\mathbf{R}^s$ whose components are $\mu([\alpha])$, $\alpha\in A$, is the normalized right eigenvector corresponding to the Perron-Frobenius eigenvalue.*

5.4.3 Matrices M_ℓ

For $\ell \geq 2$, we denote by M_ℓ the composition matrix of the substitution ζ_ℓ, defined in the foregoing subsection 5.4.1. When ζ is primitive, we are now able to compute the occurrence frequency of any word of length ℓ in $\mathscr{L}_\zeta$, by determining the normalized eigenvector of M_ℓ corresponding to the Perron-Frobenius eigenvalue (last corollary). We shall outline simple algebraic considerations leading to a rather faster method to carry out these calculations. We shall see that all the informations on words are, in some sense, contained in the two-words; in fact, we shall show how to deduce the occurrence frequency of words in Ω_ℓ from the distribution of the two-words. We already proved :

Proposition 5.10. *If $M := M(\zeta)$ is a primitive matrix with the dominant positive eigenvalue θ, then, for every $\ell \geq 2$, M_ℓ is a primitive matrix with same dominant eigenvalue.*

We now fix $\ell \geq 2$. Recall that, for every $p \geq 1$, $(\zeta_\ell)^p = (\zeta^p)_\ell$ so that the iterate $(\zeta_\ell)^p$ is defined on Ω_ℓ as follows : if $\omega = \omega_0 \cdots \omega_{\ell-1}$ and

$$\zeta^p(\omega) = \zeta^p(\omega_0)\cdots\zeta^p(\omega_{\ell-1}) = y_0 y_1 \cdots y_{|\zeta^p(\omega)|-1}$$

then

$$\zeta_\ell^p(\omega) = (y_0 \cdots y_{\ell-1})(y_1 \cdots y_\ell)\cdots(y_{|\zeta^p(\omega_0)|-1} \cdots y_{|\zeta^p(\omega_0)|+\ell-2}).$$

An immediate observation is the following : ζ_ℓ^p is entirely determined on $\omega \in \Omega_\ell$ by the knowledge of the two first letters $\omega_0\omega_1$ of ω, provided that p is large enough; more precisely, as soon as p is larger than ℓ in such a way that

$$|\zeta_\ell^p(\omega_0)| + \ell - 2 < |\zeta_\ell^p(\omega_0)| + |\zeta_\ell^p(\omega_1)|. \tag{5.17}$$

By proposition 5.9, $|\zeta^p(\alpha)| \sim \theta^p \|P_\theta(L(\alpha))\|$ as $p \to \infty$, so that the condition (5.17) can be formulated in terms of θ :

$$\theta^p > C\ell \quad \text{for some positive constant } C. \tag{5.18}$$

We fix now p and ℓ in view of (5.18) and we consider $\pi_2 : \Omega_\ell \to \Omega_2$, the restriction to the first two letters :

$$\pi_2(\omega_0 \cdots \omega_{\ell-1}) = \omega_0\omega_1.$$

We then define a map $\pi_{2,\ell,p} : \Omega_2 \to \Omega_\ell^*$ by setting

$$\pi_{2,\ell,p}(\omega_0\omega_1) = (y_0 \cdots y_{\ell-1})(y_1 \cdots y_\ell)\cdots(y_{|\zeta^p(\omega_0)|-1} \cdots y_{|\zeta^p(\omega_0)|+\ell-2})$$

for every $\omega_0\omega_1 \in \Omega_2$ with $\zeta^p(\omega_0\omega_1) = y_0 \cdots y_{|\zeta^p(\omega_0)|-1} y_{|\zeta^p(\omega_0)|} \cdots y_{|\zeta^p(\omega_0\omega_1)|-1}$.

It is obviously checked that

$$\pi_{2,\ell,p} \circ \pi_2 = \zeta_\ell^p$$
$$\pi_2 \circ \pi_{2,\ell,p} = \zeta_2^p$$

and

$$\zeta_\ell \circ \pi_{2,\ell,p} = \pi_{2,\ell,p} \circ \zeta_2.$$

If we naturally extend $\pi_{2,\ell,p}$ by juxtaposition to a map from Ω_2^* into Ω_ℓ^*, and π_2 to a map from Ω_ℓ^* into Ω_2^*, we may draw the following commutative diagram :

$$\begin{array}{ccccc}
\Omega_\ell^* & \xrightarrow{\zeta_\ell^p} & \Omega_\ell^* & \xrightarrow{\zeta_\ell} & \Omega_\ell^* \\
\downarrow{\scriptstyle \pi_2} & \nearrow{\scriptstyle \pi_{2,\ell,p}} & & \nearrow{\scriptstyle \pi_{2,\ell,p}} & \downarrow{\scriptstyle \pi_2} \\
\Omega_2^* & \xrightarrow[\zeta_2]{} & \Omega_2^* & \xrightarrow[\zeta_2^p]{} & \Omega_2^*
\end{array}$$

Denoting by L_ℓ (in the following only) the composition function in "letters" of Ω_ℓ, then we have

$$L_\ell(\zeta_\ell(\omega)) = M_\ell L_\ell(\omega) \quad \text{if} \quad \omega \in \Omega_\ell^*$$

and

$$M_{2,\ell,p} L_2(\pi_2(\omega)) = L_\ell(\zeta_\ell^p(\omega)),$$

where $M_{2,\ell,p}$ is the matrix of the map $\pi_{2,\ell,p}$. This leads to the following commutative diagram, with $\xi = \pi_2(\omega)$ and A, the matrix of the projection π_2.

$$\begin{array}{ccccc}
L_\ell(\omega) & \xrightarrow{M_\ell^p} & L_\ell(\zeta_\ell^p(\omega)) & \xrightarrow{M_\ell} & L_\ell(\zeta_\ell^{p+1}(\omega)) \\
\downarrow{\scriptstyle A} & \nearrow{\scriptstyle M_{2,\ell,p}} & & \nearrow{\scriptstyle M_{2,\ell,p}} & \downarrow{\scriptstyle A} \\
L_2(\xi) & \xrightarrow[M_2]{} & L_2(\zeta_2(\xi)) & \xrightarrow[M_2^p]{} & L_2(\zeta_2^{p+1}(\xi))
\end{array}$$

We deduce from these diagrams the description of the spectrum of M_ℓ.

Corollary 5.5. *The eigenvalues of M_ℓ are those of M_2, with possibly the additional eigenvalue zero.*

Proof. From the identity : $M_{2,\ell,p}\, M_2 = M_\ell\, M_{2,\ell,p}$ we get, for every algebraic polynomial Q,

$$M_{2,\ell,p}\, Q(M_2) = Q(M_\ell)\, M_{2,\ell,p}.$$

It follows that, using once more the diagram,

$$\begin{aligned} M_{2,\ell,p}\, Q(M_2)\, A &= Q(M_\ell)\, M_{2,\ell,p}\, A \\ &= Q(M_\ell)\, M_\ell^p \end{aligned}$$

Suppose now that $Q(M_2)=0$; then, the polynomial $X \to Q(X)\cdot X^p$ annihilates the matrix M_ℓ. In the same way, we deduce from the identity

$$M_2^p\, Q(M_2) = AQ(M_\ell)\, M_\ell^p$$

that, in turn, $X \to Q(X)\cdot X^p$ annihilates the matrix M_2 as soon as $Q(M_\ell)=0$. We have proved that the matrices M_2 and M_ℓ have the same non zero eigenvalues. □

Corollary 5.6. *If V_2 is an eigenvector of M_2 corresponding to the dominant eigenvalue θ, then $M_{2,\ell,p}(V_2)$ is an eigenvector of M_ℓ corresponding to the dominant eigenvalue θ.*

Proof. Immediate from the identity $M_{2,\ell,p}\, M_2 = M_\ell\, M_{2,\ell,p}$.

Application : We shall now derive, from this analysis, a calculation algorithm for the occurrence frequency of the different factors of the fixed point u. Suppose that one intends to compute the occurrence frequency of the ℓ-words $\omega \in \Omega_\ell$, for a fixed $\ell \geq 2$. For every two-word $\alpha\beta \in \Omega_2$, one enumerates the occurrences of ω in $\zeta^p(\alpha\beta)$, with its first letter in $\zeta^p(\alpha)$ (p determined by (5.18)). This gives the $(\omega,\alpha\beta)$-entry of the matrix $M_{2,\ell,p}$, and once the matrix is known, one applies the last corollary.

Example : We wish to calculate the occurrence frequency of the factors of length 5 in the Thue-Morse sequence. There are 12 distinct words of length 5 :

00101 00110 01001 01011 01100 01101

11010 11001 10110 10100 10011 10010

It is necessary to choose $p=3$ to implement the algorithm and we expand $\zeta^3(\alpha\beta)$ for $\alpha\beta \in \Omega_2$.

$$\zeta^3(00) = 01101001\cdot 01101001$$

$$\zeta^3(01) = 01101001\cdot 10010110$$

$$\zeta^3(10) = 10010110\cdot 01101001$$

$$\zeta^3(11) = 10010110\cdot 10010110$$

The 5-words arising from $\zeta^3(00)$ are

$$(01101)(11010)(10100)(01001)(10010)(00101)(01011)(10110)$$

and the specific word $\omega = 00101$ occurs once in this enumeration.

Continuing in this way, we find the different $(\omega,\alpha\beta)$-entries of the rectangular matrix $M_{2,5,3}$, with 12 rows and 4 columns.

$$\begin{pmatrix} 1 & 0 & 1 & 1 \\ 0 & 1 & 1 & 0 \\ 1 & 1 & 0 & 1 \\ 1 & 0 & 1 & 1 \\ 0 & 1 & 1 & 0 \\ 1 & 1 & 0 & 1 \\ 1 & 1 & 0 & 1 \\ 0 & 1 & 1 & 0 \\ 1 & 0 & 1 & 1 \\ 1 & 1 & 0 & 1 \\ 0 & 1 & 1 & 0 \\ 1 & 0 & 1 & 1 \end{pmatrix}$$

Since $V_2 = (1,2,2,1)$, we find that $V_5 = M_{2,5,3}(V_2)$ has identical components, all equal to 4; it follows that the factors of length 5 in the Thue-Morse sequence have the same occurrence frequency, thus equal to $1/12$.

5.5 Combinatorial Aspects

In order to describe the structure of substitution dynamical systems, additional combinatorial properties of substitution maps and substitutive sequences will be needed.

5.5.1 *Complexity Function and Topological Entropy*

We leave substitutions for a moment and we turn back to the general framework of sequences taking values in a finite alphabet A. A natural way of measuring the disorder of such a sequence is to count, for every $n \geq 1$, the number of different words of length n occurring in the sequence.

Definition 5.11. Let y be a finite-valued sequence and, for every $n \geq 1$, let Ω_n be the set of different words of length n occurring in the sequence y. We call *complexity function* of y and denote by p_y, or simply p, the function

$$p : n \mapsto p(n) = \text{Card } \Omega_n.$$

It is clear that p is nondecreasing; also, $2 \leq p(n) \leq (\text{Card } A)^n$ and the larger $p(n)$ is, the more random the sequence must be. In the other direction, the complexity function of a periodic sequence is ultimately constant. Actually, Coven and Hedlund have proved the following characterization [62]:

Proposition 5.11 (Coven & Hedlund). *Let y be a sequence in $A^{\mathbb{N}}$. The following assertions are equivalent :*

a) y is ultimately periodic.

b) There exists $n \geq 1$ such that $p(n) \leq n$.

c) There exists $n \geq 1$ such that $p(n+1) = p(n)$.

Proof. ▷ Suppose that $y = y_0 y_1 \cdots y_{k-1}\, y_k y_{k+1} \cdots$ where $y_k y_{k+1} \cdots$ is periodic with period length d. Obviously, $p(n) \leq d+k$ for every n, and b) holds.

▷ Assume now b) and let m be the least integer such that $p(m) \leq m$. If $m = 1$, c) is clearly true. Suppose then $m > 1$ and $p(n+1) > p(n)$ for every $n \geq 1$; we get a contradiction since

$$p(m) = p(1) + \sum_{j=2}^{m} \big(p(j) - p(j-1)\big)$$

must be $\geq m+1$. And b) implies c).

▷ Assume c) and put $k = p(n) = p(n+1)$ for such an $n \geq 1$. We consider

$$Y = \{y_j \cdots y_{j+n-1} =: y_{[j,\, j+n-1]},\ \ 1 \leq j \leq k+1\}.$$

$Y \subset \Omega(n)$ and thus, Card $Y \leq k$. Since Y consists in $k+1$ words of length n, by the Pigeon Hole principle, two of them must be identical and there exist p, q, $1 \leq p < q \leq k+1$ such that

$$y_{[p,\, p+n-1]} = y_{[q,\, q+n-1]}.$$

Remember now that $p(n) = p(n+1)$, which means that each word in Ω_n has a unique extension to Ω_{n+1}; this imposes that $y_{p+n} = y_{q+n}$ and, in turn,

$$y_{[p+1,\, p+n]} = y_{[q+1,\, q+n]} \in \Omega_n.$$

By the same argument, $y_{p+n+1} = y_{q+n+1}$ and finally by iteration, $y_{p+j} = y_{q+j}$ for every $j \geq 0$. It follows that $y_p y_{p+1} \cdots$ is periodic, with period length $q - p \leq k$. □

If y, now, is not ultimately periodic, then $p(n) \geq n+1$ for every $n \geq 1$. It is remarkable that such non-periodic sequences, for which $p(n) = n+1$ for all n, do exist. These are the *sturmian sequences*. Note that $p(1) = 2$ which is the cardinal of the alphabet. This very interesting class of sequences with extremal complexity function has been intensively studied and equivalent characterizations or properties can be found in chapter 6 of [88]. The most famous example of such a sequence is given by the Fibonacci sequence, this can be proved directly by induction on n. Actually, very few Sturmian sequences can be generated by substitutions since the set of substitutive sequences on a given alphabet is countable whereas the set of sturmian sequences is not. Sequences inside both classes have been described in [66].

Definition 5.12. If $u \in A^{\mathbb{N}}$ with Card $A = s$, $h = \lim_{n\to\infty} \dfrac{\log_s p(n)}{n}$ is called the *topological entropy* of the sequence u.

Note that the limit exists since $p(n+m) \leq p(n)p(m)$ for every n and m. Also, $0 \leq h \leq 1$ and the sequences with zero entropy are said to be *deterministic*.

Turning back to the fixed point u of some primitive substitution ζ on A with Card $A = s$, we are able to give bounds for the complexity function of u.

Proposition 5.12. *Let p be the complexity function of the sequence $u = \zeta(u)$. Then, there exists C, a positive constant, such that*

$$p(n) \leq Cn \quad \textit{for every } n \geq 1. \tag{5.19}$$

Proof. For every $n \geq 1$, one can find $p \geq 1$ such that

$$\inf_{\alpha \in A} |\zeta^{p-1}(\alpha)| \leq n \leq \inf_{\alpha \in A} |\zeta^p(\alpha)| \tag{5.20}$$

since the sequence of integers $(\inf_{\alpha \in A} |\zeta^p(\alpha)|)_{p \geq 1}$ is non-decreasing.

Remember, with notations of proposition 5.8, that $|\zeta^p(\alpha)|/\theta^p$ tends to $< v(\alpha), \mathbf{1} >$ as $p \to \infty$. One can thus find positive constants A and B, such that, for every $p \geq 1$,

$$A\theta^p \leq \inf_{\alpha \in A} |\zeta^p(\alpha)| \leq \sup_{\alpha \in A} |\zeta^p(\alpha)| \leq B\theta^p. \tag{5.21}$$

Let $n \geq 1$ and $\omega \in \Omega_n$. As $u = \zeta^p(u)$, u is the juxtaposition of words $\zeta^p(u_0)\zeta^p(u_1)\cdots$ and every word $\omega \in \Omega_n$ is a factor of some $\zeta^p(\alpha)$ or some $\zeta^p(\alpha\beta)$ where $\alpha\beta$ is a two-word in u. There exist at most s^2 different two-words in u, and at most $\sup_{\alpha \in A} |\zeta^p(\alpha)|$ different words of length n with first letter in $\zeta^p(\alpha)$, thus whole contained in some $\zeta^p(\alpha\beta)$. As a consequence, for every $n \geq 1$,

$$\begin{aligned} p(n) &\leq s^2 \sup\nolimits_{\alpha \in A} |\zeta^p(\alpha)| \leq B\theta^p s^2 \\ &\leq \tfrac{B}{A} s^2 \theta \cdot n. \end{aligned}$$

The proposition follows. □

Corollary 5.7. *Every sequence $x \in X = \overline{O(u)}$ is deterministic.*

Remark 5.4. Explicit value of certain complexity functions has been computed with the help of the following beautiful result due to B.Mossé [191, 192].

Theorem 5.7. *Let u be the fixed point of a primitive substitution of constant length q. Then, the sequence*

$$(p(n+1) - p(n))_{n \geq 1} \quad \textit{is } q\textit{-automatic}$$

As a consequence of the constructive proof of this result, one gets, for instance :

1. Let ζ be defined on the alphabet $\{0, 1\}$ by $\zeta(0) = 001$ and $\zeta(1) = 101$ (Chacon's substitution). Then

$$p(1) = 2 \text{ and } p(n) = 2n - 1, \text{ for } n > 1.$$

2. Let ζ be the Thue-Morse substitution. Then $(p(n+1) - p(n))_{n\geq 1}$ is 2-automatic, and for $a \geq 1, 0 \leq b < 2^{a-1}$,

$$p(n+1) = \begin{cases} 4n - 2^a & \text{if } n = 2^a + b \\ 4n - 2^a - 2b & \text{if } n = 2^a + 2^{a-1} + b, \end{cases}$$

Other computations of complexity functions can be found in [14, 126], and, to the opposite direction, minimal sequences with a given complexity function have been described (up to isomorphism) [17,53,88].

The spectral study of a uniquely ergodic system (X, μ, T) consists of the spectral study of $U := U_T$, the operator on $L^2(X, \mu)$ defined by $Uf = f \circ T$, μ being the unique T-invariant probability measure on X. U may fail to be unitary; but, U being an isometry, it satisfies $U^*U = I$ while UU^* is the orthogonal projection onto Im U.

Consider now the system associated with some primitive substitution. Most of the time, the one-sided shift on $X := X_\zeta$ is not one-to-one but, the system (X, T) being minimal, T must be onto (otherwise, $T(X)$ would be a non trivial compact subset of X). We shall see, actually, that T is not far from being an homeomorphism.

Proposition 5.13. *There exists a countable subset $\mathscr{D}$ of X such that T is an homeomorphism of $X \setminus \mathscr{D}$.*

Proof. Let us consider

$$Z = \{x \in X, \text{ Card } T^{-1}\{x\} \geq 2\}.$$

Z consists in $x \in X$ which can be extended on the left by some letter, in at least two ways; this means, for such an x, that one can find $\alpha \neq \beta \in A$ such that $\alpha x_0 x_1 \cdots$ and $\beta x_0 x_1 \cdots$ still belong to X. We shall prove that Z has finitely many elements.

We need the following interpretation of the previous proposition. Let K be an integer, $K > C$, the constant given by (5.19) :

There exist infinitely many n such that, at most K words in Ω_n admit at least two left extensions to Ω_{n+1}.

Actually, in the contrary, there would exist N such that $p(n+1) > p(n) + K$ for every $n > N$; then $p(N+m) > p(N) + mK$ for every $m \geq 1$ which provides a contradiction. Let now $x^{(1)}, \ldots, x^{(k)}$ be k distinct elements of Z; there exists N such that, for every $n \geq N$, the k words in Ω_n

$$x^{(j)}_{[0,\, n-1]}, \quad 1 \leq j \leq k,$$

are distinct, and admit at least two left extensions to Ω_{n+1}; this imposes $k \leq K$ and Z is a finite set. By definition of Z, $x \in X \setminus Z$ admits a unique left extension to X by some letter. Take now $\mathscr{D} = \cup_{n \in \mathbf{Z}} T^n Z$; $\mathscr{D}$ is at most countable; $x \in X \setminus \mathscr{D}$ admits a unique left extension to X by some arbitrarily long word and T is a one-to-one map from $X\backslash\mathscr{D}$ onto $X\backslash\mathscr{D}$. □

Corollary 5.8. *The operator $U := U_T$ acting on $L^2(X,\mu)$, where $X = X_\zeta$ for some primitive substitution, is unitary.*

Proof. The unique T-invariant probability measure on X is either discrete or continuous by T-ergodicity. If μ is discrete, μ must be a finite discrete measure, which imply that X itself is finite and u shift-periodic. U is unitary in this case.

Otherwise, μ is continuous with its closed support equal to X, and $\mu(\mathscr{D}) = 0$. It follows that, if $f,g \in L^2(X,\mu)$,

$$
\begin{aligned}
< UU^*f,g > &= \int_{X\backslash\mathscr{D}} f(T^{-1}x)\overline{g(T^{-1}x)}\,d\mu(x) \quad \text{by definition of } U^* \\
&= \int_{X\backslash\mathscr{D}} f(x)\overline{g(x)}\,d\mu(x) \quad \text{by T-invariance of } \mu \text{ and } \mathscr{D} \\
&= < f,g > .
\end{aligned}
$$

The corollary is proved. □

This result allows one to consider one-sided sequences, as they naturally appear, and one-sided shift systems, without any disadvantage in the spectral analysis. But bilateral substitutive sequences enjoy an important combinatorial property which is not shared by one-sided ones, as we shall see.

5.5.2 Recognizability Properties

A very important problem, related to the shift-periodicity of elements of X_ζ, is the uniqueness of the decomposition of any word $\omega \in \mathscr{L}_\zeta$ in consecutive elementary words of the form $\zeta^k(\alpha)$, $k \geq 1$, $\alpha \in A$ (namely *substituted words*). Such a decomposition has already been useful in the proof of unique ergodicity (lemma 5.5). But the uniqueness of the decomposition will make a "desubstitution" of ζ possible.

In what follows, we only consider primitive substitutions so that $\mathscr{L}_\zeta = \mathscr{L}(u)$, and every word of $\mathscr{L}_\zeta$ occurs in u with bounded gaps.

Since $u = u_0u_1\cdots = \zeta^k(u_0)\zeta^k(u_1)\cdots$ for every $k \geq 0$, the integers $|\zeta^k(u_{[0,\,p-1]})|$, $p > 0$, are of great help to locate the cutting of u in substituted words; whence the definitions :

Definition 5.13. For every $k \geq 1$, $E_k = \{0\} \cup \{|\zeta^k(u_{[0,\,p-1]})|,\ p > 0\}$ is the set of *cutting bars* of order k.

Notice that $r \in E_k$ if and only if $u_r = \zeta^k(u_p)_0$ for some $p > 0$, and that $E_k \supset E_{k+1}$. For example, $E_k = q^k\mathbf{N}$ for any ζ of constant length q.

Definition 5.14. ζ is said to be *recognizable* if there exists an integer $K > 0$ such that

$$n \in E_1 \text{ and } u_{[n,\, n+K]} = u_{[m,\, m+K]} \quad \text{imply } m \in E_1.$$

In this case, it is possible to decide whether u_m is the first letter of some substituted word $\zeta(u_p)$ by inspecting K terms following u_m. The smaller integer K enjoying this property will be called *recognizability index* of ζ.

Examples

1. ζ defined on $\{0,1\}$ by $\zeta(0) = 010$, $\zeta(1) = 101$ is not recognizable and more generally, so is no shift-periodic sequence.
2. The Fibonacci substitution defined by $\zeta(0) = 01$, $\zeta(1) = 0$ is easily seen to be recognizable since each 0 in $u = \zeta^\infty(0)$ is the beginning of some substituted word.
3. We shall determine the recognizability index of the Thue-Morse substitution. Recall that $\zeta(0) = 01$, $\zeta(1) = 10$ and

$$u = \zeta^\infty(0) = 0110100110010110\cdots$$

The question is to decide whether some 0 (or 1) is the first letter of $\zeta(0)$ (or $\zeta(1)$). We already know that 00 and 11 are not substituted words and that no word with more than two consecutive 0, or two consecutive 1, does occur in u. It follows that 011 is the beginning of a substituted word (otherwise, 11 must be one) while 001 is not. As for 010, we cannot conclude at once, since this word, prefix of $\zeta(00)$, may occur at the end of $\zeta(11)$ as well. We must continue :

0100 is never a substituted word since 00 is not.

0101 always occurs as $\zeta(00)$. Suppose, in the contrary, that 0101 admits a left extension to a substituted word $\alpha0101$. The only possibility would be 10101 since $\alpha = 0$ is excluded hence 10101 is a prefix of $\zeta(11\beta)$. But β must be 0 since 111 does not appear; now $\zeta(110) = 101001$ and we get a contradiction.

So we proved that 0 is the first letter of $\zeta(0)$ in u if and only if it is the first letter of 011 or 0101. By symmetry, we conclude that $K(\zeta) = 3$.

Remark 5.5. The Thue-Morse sequence satisfies a strong aperiodicity property : if $\omega = \omega_0 \cdots \omega_{\ell-1}$ is a factor of u, then $\omega\omega\omega_0$ does not occur in u [105].

Definition 5.15. We say that the substitution ζ is *aperiodic* if there exists $x \in X_\zeta$ which is not shift-periodic.

If ζ is primitive, it is aperiodic if and only if X_ζ is infinite.

We quote (without proof) the following result on substitutions of constant length [88, 173].

Proposition 5.14. *A primitive, aperiodic and constant-length substitution, which is one-to-one on the alphabet, is recognizable.*

But the notion of recognizability is not quite satisfactory in the general setting since very simple primitive and aperiodic sequences may fail to share this property. This is the case for the following ζ, defined on $\{0,1\}$ by $\zeta(0)=010$, $\zeta(1)=10$. Moreover, an iterate of a recognizable substitution may fail to be recognizable itself. B. Mossé in [191] suggests the weaker definition

Definition 5.16. ζ is said to be *bilaterally recognizable* if there exists an integer $L>0$ such that

$$n \in E_1 \text{ and } u_{[n-L,\, n+L]} = u_{[m-L,\, m+L]} \quad \text{imply} \quad m \in E_1.$$

The following now holds

Theorem 5.8. *Every primitive aperiodic substitution is bilaterally recognizable.*

In addition, an injectivity property can be deduced from theorem 5.8 :

Theorem 5.9. *If u is a fixed point of a primitive aperiodic substitution, there exists an integer $M>0$ such that*

$$r,t \in E_1 \text{ and } u_{[r-M,\, r+M]} = u_{[t-M,\, t+M]} \quad \text{imply} \quad u_r = u_t.$$

Any word in $\mathscr{L}_\zeta$ can be "desubstituted", up to some prefix and some suffix at the ends of the word [88]. A very recent contribution provides a complete answer to this decomposition problem in case of a two-letters alphabet [65].

5.6 Structure of Substitution Dynamical Systems

The recognizable property has an important consequence on the topological and geometrical structures of the system (X_ζ, T) as we shall see : ζ acting on X_ζ transforms this set into a similar one, providing a geometrical description of X_ζ.

In this section, we fix ζ, a **primitive, aperiodic** substitution with a fixed point u and put $X=X_\zeta$. We deal first with the one-sided shift and one-sided sequences, assuming recognizability if necessary, and we give the more general result, valid for two-sided substitutive sequences at the end the section.

5.6.1 Consequences of the Recognizability

ζ defines a map from X into $A^{\mathbf{N}}$ by the formula

$$\zeta(x_0x_1\cdots) = \zeta(x_0)\zeta(x_1)\cdots$$

and it is a continuous map, more precisely :

Proposition 5.15. *ζ defines a contraction on X.*

Proof. It is clear, by using the definition of a subshift, that $\zeta(X) \subset X$. If now x and y are close elements of X, there exists $n \geq 0$ such that $x_{[0,\,n]} = y_{[0,\,n]}$ and thus

$$\zeta(x)_j = \zeta(y)_j \quad \text{for every } 0 \leq j < |\zeta(x_{[0,\,n]})|.$$

□

We shall be needing, throughout this chapter, the following useful identity

$$\zeta^k(T^n x) = T^{|\zeta^k(x_{[0,\,n[})|}\zeta^k(x), \tag{5.22}$$

for every $x \in X$ and $k, n \geq 1$, which is immediate from $\zeta^k(T^n x) = \zeta^k(x_n)\zeta^k(x_{n+1})\cdots$.

Let E_k be the set of k-cutting bar defined in subsection 5.5.2. Note the first observation.

Proposition 5.16. *Let $y \in X$. Then $y \in \zeta^k(X)$ if and only if $y = \lim_{m_i \to \infty} T^{m_i} u$ with $m_i \in E_k$ for i large enough.*

Proof. ▷ If $y \in \zeta^k(X)$, $y = \zeta^k(x)$ where $x = \lim_{i\to\infty} T^{n_i} u$, and $y = \lim_{i\to\infty} \zeta^k(T^{n_i} u)$ by continuity of ζ. Applying (5.22) we get readily $y = \lim_{i\to\infty} T^{m_i} u$ with $m_i = |\zeta^k(u_{[0,\,n_i[})| \in E_k$.

◁ Conversely, suppose that $y = \lim_{m_i\to\infty} T^{m_i} u$, with $m_i \in E_k$ for $i \geq i_0$. Then, for every $i \geq i_0$, one can find n_i such that $m_i = |\zeta^k(u_{[0,\,n_i[})|$. Let x be a cluster point in X of the sequence $(T^{n_i} u)_i$; $\zeta^k(x)$ is a cluster point in X of the sequence $(T^{m_i} u)_i$ and $\zeta^k(x) = y$. □

From now on, we suppose ζ to be *recognizable*, with K as recognizability index (subsection 5.5.2). We shall give a characterization of the range $\zeta(X)$.

Proposition 5.17. *$\zeta(X)$ is a clopen set in X. In other words, $\lim_{i\to\infty} T^{n_i} u$ is in $\zeta(X)$ if and only if $n_i \in E := E_1$ for i large enough.*

Proof. Let $y \in \zeta(X)$; by the previous remark, $y = \lim_{i\to\infty} T^{m_i} u$ with $m_i \in E$. Consider now $x = \lim_{i\to\infty} T^{n_i} u \in X$ and suppose that $x_{[0,\,K]} = y_{[0,\,K]}$; for i large enough, $(T^{n_i} u)_j = x_j$ and $(T^{m_i} u)_j = y_j$ for every $0 \leq j \leq K$ so that

$$u_{[n_i,\,n_i+K]} = u_{[m_i,\,m_i+K]}$$

and n_i in turn belongs to E (by recognizability property). This proves that $x \in \zeta(X)$ and $\zeta(X)$ is open in X. □

The analogue for $k > 1$ requires the recognizability property for the substitution ζ^k, which is not automatically satisfied. But this is true under an additional assumption on ζ.

Lemma 5.6. *Suppose in addition that ζ is one-to-one on X. Then ζ^k is recognizable for every $k > 1$.*

Proof. We proceed by induction on k. Obviously, ζ^k is one-to-one for every $k > 1$. Suppose that K_j, the recognizability index of ζ^j, exists for some $j \geq 1$. Since ζ^j is an homeomorphism from X onto $\zeta^j(X)$, one can find $K \geq K_j$ such that

$$\zeta^j(x)_{[0,\,K]} = \zeta^j(x)_{[0,\,K]} \quad \text{implies} \quad x_{[0,\,K_1]} = y_{[0,\,K_1]}.$$

We shall prove that $K_{j+1} \leq K$. Suppose that $u_{[n,\,n+K]} = u_{[m,\,m+K]}$ with $n \in E_{j+1}$; since $K \geq K_j$, $m \in E_j$ already. Thus,

$$T^m u = \zeta^j(T^p u) \quad \text{for some} \quad p > 0$$

and

$$T^n u = \zeta^{j+1}(T^q u) = \zeta^j(\zeta(T^q u)) \quad \text{for some} \quad q > 0.$$

It follows from above that $(T^p u)_{[0,\,K_1]} = (\zeta(T^q u))_{[0,\,K_1]}$ and p must belong to E. Finally, $T^p u = \zeta(T^r u)$ for some $r > 0$ and $T^m u = \zeta^{j+1}(T^r u)$, which means $m \in E_{j+1}$. □

We summarize these observations below :

Corollary 5.9. *Suppose ζ to be a recognizable and one-to-one substitution. Then, for every $k \geq 1$,*

(i) $\zeta^k(X)$ is a clopen set in X.
(ii) ζ^k is an homeomorphism from X onto $\zeta^k(X)$.
(iii) Let $x = \lim_i T^{n_i} u \in X$; $x \in \zeta^k(X)$ if and only if $n_i \in E_k$ for i large enough.

5.6.2 $\zeta(X)$-Induced Map

We recall a few general facts on induced transformations. Let $(X, \mathscr{B}, \mu, T)$ be a dynamical system, and let $Y \in \mathscr{B}$ be a set of positive measure : $\mu(Y) > 0$.

Definition 5.17. The transformation induced by T on Y, T_Y, is defined on $y \in Y$ by

$$T_Y(y) = T^{n(y)}(y),$$

where $n(y)$, called the *return time* in Y, is

$$n(y) = \inf\{j > 0,\ T^j y \in Y\}.$$

This function $y \mapsto n(y)$ is $\mathscr{B}_Y$-measurable, where $\mathscr{B}_Y = \mathscr{B} \cap Y$, and finite for μ-almost every $y \in Y$ (from the Poincaré recurrence theorem). If, in addition, T is an ergodic automorphism,

$$\int_Y n(y)\, d\mu_Y(y) = 1/\mu(Y),$$

denoting by μ_Y the induced probability measure on $(Y, \mathscr{B}_Y)$, defined on $B \in \mathscr{B}_Y$ by $\mu_Y(B) = \mu(B)/\mu(Y)$ (Kac's lemma).

Let us turn back to the substitution dynamical system (X, μ, T). By strict ergodicity of the system, μ gives positive measure to any open subset of X (proposition 4.4), and, when ζ is recognizable, $\mu(\zeta(X)) > 0$.

Notation : We denote by S the transformation induced by T on $\zeta(X)$ and more generally, for $k > 1$, we denote by S_k, the induced transformation on $\zeta^k(X)$ (if $\mu(\zeta^k(X)) > 0$).

Proposition 5.18. *If ζ is recognizable and $y = \zeta(x)$ for some $x \in X$, then*

$$Sy = T^{|\zeta(x_0)|}y.$$

Proof. If $y = \zeta(x)$, $T^{|\zeta(x_0)|}y = \zeta(Tx)$ belongs to $\zeta(X)$. We have just to prove that $|\zeta(x_0)| = \inf\{\ell > 0,\ T^\ell y \in \zeta(X)\}$. Let us write $x = \lim_{i \to \infty} T^{n_i}u$ so that $y = \lim_{i \to \infty} T^{m_i}u$ where $m_i = |\zeta(u_{[0,\, n_i[})| \in E$ (proposition 5.16). If now $T^\ell y \in \zeta(X)$, $T^\ell y = \lim_{i \to \infty} T^{m_i+\ell}u$ and, by proposition 5.17, $m_i + \ell \in E$ for i large enough. Thus

$$m_i + \ell = |\zeta(u_{[0,\, n_i+p[})| \quad \text{for some } p > 0$$

$$= m_i + |\zeta(u_{[n_i,\, n_i+p[})|.$$

It follows that $\inf\{\ell > 0,\ T^\ell y \in \zeta(X)\} = |\zeta(u_{n_i})| = |\zeta(x_0)|$ for sufficiently large i. □

One deduces from what precedes the following commutative diagram :

$$\begin{array}{ccc} X & \xrightarrow{T} & X \\ \zeta \downarrow & & \downarrow \zeta \\ \zeta(X) & \xrightarrow{S} & \zeta(X) \end{array}$$

Same difficulties as before arise when we try to describe S_k for $k > 1$, since we need the recognizability property for ζ^k. From the previous proposition, we derive a sufficient condition for ζ to be one-to-one on X.

Lemma 5.7. *If ζ is recognizable and* one-to-one *on letters, then, for every $k > 1$, ζ^k is recognizable.*

Proof. In regard of lemma 5.6, it is sufficient to prove that ζ is one-to-one on X as soon as it is one-to-one on letters. Let thus $x, x' \in X$ be such that $\zeta(x) = \zeta(x') = y = y_0 y_1 \cdots$; by proposition 5.18, $|\zeta(x_0)| = |\zeta(x'_0)| = \ell$, so that

$$\zeta(x_0) = \zeta(x'_0) = y_0 y_1 \cdots y_{\ell-1}.$$

But ζ being one-to-one on letters, then $x_0 = x'_0$ in turn, and we apply a similar argument to $T^\ell y \in \zeta(X)$ to obtain $x_1 = x'_1, x_2 = x'_2 \cdots$. Finally, $x = x'$, and the claim is proved. □

We shall say that ζ is *admissible* if it is recognizable and one-to-one on letters.

Corollary 5.10. *Let ζ be an admissible substitution. Then, for every $k \geq 1$, the induced transformation S_k on $\zeta^k(X)$ is defined on $y = \zeta^k(x)$ by*

$$S_k(y) = T^{|\zeta^k(x_0)|}y;$$

in particular, the systems (X,T) and $(\zeta^k(X), S_k)$ are topologically isomorphic, and the following diagram is commutative :

$$\begin{array}{ccc} X & \xrightarrow{T} & X \\ \zeta^k \downarrow & & \downarrow \zeta^k \\ \zeta^k(X) & \xrightarrow{S_k} & \zeta^k(X) \end{array}$$

For every $k \geq 1$, we denote by $\zeta^k(\mu)$ the pull-back of μ under ζ^k. The previous corollary, moreover, says that the system $(\zeta^k(X), S_k)$ is uniquely ergodic, with $\zeta^k(\mu)$ as its unique S_k-invariant probability measure. We shall identify this measure later.

Note the following, for which we refer to [52]:

Proposition 5.19. *A Pisot substitution is always admissible.*

5.6.3 Metric Properties

We begin with a geometrical description of X which makes clear the dynamics on X, and will play a key role in the spectral study. Recall that, for an explicit countable subset $\mathscr{D}$ of X, T is an homeomorphism of $X \backslash \mathscr{D}$ (proposition 5.13), whence a metric automorphism of X. This description makes useless the (strong) recognizability assumption. Actually, for every $k \geq 1$, ζ^k is always bilaterally recognizable (theorem 5.8) which appears to be sufficient for our purpose [87] as we shall see.

Lemma 5.8. *Let $x = \lim_{i\to\infty} T^{n_i}u \in X \backslash \mathscr{D}$. Then $x \in \zeta^k(X)$ if and only if $n_i \in E_k$ for i large enough.*

Proof. If $x \in \zeta^k(X)$, $x = \lim_{i\to\infty} T^{m_i}u$ with $m_i \in E_k$ (proposition 5.16). Suppose, in addition, that $x = \lim_{i\to\infty} T^{n_i}u \in X \backslash \mathscr{D}$. If $L := L_k$ is the bilateral recognizability index of ζ^k, then in turn, $T^{-L}x = \lim_{i\to\infty} T^{n_i - L}u$. Let now i be large enough in such a way that

$$(T^{m_i - L}u)_{[0,\,2L]} = (T^{-L}x)_{[0,\,2L]} = (T^{n_i - L}u)_{[0,\,2L]}$$

hence

$$u_{[m_i-L,\, m_i+L]} = u_{[n_i-L,\, n_i+L]};$$

by definition of L, n_i as m_i, must be in E_k, and the lemma is proved. □

Proposition 5.20. *For every $n > 0$, we set*

$$\mathscr{P}_n = \{T^k(\zeta^n[\alpha]),\ \ \alpha \in A,\ 0 \le k < |\zeta^n(\alpha)|\}.$$

Then $\mathscr{P}_n$ is a metric partition of X.

Proof. ⋆ We remark first that $\mathscr{P}_n$ is a covering of X : indeed, let

$$Y = \bigcup_{\alpha \in A,\ 0 \le k < |\zeta^n(\alpha)|} T^k(\zeta^n[\alpha]).$$

Clearly, Y is a closed subset of X and we just have to show that Y contains the orbit $O(u)$. Fix $r \ge 0$. There exists $j \ge 0$ such that $m_j \le r < m_{j+1}$, where m_j runs over E_n the set of n-cutting bars.

If $m_j = |\zeta^n(u_{[0,\, p[})|$ for some $p > 0$, then $m_{j+1} = m_j + |\zeta^n(u_p)|$ and we write

$$T^r u = T^{r-m_j}(T^{m_j} u)$$

But, $T^{m_j} u \in \zeta^n[u_p]$ and $0 \le r - m_j < |\zeta^n(u_p)|$. It follows that

$$T^r u \in T^k \zeta^n[u_p] \ \text{ for some } k,\ 0 \le k < |\zeta^n(u_p)|,$$

and $T^r u \in Y$.

⋆ We now prove that the sets in $\mathscr{P}_n$ are non-intersecting μ-a.e., more precisely, we claim that :

Lemma 5.9. *For $\alpha, \beta \in A$ and $0 \le k < |\zeta^n(\alpha)|$, $0 \le j < |\zeta^n(\beta)|$, we have*

$$T^k(\zeta^n[\alpha]) \cap T^j(\zeta^n[\beta]) \cap (X \backslash \mathscr{D}) = \emptyset,$$

except if $\zeta^n(\alpha) = \zeta^n(\beta)$ and $j = k$.

Proof. Suppose first that there exists $x \in T^k(\zeta^n[\alpha]) \cap T^j(\zeta^n[\beta]) \cap (X \backslash \mathscr{D})$ with $k > j$. Then, $x = T^k y = T^j z$ for some $y \in \zeta^n[\alpha]$ and $z \in \zeta^n[\beta]$. Since $x \notin \mathscr{D}$, we have $z = T^{-j} x = T^{k-j} y$; but both y and z are in $\zeta^n(X)$ and $k - j$ must be a multiple of $|\zeta^n(\alpha)|$ (lemma 5.8) which is impossible since $0 \le k < |\zeta^n(\alpha)|$.

If now $k = j$, this provides $y = z$ with $y \in \zeta^n[\alpha]$ and $z \in \zeta^n[\beta]$, and a fortiori, $\zeta^n(\alpha) = \zeta^n(\beta)$. □

This completes the proof of the proposition. □

The partition $\mathscr{P}_n$ is a *Kakutani-Rokhlin* partition, with base $\bigcup_{\alpha\in A}\zeta^n[\alpha] = \zeta^n(X)$. A very important property of these partitions is the following [119]

Proposition 5.21. *Let $\mathscr{B}_n = \sigma(\mathscr{P}_n)$ be the σ-algebra generated by $\mathscr{P}_n$. Then $\mathscr{B}_n$ increases to $\mathscr{B}$, the σ-algebra on X.*

Proof. Let $\omega \in \mathscr{L}_\zeta$, $f = \mathbf{1}_{[\omega]}$ and $f_n = E^{\mathscr{B}_n}(\mathbf{1}_{[\omega]})$; we shall prove that $\lim_{n\to\infty} ||f - f_n||_{L^2} = 0$. Consequently, $[\omega]$ must belong to $\sigma(\cup_n \mathscr{B}_n)$ and both σ-algebras coincide.

By definition of f_n, $f_n = f$ on each stack $T^k(\zeta^n[\alpha])$ included into $[\omega]$, and $f_n = 0 = f$ on the other $T^k(\zeta^n[\alpha])$ disjoint from $[\omega]$. It follows that

$$\{f_n \neq f\} \subset \{(k,\alpha),\ T^k(\zeta^n[\alpha]) \cap [\omega] \neq \emptyset \text{ and } T^k(\zeta^n[\alpha]) \cap [\omega]^c \neq \emptyset\}$$

$$\subset \{T^k(\zeta^n[\alpha]),\ |\zeta^n(\alpha)| - |\omega| < k < |\zeta^n(\alpha)| \text{ and } a \in A\}.$$

Since each $\mu(T^k(\zeta^n[\alpha])) = \theta^{-n}\mu[\alpha] \leq \theta^{-n}$, we see that

$$\mu\{f_n \neq f\} \leq s(|\omega| - 1)\theta^{-n} \to 0$$

as n tends to infinity. We deduce that

$$||f - f_n||^2_{L^2} \leq 2||f||^2_\infty \mu\{f_n \neq f\} \leq C|\omega|\theta^{-n}$$

and $E^{\mathscr{B}_n}(f)$ tends to f in $L^2(X,\mu)$ for every $f \in L^2(X,\mu)$. □

Proposition 5.22. *Let θ be the Perron-Frobenius eigenvalue of ζ. Then*

$$\zeta(\mu) = \theta \cdot \mu_{|\zeta(X)}$$

where $\mu_{|\zeta(X)}$ is the restriction of μ to $\zeta(X)$.

Proof. If the measure $\nu = \mu_{|\zeta(X)}$ is S-invariant too, by unique ergodicity, $\nu = C\zeta(\mu)$ for some positive constant C, which is easily seen to be $C = \mu(\zeta(X))$.

$\star$ We first prove that $S(\nu) = \nu$. Let $\omega \in \mathscr{L}_\zeta$ and consider the cylinder set $\zeta[\omega] \subset \zeta(X)$. By the commutativity relation and the definition of ν

$$\nu(S(\zeta[\omega])) = \nu(\zeta(T[\omega])) = \mu(\zeta(T[\omega])).$$

But, if $\omega = \omega_0 \cdots \omega_{\ell-1}$,

$$T[\omega] = \{y \in X,\ y_{[0,\,\ell-2]} = \omega_{[1,\,\ell-1]}\}$$

so that

$$\zeta(T[\omega]) = T^{|\zeta(\omega_0)|}\zeta[\omega],$$

and, T being a metric isomorphism,

$$\mu(\zeta(T[\omega])) = \mu(T^{|\zeta(\omega_0)|}\zeta[\omega]) = \mu(\zeta[\omega]).$$

This means $\nu(S(\zeta[\omega])) = \nu(\zeta[\omega])$ and the measure ν is S-invariant.

⋆ It remains to compare $\mu(\zeta(X))$ with θ. We make use of proposition 5.20 under the weaker form

$$[\alpha] = \bigcup_{\substack{k,\beta \\ \alpha=\zeta(\beta)_k}} T^k(\zeta[\beta]),$$

for each letter α. We obtain, taking the μ-measure,

$$\mu([\alpha]) = \sum_{\substack{k,\beta \\ \alpha=\zeta(\beta)_k}} \mu(T^k(\zeta[\beta])) = \sum_{\substack{k,\beta \\ \alpha=\zeta(\beta)_k}} \mu(\zeta[\beta])$$

$$= \Sigma_\beta \ell_{\alpha\beta}\mu(\zeta[\beta])$$

where $\ell_{\alpha\beta}$ denotes the occurrence number of α in $\zeta(\beta)$; but this is exactly the $\alpha\beta$-entry of the composition matrix $M := M(\zeta)$, and, remembering the notation d for the normalized right eigenvector $(\mu([\alpha]))_{\alpha\in A}$, this identity can be written $d = Mw$ with $w = (\mu(\zeta[\alpha]))_{\alpha\in A}$. Since $\mu([\alpha]) = \zeta(\mu)(\zeta[\alpha]) = C^{-1}\mu(\zeta[\alpha])$, d and w are proportional and $d = M(Cd) = C\theta d$.

Finally $C = \theta^{-1} = \mu(\zeta(X))$ and in the same way, $\zeta^n(X) = \theta^{-n}$ for every $n \geq 1$. The claim is proved. □

Remark 5.6. From the construction of μ, $\mu(\zeta^n(X))$ is nothing else than the density of the set $E_n \subset \mathbf{N}$ of n-cutting bars. We thus have

$$\text{dens}\,(E_n) = 1/\theta^n.$$

5.6.4 Bilateral Substitution Sequences

In view of the fundamental role plaid by this property of recognizability in what precedes, we aim to relax our statements by considering two-sided substitution subshifts; applying then Mossé's theorems if possible, we could expect more general results. We refer to [122] for complements.

We consider the full shift $(A^{\mathbf{Z}}, T)$ on the finite alphabet A, T being invertible; a subshift on A, as before, is the dynamical system (X, T) where T denotes now the bilateral shift restricted to a closed and non-empty shift-invariant subset of $A^{\mathbf{Z}}$. If ζ is a substitution on A with language $\mathscr{L}_\zeta$, the subshift associated to ζ at present is

$$X_\zeta = \{x \in A^{\mathbf{Z}},\ \mathscr{L}(x) \subset \mathscr{L}_\zeta\} = \{x \in A^{\mathbf{Z}},\ x_{[-n,n]} \in \mathscr{L}_\zeta,\ \text{ for } n \geq 0\}$$

(the inclusion "$\supset$" being clear since any factor of $\omega \in \mathscr{L}$ must also be in $\mathscr{L}$).

If there exist α and β in A such that

(i) α is the first letter of $\zeta(\alpha)$ and β, the last letter of $\zeta(\beta)$,

(ii) $|\zeta^n(\alpha)|$ and $|\zeta^n(\beta)|$ tend to infinity,

we get a bilateral fixed point of ζ by iterating ζ on the two-word $\beta \cdot \alpha$, that we write, emphasizing on the 0-index,

$$u = \zeta^\infty(\beta) \cdot \zeta^\infty(\alpha) = \cdots u_{-2}u_{-1} \cdot u_0 u_1 \cdots$$

with $u_0 = \alpha$ and $u_{-1} = \beta$. One will easily be convinced that such a pair does exist for a possible iterate of ζ. When ζ is primitive, $\mathscr{L}_{\zeta^n} = \mathscr{L}_\zeta$ for every $n \geq 1$ and one may assume that ζ admits a fixed point u.

The system (X_ζ, T) is strictly ergodic for a primitive ζ. If we consider now, for every $k \geq 1$, *the set of k-cutting bars*

$$E_k = \{0\} \cup \{|\zeta^k(u_{[0,\, p[})|, |\zeta^k(u_{[-p,\, 0[})|,\ p > 0\},$$

Mossé's theorems extend in the following similar form :

Theorem 5.10. *Let ζ be a primitive aperiodic substitution.*

a) There exists an integer $L > 0$ such that

$$n \in E_1 \ \text{ and } \ u_{[n-L,\, n+L]} = u_{[m-L,\, m+L]} \quad \text{imply } \ m \in E_1.$$

b) There exists an integer $M > 0$ such that

$$r, t \in E_1 \ \text{ and } \ u_{[r-M,\, r+M]} = u_{[t-M,\, t+M]} \quad \text{imply } \ u_r = u_t.$$

Now, the key results of the previous subsections remain valid without any additional assumption on ζ, since, for $k \geq 1$, ζ^k in turn is primitive and aperiodic, and thus, satisfy theorem 5.10.

We summarize below the results for bilateral substitutions :

Corollary 5.11. *For every $n \geq 1$,*

(i) ζ^n is a homeomorphism from X onto the clopen set $\zeta^n(X)$.

(ii) Every $x \in X$ can be written in a unique way $x = T^k(\zeta^n(y))$ with $y \in X$ and $0 \leq k < |\zeta^n(y_0)|$. In particular, every x can be desubstituted.

(iii) The systems (X, T) and $(\zeta^n(X), S_n)$ are topologically isomorphic, where the induced transformation S_n is defined by

$$S_n(y) = T^{|\zeta^n(x_0)|}y, \ \text{ if } y = \zeta^n(x) \in \zeta^n(X).$$

(iv) The systems (X, μ, T) and $(\zeta^n(X), \zeta^n(\mu), S_n)$ are metrically isomorphic, where $\zeta^n(\mu)(B) = \theta^{-n}\mu(B)$ for $B \in \mathscr{B}_{\zeta^n(X)}$.

Chapter 6
Eigenvalues of Substitution Dynamical Systems

This chapter is devoted to the description of eigenvalues of the dynamical system (X, μ, T) arising from a **primitive** and **aperiodic** substitution ζ. We distinguish the case of constant-length substitutions from the case of nonconstant-length ones. In the first case, the result is due to M. Dekking [68] and J.C. Martin [174] who made use of a different approach. The description of the eigenvalues in the general case has been essentially established by B. Host [119]. Another reformulation and a new proof, in somewhat more geometric terms, have been given ten years later [87] and we just outline the ideas. The third part raises the problem of pure point spectrum for substitution dynamical systems, with emphasize on the emblematic Pisot case.

6.1 Eigenvalues of a Constant-Length Substitution

Let ζ be a substitution of length $q \geq 2$ and u be a fixed point of ζ. We shall describe the eigenvalues of the system (X, T) where $X := X_\zeta$, (in short, the eigenvalues of ζ– if it cannot lead to some ambiguity), namely the discrete part of the maximal spectral type of the system.

6.1.1 Continuous Eigenvalues

We start by studying the topological system (X_ζ, T) and by investigating the eigenvalues of ζ admitting a continuous associated eigenfunction (called *continuous eigenvalues*). We first observe

Proposition 6.1. *All the possible continuous eigenvalues of ζ are roots of unity.*

Proof. Suppose that $f_\lambda \in C(X)$ is the continuous eigenfunction corresponding to $\lambda \in \mathbf{U}$, normalized by $f_\lambda(u) = 1$ (thanks to ergodicity, eigensubspaces are one-dimensional). Let $a \geq 1$ be an index such that $u_a = u_0$. Since $u = \zeta^n(u)$, $\zeta^n(u_a) =$

M. Queffélec, *Substitution Dynamical Systems – Spectral Analysis: Second Edition*, Lecture Notes in Mathematics 1294, DOI 10.1007/978-3-642-11212-6_6,

© Springer-Verlag Berlin Heidelberg 2010

$\zeta^n(u_0)$ is a prefix of $T^{aq^n}u$ for every $n \geq 1$. Now, $|\zeta^n(u_0)|$ increases to infinity and $T^{aq^n}u$ tends to u, so that, f_λ being continuous,

$$\lim_{n\to\infty} f_\lambda(T^{aq^n}u) = \lim_{n\to\infty} \lambda^{aq^n} = 1.$$

Writing $\lambda = e^{2i\pi t}$, and considering the q-adic expansion of t, we see that taq^n must be an integer for large n, and $\lambda = e^{2i\pi k/aq^{n_0}}$. □

Corollary 6.1. *Suppose that $\lambda = e^{2i\pi/p}$ is a continuous eigenvalue where $(p,q) = 1$. Then, p divides every integer $a \geq 1$ such that $u_a = u_0$.*

Proof. From the above, $\frac{aq^{n_0}}{p} \in \mathbf{Z}$ if $a \geq 1$ satisfies $u_a = u_0$. Since $(p,q) = 1$, p must divide a. □

We are thus led to define the following number which measures how far the sequence u can be from a (shift)-periodic sequence.

Let us consider, for every $k \geq 0$,

$$S_k = \{a \geq 1,\ u_{a+k} = u_k\},$$

and

$$g_k = \gcd S_k.$$

Definition 6.1. We call the *height* of ζ and denote by $h := h(\zeta)$, the number

$$h = \max\{n \geq 1,\ (n,q) = 1,\ n \text{ divides } g_0\}.$$

We list some properties of h :

1. $h = \max\{n \geq 1,\ (n,q) = 1,\ n$ *divides* $g_k\}$ *for every* $k \geq 0$.

   Proof. We only prove that $\{n \geq 1,\ (n,q) = 1,\ n \text{ divides } g_0\} \subset \{n \geq 1,\ (n,q) = 1,\ n \text{ divides } g_k\}$; the reverse inclusion will be established in the same way. Let us choose $a_1,\ldots,a_p \in S_k$ such that $g_k = gcd\{a_1,\ldots,a_p\}$. Now let N be such that u_0 occurs in $\zeta^N(u_k)$, for instance $u_0 = \zeta^N(u_k)_j$, in other words, $u_0 = u_{kq^N+j}$. Since $u_{a_i+k} = u_k$, we have also $u_0 = u_{(a_i+k)q^N+j}$ for $1 \leq i \leq p$. It follows that $kq^N + j, (a_1+k)q^N + j, \ldots, (a_p+k)q^N + j$ are in S_0, and $a_1q^N, \ldots, a_pq^N$ are in $S_0 - S_0$. Now, if $n \geq 1$ divides g_0 and $(n,q) = 1$, then n divides $\gcd(S_0 - S_0)$, henceforth divides $a_1q^N, \ldots, a_pq^N$. This implies that n divides $\gcd(a_1,\ldots,a_p) = g_k$ and gives the first inclusion. □

   As a consequence, observe that $1 \leq h \leq s$ if Card $A = s$.
   Indeed, among the s numbers g_k, corresponding to the occurrences of each letter of A, one at least is less than or equal to s.

2. *If $h = s$, u is periodic.*

   Proof. If $h = s$, then $g_k \geq s$ for every $k \geq 0$ and the s first components $u_0u_1\cdots u_{s-1}$ of u must be distinct. Thus, the s letters of A occur exactly in $u_0u_1\cdots u_{s-1}$. Now, u_s in turn must be different from $u_1\cdots u_{s-1}$ unless $h < s$. So $u_s = u_0$, then $u_{s+1} = u_1$ in the same way, and so on; u is the sequence $u_0u_1\cdots u_{s-1}u_0u_1\cdots u_{s-1}\cdots$ □

3. We consider, for $j = 0, 1, \ldots, h$, the class

$$C_j = \{u_a,\ a \equiv j \ (\bmod h)\}.$$

Note that these classes form a partition of A. *If we identify, in u, the letters in a same class C_j, we thus obtain a periodic sequence, and h is the smallest integer $\leq s$, prime to q, with this property.*

Proof. This is clear from the definition of h. □

We are now in a position to describe the continuous eigenvalues of ζ. In the sequel, we shall identify $\mathbf{Z}(q)$, the group of q-adic rational numbers, with $\{e^{2i\pi k/q^n},\ n \geq 0, k \in \mathbf{Z}\}$ and $\mathbf{Z}/h\mathbf{Z}$, with the h-roots of unit.

Theorem 6.1. *Assume ζ to be primitive aperiodic and one-to-one on letters, and let G denote the subgroup of $\mathbf{U}$ consisting in the continuous eigenvalues. Then*

$$G = \mathbf{Z}(q) \times \mathbf{Z}/h\mathbf{Z}.$$

Proof. ▷ The first inclusion $G \subset \mathbf{Z}(q) \times \mathbf{Z}/h\mathbf{Z}$ is just a reformulation of corollary 6.1 which can be stated in the following form : each continuous eigenvalue of ζ may be decomposed into $\lambda_1\lambda_2$, where $\lambda_1 \in \mathbf{Z}(q)$ and $\lambda_2 \in \mathbf{Z}/h\mathbf{Z}$.

◁ Remember that such a substitution is recognizable (proposition 5.14) and that we have in this particular case

$$\zeta \circ T = T^q \circ \zeta$$

whence, a simple topological description of X :

Lemma 6.1. $X = \bigcup_{0 \leq j < q} T^j \zeta(X)$ *where the "stacks" are disjoint clopen sets.*

If $\lambda = e^{2i\pi/q^n}$ we consider f_λ defined by

$$f_\lambda(x) = \begin{cases} 1 & \text{if } x \in \zeta(X) \\ e^{2i\pi j/q^n} & \text{if } x \in T^j\zeta(X). \end{cases}$$

From the lemma, f_λ is continuous and $f_\lambda(Tx) = \lambda f_\lambda(x)$ in an obvious way.

If $\lambda = e^{2i\pi/h}$ we consider f_λ defined by

$$f_\lambda(x) = e^{2i\pi j/h} \text{ if } x_0 \in C_j,\ 0 \leq j < h.$$

Once more f_λ is continuous on $X = \bigcup_{0 \leq j < h}\{x,\ x_0 \in C_j\}$. Besides, the sequence deduced from u through this identification is periodic : $C_0C_1 \cdots C_{h-1}C_0C_1 \cdots$ so that, $f_\lambda(Tu) = e^{2i\pi/h} = \lambda f_\lambda(u)$; f_λ is the normalized eigenfunction corresponding to λ.

□

Examples

1. The continuous eigenvalues of the Thue-Morse substitution are the dyadic rational numbers, $\mathbf{Z}(2)$, since $h = 1$.
2. Consider ζ defined on $\{0,1,2,3,4\}$ by $\zeta(0) = 0213, \zeta(1) = 1341, \zeta(2) = 4104$, $\zeta(3) = 0413, \zeta(4) = 2134$ and consider $u = \zeta^{\infty}(0)$. By definition, h divides a whenever $u_a = 0$, thus, obviously, h divides 6 and $h = 1,2$ or 3. But $u_{3n} \in \{0,3\}$, $u_{3n+1} \in \{2,4\}$ and $u_{3n+2} = 1$ for every $n \geq 0$. Then $h = 3$ and $G = \mathbf{Z}(4) \times \mathbf{Z}/3\mathbf{Z}$.

6.1.2 L^2-Eigenvalues

We now wish to describe **all** the eigenvalues of the dynamical system (X_ζ, μ, T) where the substitution ζ is always assumed to be primitive aperiodic and one-to-one on letters. We shall see in passing that they are continuous, more precisely admitting a L^2-eigenfunction whose class contains a continuous representative, and thus they are rational.

Theorem 6.2. *Let λ be a unimodular complex number. The following assertions are equivalent*
(a) λ is an eigenvalue of ζ.
(b) There exists an integer $j \geq 1$ such that

$$z = \lim_{n\to\infty} \lambda^{q^{jn}} \text{ exists and satisfies } z^h = 1,$$

where $h = h(\zeta)$ is the height of ζ.
(c) λ is a continuous eigenvalue of ζ.

Remark 6.1. J.C. Martin [173] and F.M. Dekking [68] both proved that the set of eigenvalues of such a substitution is exactly $\mathbf{Z}(q) \times \mathbf{Z}/h\mathbf{Z}$, h being the height of ζ. We propose a somewhat different proof, more elementary but longer, and which naturally leads to the description of the eigenvalues in the nonconstant-length case, obtained by B. Host, and only partially discovered by J.C. Martin [174].

Proof. $\triangleright$ We first prove that (b) implies (c). Suppose that the unimodular complex number λ satisfies (b). Writing $\lambda = e^{2i\pi t}$ and $z = e^{2i\pi\tau}$, (b) implies

$$\lim_{n\to\infty} q^{jn} t \equiv \tau \pmod 1,$$

so that, the decomposition of t in the q^j-adic expansion must be ultimately stationary. Whence, the following expression for t:

$$t = \frac{a}{q^{jn_0}} + \frac{b}{q^{jn_0}(q^j - 1)}, \quad \text{for some } a \in \mathbf{Z},\ 0 \leq b < q^j;$$

turning back to our assumption (b), we deduce that $\tau = \frac{b}{(q^j-1)}$. Now, taking into account the second condition $e^{2i\pi h\tau} = 1$, we see that

$$t = \frac{a}{q^{jn_0}} + \frac{c}{h}, \quad \text{for some } a, c \in \mathbf{Z}$$

and $\lambda \in \mathbf{Z}(q) \times \mathbf{Z}/h\mathbf{Z}$; in particular λ is a continuous eigenvalue by theorem 6.1, and (c) holds.

$\triangleright$ We assume that λ is an eigenvalue of ζ, and let $f := f_\lambda$ the associated L^2-eigenfunction, normalized by $|f| = 1$. By using the Kakutani-Rokhlin partitions $(\mathscr{P}_n)$ (proposition 5.17), we have

$$\zeta^n[\alpha] = \bigcup_{\substack{k,\beta \\ \alpha=\zeta(\beta)_k}} \zeta^n(T^k(\zeta[\beta])) = \bigcup_{\substack{k,\beta \\ \alpha=\zeta(\beta)_k}} T^{kq^n}(\zeta^{n+1}[\beta]) \;\; \mu - a.e,$$

so that, by invoking proposition 5.13, we can write

$$\begin{aligned} \int_{\zeta^n[\alpha]} f\, d\mu &= \int_{\bigcup_{\substack{k,\beta \\ \alpha=\zeta(\beta)_k}} T^{kq^n}(\zeta^{n+1}[\beta])} f\, d\mu \\ &= \sum_{\substack{k,\beta \\ \alpha=\zeta(\beta)_k}} \lambda^{kq^n} \int_{\zeta^{n+1}[\beta]} f\, d\mu. \end{aligned} \tag{6.1}$$

We are thus led to consider V_n, the s-dimensional vector whose α^{th} component is

$$V_n(\alpha) = \frac{1}{\mu(\zeta^n[\alpha])} \int_{\zeta^n[\alpha]} f\, d\mu = \frac{q^n}{\mu([\alpha])} \int_{\zeta^n[\alpha]} f\, d\mu.$$

Equation (6.1) has the following interpretation in terms of V_n:

$$V_n = M(\lambda^{q^n})\, V_{n+1}, \tag{6.2}$$

where $M(z)$ is the $s \times s$ complex matrix whose entries are

$$m_{\alpha\beta}(z) = \frac{\mu([\beta])}{\mu([\alpha])} \frac{1}{q} \Big(\sum_{\substack{k, \\ \alpha=\zeta(\beta)_k}} z^k \Big). \tag{6.3}$$

If $|z| = 1$, $|m_{\alpha\beta}(z)| \leq m_{\alpha\beta}(1) = \frac{\mu([\beta])}{\mu([\alpha])} \frac{1}{q} \ell_{\alpha\beta} =: s_{\alpha\beta}$, where $\ell_{\alpha\beta} = L_\alpha(\zeta(\beta))$ denotes as in chapter 5 the occurrence number of α in $\zeta(\beta)$. Remember that $(\mu[\alpha])_\alpha$ is the normalized eigenvector associated to $\theta = q$, the Perron-Frobenius eigenvalue :

$$\sum_{\beta \in A} \ell_{\alpha\beta} \mu([\beta]) = q\mu([\alpha])$$

so that, for every $\alpha \in A$, $\sum_{\beta \in A} s_{\alpha\beta} = 1$. The matrix $S = (s_{\alpha\beta})$ is row-stochastic and the matrix $M(z)$, in turn, is row-sub-stochastic for $|z| = 1$.

We now aim to let n go to infinity in (6.2) and this will require two steps:

Lemma 6.2. *For every $\alpha \in A$, $\lim_{n\to\infty} |V_n(\alpha)| = 1$.*

Lemma 6.3. *Replacing if necessary ζ by ζ^j for some suitable j, we have, for every $\alpha \in A$, $\lim_{n\to\infty} |V_{n+1}(\alpha) - V_n(\alpha)| = 0$.*

Proof of lemma 6.2. After p iterations of identity (6.2), we derive

$$|V_n(\alpha)| \leq \frac{1}{q^p} \sum_{\beta \in A} \frac{\mu([\beta])}{\mu([\alpha])} L_\alpha(\zeta^p(\beta))|V_{n+p}(\beta)|.$$

Now $\lim_{p\to\infty} \frac{1}{q^p} L_\alpha(\zeta^p(\beta)) = \mu([\alpha])$, so that, taking the liminf on p, we obtain

$$|V_n(\alpha)| \leq \sum_{\beta \in A} \mu([\beta]) \liminf_{n\to\infty} |V_n(\beta)|$$

which in turn implies

$$\begin{aligned}\limsup_{n\to\infty} |V_n(\alpha)| &\leq \sum_{\beta \in A} \mu([\beta]) \liminf_{n\to\infty} |V_n(\beta)| \leq \sum_{\beta \in A} \mu([\beta]) \limsup_{n\to\infty} |V_n(\beta)| \\ &\leq \max_\beta \limsup_{n\to\infty} |V_n(\beta)|.\end{aligned}$$

Let us now choose α such that $\limsup_{n\to\infty} |V_n(\alpha)| = \max_\beta \limsup_{n\to\infty} |V_n(\beta)|$; the inequalities become equalities and we deduce from these that $\lim_{n\to\infty} |V_n(\alpha)| = \kappa$ exists for every α and is independent of α.

This limit κ must be one : indeed, since $|f| = 1$, for every $0 < \varepsilon < 1/4$ one can find a set $B \in \mathscr{B}$ such that

$$\frac{1}{\mu(B)} \left| \int_B f \, d\mu \right| \geq 1 - \varepsilon.$$

Now, thanks to proposition 5.20, one can find $B_n \in \mathscr{B}_n = \sigma(\mathscr{P}_n)$ such that $\mu(B \Delta B_n) \leq \varepsilon$ and

$$\frac{1}{\mu(B_n)} \left| \int_{B_n} f \, d\mu \right| \geq 1 - 2\varepsilon. \tag{6.4}$$

But B_n is a union of disjoint sets $T^k\zeta^n[\alpha]$ whence

$$\begin{aligned}\left| \int_{B_n} f \, d\mu \right| &= \left| \sum_{k,\alpha} \lambda^k \int_{\zeta^n[\alpha]} f \, d\mu \right| \leq \sum_{k,\alpha} |V_n(\alpha)| \mu(\zeta^n[\alpha]) \\ &\leq (\kappa + \varepsilon) \sum \mu(\zeta^n[\alpha]) = (\kappa + \varepsilon) \mu(B_n).\end{aligned}$$

The lemma then follows from equation (6.4). □

We need to replace ζ by a suitable ζ^j in order to fulfill the three requirements

(i) $g_0 = gcd\{k \geq 1,\ \zeta^j(u_0)_k = u_0\}$ (Recall that $g_0 = gcd\ S_0$ where $S_0 = \{k \geq 1,\ u_k = u_0\}$).

(ii) h divides $q^j - 1$ (Recall that $(h,q) = 1$).

(iii) For every $\alpha \in A$, $\zeta^j(\alpha)$ and $\zeta^{2j}(\alpha)$ begin with the same letter.

For this last condition with some fixed α, observe that the sequence $(\zeta^n(\alpha)_0)_{n\geq 0}$ consisting of the first letter in the different words $\zeta^n(\alpha)$ is ultimately periodic. If $r := r(\alpha)$ is the period length of this sequence, $\zeta^{mr}(\alpha)$ begins with the same letter for every $m \geq 1$, and $j =$scm $\{r(\alpha), \alpha \in A\}$ is a good choice. Finally, the scm of the j's satisfying (i), (ii) and (iii) will do the job, since these conditions remain valid for multiples of j.

We then replace ζ by ζ^j of constant length q^j but we omit the "j" to save notations.

Proof of lemma 6.3. As a consequence of lemma 6.2, note that

$$\lim_{n\to\infty} \frac{1}{\mu(\zeta^n[\alpha])} \int_{\zeta^n[\alpha]} |f - V_n(\alpha)|\, d\mu = 0.$$

It now follows from (iii) that $\zeta^{n+1}[\alpha] \subset \zeta^n[\alpha]$ and thus

$$\begin{aligned} |V_{n+1}(\alpha) - V_n(\alpha)| &\leq \frac{1}{\mu(\zeta^{n+1}[\alpha])} \int_{\zeta^{n+1}[\alpha]} |V_{n+1}(\alpha) - f| + |f - V_n(\alpha)|\, d\mu \\ &\leq \frac{1}{\mu(\zeta^{n+1}[\alpha])} \int_{\zeta^{n+1}[\alpha]} |V_{n+1}(\alpha) - f|\, d\mu \\ &+ \frac{q}{\mu(\zeta^n[\alpha])} \int_{\zeta^n[\alpha]} |f - V_n(\alpha)|\, d\mu \end{aligned}$$

tends to zero, whence the lemma. □

We shall finally prove that (λ^{q^n}) is convergent and end the proof of the theorem. Let z be a cluster-point of the sequence (λ^{q^n}); we choose V, a cluster-point of the sequence of s-dimensional vectors (V_n), in such a way that, taking the limit in (6.2), we get, with lemma 6.5,

$$V = M(z)V. \tag{6.5}$$

But $|V(\alpha)| = 1$ for every α (lemma 6.2) and this gives rise to the inequalities

$$1 \leq \sum_{\beta\in A} |m_{\alpha\beta}(z)| \leq \sum_{\beta\in A} s_{\alpha\beta} = 1.$$

It follows that

$$|m_{\alpha\beta}(z)| = s_{\alpha\beta} = m_{\alpha\beta}(1)$$

and, by (6.3),

$$\Big| \sum_{\substack{k,\\ \alpha=\zeta(\beta)_k}} z^k \Big| = \sum_{\substack{k,\\ \alpha=\zeta(\beta)_k}} 1.$$

All numbers z^k, for k such that $\alpha = \zeta(\beta)_k$, must have the same argument and, setting $V(\alpha) = e^{iv(\alpha)}$ in (6.5), we get

$$z^k = e^{iv(\alpha)}e^{-iv(\beta)} \quad \text{if} \quad \zeta(\beta)_k = \alpha.$$

In particular, $z^k = 1$ if $\zeta(u_0)_k = u_0$ and $z^h = 1$ thanks to condition (i).

It remains to prove that (λ^{q^n}) has a unique cluster-point in $\mathbf{U}$. This is clear when $h = 1$. Otherwise, let z, z' be two distinct cluster-points and $n > m$ two integers such that λ^{q^n} is close to z, and λ^{q^m} is close to z'. Then $\lambda^{q^n - q^m} =: \lambda^{q^m(q-1)d}$ is close to both z/z' and $z'^{(q-1)d}$. But, $z'^h = 1$, and remembering (ii), h divides $q-1$ (so written instead of $q^j - 1$); one concludes that $z = z'$ and the sequence (λ^{q^n}) converges.

Re-introducing the integer j, we have obtained the characterization (b) of the theorem. □

Remark 6.2. This proof can be shortened, in particular one can prove the convergence of $(\lambda^{q^{jn}})$ when j is only supposed to check (iii), by analyzing the properties of h. This improvement will also result from the forthcoming theorem, established by B. Host for general substitutions. The link between j, such that $(\lambda^{q^{jn}})$ converges, and h, has to be precised. Note that h must divide $q^j - 1$ for all such j : actually, since

$$\lim_{n\to\infty} \lambda^{q^{jn+j} - q^{jn}} = \lim_{n\to\infty} \lambda^{q^{jn}(q^j - 1)} = 1$$

and $(h, q) = 1$, we get the remark by taking $\lambda = e^{2\pi i/h}$.

Definition 6.2. Following Dekking, we say that ζ is *pure* if $h(\zeta) = 1$.

Note the following observation.

Corollary 6.2. *a) If spectrum of the system (X, T) is purely discrete, then*

$$\lim_{n\to\infty} ||U^{q^{jn}} f - f||_{L^2} = 0 \quad \textit{for every } f \in \mathscr{F} \tag{6.6}$$

where $\mathscr{F}$ is a total set of functions in $L^2(X, \mu)$.
b) The spectrum of the system (X, T) is purely discrete if the rate of convergence in (6.6) is geometric.

Proof. The system is discrete if and only if the eigenfunctions span $L^2(X, \mu)$.

a) Suppose the system to be discrete and consider $\mathscr{F}$ to be the set of all normalized eigenfunctions. If $f = f_\lambda \in \mathscr{F}$,

$$||U^{q^{jn}} f - f||_{L^2}^2 = \int_X |f_\lambda \circ T^{q^{jn}} - f_\lambda|^2 \, d\mu = |\lambda^{q^{jn}} - 1|^2$$

and the lemma follows from the previous theorem.

b) Conversely, if (6.6) holds geometrically, then, according to proposition 2.16, U has purely discrete spectrum. □

6.2 Eigenvalues of a Nonconstant-Length Substitution

Let ζ be a primitive and aperiodic substitution over A with u as a fixed point. We shall give two formulations of the description of eigenvalues, the first original one by B. Host and the second one due to S. Ferenczi, C. Mauduit and A. Nogueira. As we deal with unilateral subshifts, we need the recognizability property to validate the topological description of the system given by corollary 5.10; so we shall assume ζ to be **admissible**, i.e. ζ primitive, one-to-one on letters and recognizable (thus automatically aperiodic).

6.2.1 Host's Theorem

We need a tool, which will play the part of the height introduced in the constant-length case.

Definition 6.3. The map $h : A \to \mathbf{U}$ is called a *coboundary of* ζ if, for every factor $u_{[m,\,n[}$ of u such that $u_n = u_m$, then

$$h(u_m)h(u_{m+1})\cdots h(u_{n-1}) = 1.$$

Lemma 6.4. *h is a coboundary of ζ if and only if there exists a map $g : A \to \mathbf{U}$ such that*

$$g(\beta) = g(\alpha)h(\alpha) \quad \textit{for every two-word} \ \ \alpha\beta \ \text{ of } \ u.$$

Proof. The sufficiency is obvious. Let now h be a coboundary of ζ and put $g(u_n) = h(u_0)h(u_1)\cdots h(u_{n-1})$. If $u_n = u_m$ for some $m \le n$, clearly

$$g(u_n) = g(u_m)h(u_m)h(u_{m+1})\cdots h(u_{n-1}) = g(u_m)$$

and g extends to a map $A \to \mathbf{U}$. Let $\alpha\beta \in \mathscr{L}_\zeta$ and let $n \ge 1$ be such that $u_{n-1}u_n = \alpha\beta$; then

$$g(\beta) = g(u_n) = g(u_{n-1})h(u_{n-1}) = g(\alpha)h(\alpha).$$

□

Note that every coboundary is constant and equal to one for a two-letters alphabet. By consistency with the uniform case, a substitution with only trivial coboundary is said to be *pure*.

B. Host proved in [119] the following result.

Theorem 6.3 (B. Host). *Let ζ be an admissible substitution over A. Then, the following assertions are equivalent.*

(a) $\lambda \in \mathbf{U}$ is an eigenvalue of ζ.

(b) There exists $j \ge 1$ such that,

$$h(\alpha) = \lim_{n\to\infty} \lambda^{|\zeta^{jn}(\alpha)|} \quad \textit{exists for every} \ \alpha \in A \tag{6.7}$$

where h is a coboundary of ζ.
(c) λ is a continuous eigenvalue of ζ.

Proof. We only sketch the main steps of the proof and we focus on the points specific to the nonconstant case, or leading to generalizations.

First of all, we interpret the assertion "$\lambda^{|\zeta^{jn}(\alpha)|}$ converges" in terms of the composition matrix. Observe that, for every α, $|\zeta^n(\alpha)| =< M^n L(\alpha), \mathbf{1} >$ with previous notations, so that, if P denotes the transpose of M, the components of the s-dimensional vector $P^n\mathbf{1}$ are $(|\zeta^n(\alpha)|, \alpha \in A)$. If $\lambda = e^{2i\pi t}$, "$\lambda^{|\zeta^n(\alpha)|}$ converges for every $\alpha \in A$" (omitting the "j") now means "$P^n\mathbf{t}$ converges modulo $\mathbf{Z}^s$ with $\mathbf{t} = (t, \dots, t)$". We shall need the following useful lemma [119].

Lemma 6.5. *Let Q be a $s \times s$ integral matrix and $y \in \mathbf{R}^s$. Then, $Q^n y$ converges modulo $\mathbf{Z}^s$ if and only if $y = y_1 + y_2 + y_3$ with*

$$\begin{aligned} &Q^n y_1 \to 0 \ \textit{in} \ \mathbf{R}^s \\ &Q^n y_2 \in \mathbf{Z}^s \ \textit{for n large enough} \\ &(Q - I) y_3 \in \mathbf{Z}^s \end{aligned}$$

and then $Q^n y$ tends to y_3 modulo $\mathbf{Z}^s$. Moreover the convergence is geometric : there exist $0 \le \rho < 1$ and a positive constant C such that

$$||Q^n y|| \le C\rho^n, \ \textit{for all} \ n \in \mathbf{N},$$

if $||\cdot||$ is the distance to the nearest element for the sup norm in $\mathbf{Z}^s$.

▷ Suppose (b) : for each $\alpha \in A$, $\lim_{n\to\infty} \lambda^{|\zeta^n(\alpha)|}$ exists and is equal to $h(\alpha)$ where h is a coboundary of ζ. We deduce from the above lemma that, for every $m \ge 0$,

$$|h(u_0)h(u_1)\cdots h(u_m) - \lambda^{|\zeta^n(u_{[0,\,m]})|}| \le K\rho^n \tag{6.8}$$

for some constants K and $0 \le \rho < 1$. We shall exhibit a continuous eigenfunction f corresponding to λ. Let g be the function associated to h by lemma 6.4. For every $n \ge 0$ and $x \in X$, set

$$r_n(x) = \min\{r \ge 0, \ T^r x \in \zeta^n(X)\},$$

and now define the function f_n on X by

$$f_n(x) = \lambda^{-r_n(x)} g(\alpha) \ \text{ if } \ T^{r_n(x)} x \in \zeta^n[\alpha].$$

Since $\zeta^n(X)$ is a clopen set in X, r_n and then f_n are continuous on X. Moreover, applying (6.8), we get

$$|f_n(T^k u) - \lambda^k f_n(u)| \leq K\rho^n$$

for every $k \geq 0$. But $f_n(u) = g(u_0)$ for any n. It follows that

$$|f_n(T^k u) - f_m(T^k u)| \leq K(\rho^n + \rho^m) \text{ for all } m,n,k,$$

and the sequence (f_n) converges uniformly to a continuous function f on $X = \overline{O(u)}$ satisfying

$$f(T^k u) = \lambda^k f(u), \text{ for every } k \geq 0.$$

Therefore $f \circ T = \lambda f$ and (c) holds.

▷ Suppose (a) and let f be the eigenfunction corresponding to λ with $|f| = 1$. We shall prove that $h(\alpha) := \lim_{n\to\infty} \lambda^{|\zeta^{jn}(\alpha)|}$ exists for every $\alpha \in A$, where j satisfies the requirement (iii) in the proof of theorem 6.2 (in other words, we assume that the period of the initial letters of ζ is one, with ζ instead of ζ^j).

Recall that we proved the following (proposition 5.21) : if $\mathscr{B}_n := \sigma(\mathscr{P}_n)$ is the σ-algebra generated by $\mathscr{P}_n$, then $\mathscr{B}_n$ increases to $\mathscr{B}$, the σ-algebra on X. It follows that the sequence $(f_n) = (E^{\mathscr{B}_n}(f))$, for every $f \in L^2(X,\mu)$, is a martingale converging to f in $L^2(X,\mu)$. In particular,

$$\lim_{n\to\infty} \frac{1}{\mu(\zeta^n[\alpha])} \int_{\zeta^n[\alpha]} |f - f_n|^2 \, d\mu = 0,$$

and, if $d_n(\alpha)$ is the constant value of f_n on $\zeta^n[\alpha]$, $\lim_{n\to\infty} |d_n(\alpha)| = 1$ for every $\alpha \in A$. One successively obtains the following limits :
If β is the first letter of $\zeta(\alpha)$,

$$\lim_{n\to\infty} d_{n+1}(\alpha)/d_n(\beta) = 1. \tag{6.9}$$

By the assumption on ζ, if α_{j+1} is the first letter of $\zeta(\alpha_j)$, there exists k such that $\alpha_k = \alpha_{k+1}$. From (6.9) $\lim_{n\to\infty} d_{n+k}(\alpha)/d_n(\alpha_k) = \lim_{n\to\infty} d_n(\alpha_{k+1})/d_{n+k+1}(\alpha) = 1$ so that, taking quotients,

$$\lim_{n\to\infty} d_{n+1}(\alpha)/d_n(\alpha) = 1 \text{ for every } \alpha \in A. \tag{6.10}$$

Finally, if $\omega = \alpha\beta$ is a two-word in $\mathscr{L}_\zeta$, $\zeta^n[\omega] \subset \zeta^n[\alpha]$ and, putting $k = |\zeta^n(\alpha)|$, one has $T^k\zeta^n[\omega] \subset \zeta^n[\beta]$. But

$$\begin{aligned} f_n &= \lambda^k d_n(\alpha) \text{ on } T^k\zeta^n[\omega] \\ &= d_n(\beta) \text{ on } \zeta^n[\beta]. \end{aligned}$$

Combined with the convergence of $||f - f_n||_{L^2}$ to zero, this leads to the following:

$$\lim_{n\to\infty} |\lambda^{|\zeta^n(\alpha)|} d_n(\alpha) - d_n(\beta)| = 0,$$

$$\lim_{n\to\infty} \lambda^{|\zeta^n(\alpha)|} d_n(\alpha)/d_n(\beta) = 1 \tag{6.11}$$

Thanks to (6.10) and (6.11), one obtains, for every α

$$\lim_{n\to\infty} \lambda^{|\zeta^{n+1}(\alpha)|-|\zeta^n(\alpha)|} = 1.$$

The convergence of $\lambda^{|\zeta^n(\alpha)|}$ now is a consequence of lemma 6.5.

If $h(\alpha)$ is the limit of $\lambda^{|\zeta^n(\alpha)|}$, h is easily seen to be a coboundary of ζ : since $\lim_{n\to\infty} d_n(\alpha)/d_n(\beta)$ exists for every pair of letters α and β, put for example $g(\alpha) = \lim d_n(\alpha)/d_n(\alpha_0)$; clearly, $h(\alpha)g(\alpha) = g(\beta)$ if $\alpha\beta$ is a two-word of u and (b) holds. The sketch of proof is complete. □

Remark 6.3. 1. For a unilateral substitution subshift we always have equivalence between (a) and (b), without extra assumptions on ζ (though primitive and aperiodic). But the recognizability property is needed for the partitions $\mathscr{P}_n$ to span the topology of X and thus for getting the continuity of eigenvalues (c). Of course, this assumption can be released by considering bilateral substitution subshifts.
2. In the constant-length case, all the eigenvalues are rational (roots of unit). Conversely, what can we say about a substitution whose eigenvalues are rational ?
3. A similar condition for a complex number to be an eigenvalue was given simultaneously by Livshits in the framework of *adic systems* [169].
4. The eigenvalues of a substitution associated with the trivial coboundary depend on the matrix of the substitution only; however, eigenvalues associated with a nontrivial coboundary do exist and depend, in addition, on the combinatorics of the substitution; we get used to call them *exotic eigenvalues* [85, 109].

A more geometric formulation of Host's equivalence condition has been given by Ferenczi, Mauduit and Nogueira [87] where they emphasize the key role of the Kakutani-Rokhlin partitions. Let us say that $\omega = \omega_0\omega_1 \cdots \omega_{\ell-1} \in \mathscr{L}_\zeta$ is a *generalized return word* if there exists $\omega_\ell \in A$ such that $\omega\omega_\ell \in \mathscr{L}_\zeta$, satisfying for n large enough,

$$\zeta^n(\omega_0) = \zeta^n(\omega_\ell) \quad \text{but} \quad \zeta^n(\omega_0) \neq \zeta^n(\omega_j) \ \text{if}\ 0 < j < \ell.$$

Observe that, for n large enough, $\zeta^n(\omega)$ is a return word over $\zeta^n(\omega_0)$ in the classical sense (remark 5.1).

Definition 6.4. If ω is a generalized return word, the associated *return time sequence* $(r_n(\omega))_n$ is defined by

$$r_n(\omega) = |\zeta^n(\omega)|.$$

Theorem 6.4. *Let ζ be a primitive and aperiodic substitution. The unimodular complex number λ is an eigenvalue of the system (X,T) if and only if*

$$\lim_{n\to\infty} \lambda^{r_n(\omega)} = 1$$

for every generalized return word ω.

Examples

1. Host's description makes use of the linearly recurrent sequence of vectors in $\mathbf{Z}^s$, $(|\zeta^n(\alpha)|)_{n\geq 0}$, $\alpha \in A$, whose matrix is M, and thus depends directly on the eigenvalues of M (not to be confused with those of the system). Suppose that ζ is a **Pisot substitution** (Subsection 5.3.3). Then, assuming the only coboundaries to be the trivial ones, one gets the following:
For every $\alpha \in A$, $\lambda = e^{2\pi i\mu[\alpha]}$ is an eigenvalue of the system. Moreover, if the determinant of M is ± 1, these numbers generate the group of eigenvalues of the system, and, except for $\lambda = 1$, all of them are irrational numbers.

   Proof. With the notations of subsection 5.3.2, we have for every α, β,

$$\lim_{n\to\infty} [(M^n)_{\alpha\beta} - \theta^n \mu[\alpha] q_\beta] = 0,$$

   since $\max\{|\lambda|,\ \lambda \text{ eigenvalue of } M,\ \lambda \neq \theta\} =: \tau < 1$. And, we noticed in (5.11) that $|\zeta^n(\beta)| \sim \theta^n q_\beta$, so that, for every α, β,

$$e^{2\pi i\mu[\alpha]|\zeta^n(\beta)|} \sim e^{2\pi i (M^n)_{\alpha\beta}} = 1;$$

   we now deduce from theorem 6.3 that $\lambda = e^{2\pi i\mu[\alpha]}$ is an eigenvalue of the system. □

2. The description of eigenvalues by means of return times (theorem 6.4) provides new examples. Consider for instance ζ defined on $\{0,1,2,3\}$ by $\zeta(0) = 0133$, $\zeta(1) = 12$, $\zeta(2) = 3$, $\zeta(3) = 0$. Among the generalized return words we find $3, 0, 301, 1200$ and this is sufficient to conclude that the eigenvalues of the associated system are the $e^{ik\pi\sqrt{2}}$, $k \in \mathbf{Z}$, [87].

Host's theorem naturally raises the following question : given a topological dynamical system, when is a L^2-eigenvalue λ associated with a continuous eigenfunction ? An complete answer has been given in the framework of *linearly recurrent Cantor systems* [42]. These systems are characterized (among technical properties) by the existence of a nested sequence of clopen Rokhlin-Kakutani partitions, the height of each tower of which increases linearly. Note that substitution dynamical systems and *sturmian systems* (generated by some sturmian sequence) with badly approximable rotation number (see appendix B) are examples of such systems. This sheds a new light on the problem.

6.2.2 Application to Mixing Property

In spite of various criteria, the problem to decide, in the nonconstant case, whether the weak mixing property holds or not, remains a difficult one. M. Keane and M. Dekking in [71] have already studied this question of mixing for dynamical systems arising from substitutions. In particular, they proved the following.

Theorem 6.5. *The dynamical systems arising from a primitive substitution is never strongly mixing.*

Proof. For sake of simplicity, we restrict our attention to the alphabet $\{0,1\}$. We may assume X to be infinite and thus $\mu([00])$ or $\mu([11])$ is non zero. Set $r=\mu([00])>0$ for example. If the system were strongly mixing, we should have, for every ω in $\mathscr{L}_\zeta$,

$$\lim_{n\to\infty}\mu([\omega]\cap T^{-s_n}[\omega])=(\mu([\omega]))^2$$

where $s_n=|\zeta^n(0)|\to\infty$. On the contrary, the authors establish that

$$\liminf_{n\to\infty}\mu([\omega]\cap T^{-s_n}[\omega])\geq C\mu([\omega]) \tag{6.12}$$

for some constant C independent of ω; applied with ω long enough, inequality (6.12) violates the strong mixing property.

Recall the notation $L_C(B)$ for the occurrence number of the word C in the word B. We fix n, $D_n=[\omega]\cap T^{-s_n}[\omega]$ and $N\gg n$. By definition, D_n occurs in $\zeta^N(0)$ if two ω's occur at distance s_n i.e. the word $\omega B\omega$ occurs in $\zeta^N(0)$ for some B of length $s_n-|\omega|$. Clearly,

$$L_\omega(\zeta^n(0))\leq L_{\omega B\omega}(\zeta^n(00))$$

and

$$\begin{aligned}L_{\omega B\omega}(\zeta^N(0))&\geq L_{\omega B\omega}(\zeta^n(00))L_{\zeta^n(00)}(\zeta^N(0))\\&\geq L_\omega(\zeta^n(0))L_{00}(\zeta^{N-n}(0))\end{aligned}$$

which implies

$$\mu(D_n)\geq\lim_{N\to\infty}\frac{1}{s_N}L_\omega(\zeta^n(0))L_{00}(\zeta^{N-n}(0)).$$

As already noted, s_N is asymptotically equal to $c\theta^N$ where θ is the Perron-Frobenius eigenvalue of ζ and c a positive constant. Therefore,

$$\lim_{N\to\infty}\frac{1}{s_N}L_{00}(\zeta^{N-n}(0))=\mu([00])\lim_{N\to\infty}\frac{s_{N-n}}{s_N}=r\theta^{-n}$$

and finally,

$$\lim_{n\to\infty} \mu(D_n) \geq r \liminf_{n\to\infty} \frac{L_\omega(\zeta^n(0))}{s_n} \frac{s_n}{\theta^n}$$

$$\geq rc\mu([\omega])$$

which proves (6.12) with $C = cr$. □

In the same article, the authors provide examples of weakly mixing dynamical systems arising from substitutions (in short, weakly mixing substitutions). The following example is an easy consequence of theorem 6.3.

Corollary 6.3. *The substitution ζ defined on $\{0,1\}$ by $\zeta(0) = 01010$ and $\zeta(1) = 011$ is weakly mixing.*

Proof. First, remark that ζ is primitive, apcriodic and that the integer j defined by (iii) is equal to one. Moreover, the unique coboundary of ζ is the function one : ζ is pure. We shall show that the unique eigenvalue of ζ is $\lambda = 1$.

As a consequence of theorem 6.3, λ is an eigenvalue of ζ if and only if

$$\lim_{n\to\infty} \lambda^{|\zeta^n(0)|} = \lim_{n\to\infty} \lambda^{|\zeta^n(1)|} = 1.$$

Setting $r_n = |\zeta^n(1)|$ and $s_n = |\zeta^n(0)|$, one easily computes

$$s_{n+1} = 3s_n + 2r_n, \quad r_{n+1} = s_n + 2r_n$$

so that

$$s_n = (4^{n+1} - 1)/3, \quad r_n = (2 \cdot 4^n + 1)/3.$$

If $\lambda = e^{2i\pi t}$, $\lim_{n\to\infty} \lambda^{s_n} = 1$ implies $4t/3 \equiv t/3$ (modulo 1) since $4^n \equiv 4 \pmod 3$, and $t = 0$. □

Remark 6.4. This corollary proves that the reduced maximal spectral type of the previous system is a continuous measure. Can we say something more about the spectrum ? Actually, we do not know how to describe $\sigma_{\max}$ for such a substitution (while this will be "possible" in the constant-length case as we shall see); but, if $u = \zeta^\infty(0)$ and if σ is the unique correlation measure of the occurrences of 0 in u defined by

$$\widehat{\sigma}(k) = \lim_{N\to\infty} \frac{1}{N} \text{Card}\,\{n + k < N,\ u_{n+k} = u_n = 0\},$$

one can establish the following:

$$\lim_{n\to\infty} \widehat{\sigma}(s_n + k) = \frac{1}{3}(2\widehat{\sigma}(k) + \widehat{\sigma}(k+1)).$$

This means that there exists a character $\chi \in \overline{\Gamma}$ such that $\chi_\sigma = \frac{2}{3} + \frac{\gamma}{3}$, $\gamma(t) = e^{it}$. Nevertheless, one conjectures that σ is not a Dirichlet measure and could even belong to the ideal $\mathscr{L}_I$ (see chapter 1).

In a very similar way, one can prove the following [84].

Corollary 6.4. *The Chacon's substitution ζ defined on $\{0,1,2\}$ by $\zeta(0) = 0012$, $\zeta(1) = 012$ and $\zeta(2) = 12$ is weakly mixing.*

6.3 Pure Point Spectrum

6.3.1 Discrete Constant-Length Substitutions

We proved in section 6.1 that the spectrum of the system associated with a constant-length substitution always possesses a discrete part, which only depends on the length q and the height h of ζ. The problem which naturally arises now is to describe, if this is possible, the substitutions leading to a system with a purely discrete spectrum. We say, in short, that ζ is discrete in this case. Such a characterization has been obtained by M. Dekking, to that effect he introduced the following.

Definition 6.5. Let ζ be a substitution of constant length q, defined on the alphabet $A = \{0,1,\dots,s-1\}$. We say that ζ admits a *coincidence* if there exist $k \geq 1$ and $j < q^k$ such that

$$\zeta^k(0)_j = \zeta^k(1)_j = \cdots = \zeta^k(s-1)_j$$

(that means : ζ^k admits a column of identical values). We say that the coincidence occurs at order k.

Example 6.1. Consider ζ defined on $\{0,1,2\}$ by $\zeta(0) = 11$, $\zeta(1) = 21$, $\zeta(2) = 10$. If we superpose

$$\begin{aligned} \zeta^2(0) &= 1\,2\,1\,2 \\ \zeta^2(1) &= 0\,1\,1\,2 \\ \zeta^2(2) &= 1\,2\,1\,1 \end{aligned}$$

the third column provides a coincidence (so $k = 2$ and $j = 2$).

6.3.1.1 Pure Case

We suppose now that ζ is a primitive substitution of length q and in addition, that $h := h(\zeta) = 1$. The following has been proved in [68] (theorem 7) with a different approach.

Theorem 6.6 (Dekking). *The pure substitution ζ is discrete if and only if it admits a coincidence.*

Proof. Since the ζ-system is uniquely ergodic, the sequence $u = \zeta(u)$ admits a unique correlation measure σ (see subsection 4.3.4).

◁ Suppose that ζ admits a coincidence. We intend to prove that the unique correlation measure of u is a discrete measure. We shall then conclude with help of proposition 4.14. We may assume that $k = 1$, by exchanging ζ and ζ^k if necessary. Denote by C_n the number of coincidences at order n, in the blocks $\zeta^n(0), \zeta^n(1), \ldots, \zeta^n(s-1)$; by our assumption, $C_1 \geq 1$. Since a coincidence at order n leads to q coincidences at order $n+1$, it is easy to establish, for $n \geq 1$, the inequality

$$C_{n+1} \geq (q^n - C_n) + qC_n. \tag{6.13}$$

The solution of the inductive equation

$$c_{n+1} = q^n + (q-1)c_n, \quad c_1 = 1$$

being $c_n = q^n - (q-1)^n$, the inequality (6.13) together with $C_1 \geq 1$ implies $C_n \geq q^n - (q-1)^n$.

We recall a definition.

Definition 6.6. A sequence of complex numbers $(v_n)_n$ is said to be *mean-almost periodic* if, for every $\varepsilon > 0$, one can find a relatively dense set E_ε of integers such that

$$\limsup_{n\to\infty} \frac{1}{N} \sum_{n<N} |v_n - v_{n+k}| \leq \varepsilon \quad \text{if} \quad k \in E_\varepsilon. \tag{6.14}$$

We are now in a position to determine the nature of the correlation measure of u by using the following classical result.

Lemma 6.6. *Let σ be the unique correlation measure of a complex sequence v; then σ is a discrete measure if and only if v is mean-almost periodic.*

Proof of the lemma. We shall make use of corollary 1.2, a direct consequence of the Dunkl-Ramirez inequality, that we recall now : σ is a discrete measure if and only if $(\widehat{\sigma}(n))_{n\in\mathbf{Z}}$ is an almost periodic sequence; or, equivalently : for every $\varepsilon > 0$ one can find a relatively dense set of integers E_ε such that

$$\sup_{n\in\mathbf{Z}} |\widehat{\sigma}(n) - \widehat{\sigma}(n+k)| \leq \varepsilon \text{ if } k \in E_\varepsilon. \tag{6.15}$$

We thus have to compare (6.14) and (6.15). Since

$$\sum_{n<N} |v_{n+k} - v_n|^2 = \sum_{n<N} (|v_{n+k}|^2 + |v_n|^2) - 2\Re \sum_{n<N} v_{n+k}\bar{v}_n,$$

we see that

$$\begin{aligned} \lim_{N\to\infty} \tfrac{1}{N} \textstyle\sum_{n<N} |v_{n+k} - v_n|^2 &= 2\Re\big(\widehat{\sigma}(0) - \widehat{\sigma}(k)\big)\big) \\ &\leq 2|\widehat{\sigma}(0) - \widehat{\sigma}(k)|. \end{aligned} \tag{6.16}$$

It follows from (6.15) and (6.16) that (v_n) must be mean-almost periodic, with, in addition, the same set E_ε.

Conversely, suppose that v is mean-almost periodic with a unique correlation measure σ. Then

$$\begin{aligned}|\widehat{\sigma}(n+k)-\widehat{\sigma}(n)|^2 &\leq \widehat{\sigma}(0)\cdot\textstyle\int_{\mathbf{T}}|e^{ikt}-1|^2\,d\sigma(t)\\ &= 2\widehat{\sigma}(0)\cdot\Re\big(\widehat{\sigma}(0)-\widehat{\sigma}(k)\big)\\ &= \widehat{\sigma}(0)\cdot\lim_{N\to\infty}\tfrac{1}{N}\textstyle\sum_{n<N}|v_{n+k}-v_n|^2\end{aligned}$$

by (6.16) again. Finally the property for σ follows from (6.14). □

Let us get back to the fixed point u of ζ, which is the juxtaposition of elementary words $\zeta^n(\alpha)$, $\alpha\in A$. If u_k runs over the word $\zeta^n(\alpha)$, u_{k+jq^n} for a fixed j, runs over some $\zeta^n(\beta)$, so that (u_k) and (u_{k+jq^n}) must take at least C_n identical values for such indices k. It follows from the estimate of C_n that

$$\sum_{\{k,\ u_k\in\zeta^n(\alpha)\}}|u_k-u_{k+jq^n}|\leq(s-1)(q^n-C_n)\leq(s-1)(q-1)^n$$

for every j. We now estimate

$$\begin{aligned}\frac{1}{N}\sum_{k<N}|u_k-u_{k+jq^n}| &= \frac{1}{N}\sum_{\ell<N/q^n}\sum_{\substack{k,\\ u_k\in B_\ell}}|u_k-u_{k+jq^n}|\\ &\leq(s-1)\Big(\frac{q-1}{q}\Big)^n\end{aligned}$$

where (B_ℓ) is an enumeration of the words $\zeta^n(u_j)$ in u. Now, let $\varepsilon>0$ and n be such that $(s-1)\left(1-\frac{1}{q}\right)^n\leq\varepsilon$; we see that $(u_n)_n$ satisfies (6.14) with the relative dense set $E_\varepsilon=\{q^n\mathbf{Z}\}$. As already observed, σ and the maximal spectral type of the system $\sigma_{\max}$ are thus discrete measures.

▷ In the opposite direction, since h is assumed to be one, we know from corollary 6.6 that

$$\lim_{k\to\infty}||U^{q^k}f-f||_{L^2}=0\ \text{ for every } f\in L^2(X,\mu)$$

(writing q instead of q^j). If $f=\pi_0$ is the projection onto the zero-component defined by $x\in X\mapsto x_0$, we get, by definition of the invariant measure μ,

$$\begin{aligned}||U^{q^k}f-f||_{L^2} &= \lim_{N\to\infty}\frac{1}{N}\sum_{n<N}|f(T^{q^k}T^nu)-f(T^nu)|^2\\ &= \lim_{N\to\infty}\frac{1}{N}|u_{n+q^k}-u_n|^2;\end{aligned}$$

applying the recalled corollary with $f = \pi_0$, we deduce that

$$\lim_{k\to\infty}\lim_{N\to\infty}\frac{1}{N}\,\mathrm{Card}\,\{n<N,\ u_{n+q^k}\neq u_n\}=0.$$

Moreover, thanks to the shift-invariance, this implies for every $j \geq 1$,

$$\lim_{k\to\infty}\,\mathrm{dens}\,\{n,\ u_{n+jq^k}\neq u_{n+(j-1)q^k}\}=0.$$

Let ℓ be chosen in such a way that every word of length ℓ contains all the letters of A (this is possible by minimality). One can find $\varepsilon > 0$ and k_0 such that

$$\mathrm{dens}\,\{n,\ u_{n+jq^k}\neq u_{n+(j-1)q^k}\}\leq\varepsilon,\quad \text{for } k\geq k_0 \text{ and } j=1,2,\ldots,\ell.$$

It follows that $\mathrm{dens}\,\{n,\ u_n = u_{n+q^k} = \cdots = u_{n+\ell q^k}\}$ is positive if $k \geq k_0$. Thus, n and k can be found such that

$$u_n = u_{n+q^k} = \cdots = u_{n+\ell q^k}.$$

But $u_n = \zeta^k(u_m)_j$ for some m and j so that

$$u_{n+q^k}=\zeta^k(u_{m+1})_j,\ldots,u_{n+\ell q^k}=\zeta^k(u_{m+\ell})_j$$

and, by the choice of ℓ, the factor $u_{[m,\ m+\ell]}$ contains all the letters of A. We have proved that ζ admits a coincidence. The proof is complete. □

Remark 6.5. Define now the *column number*

$$C(\zeta)=\min_{k\geq 1}\min_{0\leq j<q^k}\,\mathrm{Card}\,\{\zeta^k(0)_j,\zeta^k(1)_j,\ldots,\zeta^k(s-1)_j\}$$

(when $h = 1$). Dekking's theorem asserts that ζ is discrete if and only if $C(\zeta) = 1$. Obviously, $C(\zeta) = s$ means that ζ is, what we shall call, a *bijective* substitution (see chapter 10) for which a complete spectral description will be achieved. A geometric interpretation of $C(\zeta)$ has been given by Rauzy [208]. What about a spectral one when $C(\zeta) > 1$?

6.3.1.2 General Constant-Length Substitutions

Actually, the description by Dekking of discrete constant-length substitutions in [68] is complete but requires a new tool. Let ζ be a primitive substitution of length q with $h := h(\zeta) > 1$ and let $u = \zeta(u)$. If J is the collection of words of length h appearing in u at places nh, $n \geq 0$, one defines a new substitution η on J in this way : for every $w \in J$, the word $\zeta(w)$ of length qh can be split into q h-words belonging to J : $\omega_1 \omega_2 \cdots \omega_q$, $\omega_j \in J$; one naturally puts

$$\eta(w) = \omega_1 \omega_2 \cdots \omega_q,$$

and η is of constant length q. One also checks that η is primitive, so that (X_η, T') is minimal and ergodic (denoting by T' the shift on X_η). Of course, in case the initial substitution is pure, η is nothing but ζ and, in the general case, η is called "the pure base of ζ".

Let $X_0 = \overline{\{T^{nh}u,\ n \geq 0\}}$; it follows that

1. $(X_0, T^h) \simeq (X_\eta, T')$;
2. the height $h(\eta) = 1$;
3. the initial system and the *flow* $(X_\eta \times \mathbf{Z}_h, \sigma)$ are metrically isomorphic where

$$\sigma(x,k) = \begin{cases} (x, k+1) & \text{if } k < h, \\ (T'x, 0) & \text{if } k = h \end{cases}$$

In particular, the systems (X_ζ, T) and $(X_\eta \times \mathbf{Z}_h, \sigma)$ are spectrally isomorphic.

Proposition 6.2 (Dekking). *In case the height is greater than one, ζ is discrete if and only if its* pure base η *admits a coincidence.*

Example 6.2. Consider ζ defined on $A = \{0,1,2\}$ by

$$\zeta(0) = 010,\ \zeta(1) = 102,\ \zeta(0) = 201.$$

Clearly, $J = \{01, 02\}$ and η is the substitution over two letters $a \to aab$, $b \to aba$. The initial substitution is thus discrete.

6.3.2 Discrete Nonconstant-Length Substitutions

In the nonconstant-length case, there is no satisfactory condition on the substitution ζ leading to pure point spectrum as in theorem 6.6, and many questions remain unanswered. Two main different approaches have been developed since the pioneer results of P. Michel [189] and G. Rauzy [210]. The first one consists in proving directly that all the spectral measures are discrete, while the second one, more geometric, proceeds by constructing an explicit metric conjugacy from the substitution system onto an ergodic rotation. But no general study has yet been carried out.

6.3.2.1 Coincidences and Balanced Blocks

We restrict our attention to admissible substitutions on a two-letters alphabet $\mathscr{A} = \{0,1\}$ and we put $u = \zeta^\infty(0)$. For such an alphabet, recall that the unique possible coboundary in theorem 6.3 is the trivial one. Inspired by the Dekking's method of coincidences in the constant-length case, one considers $s_k = |\zeta^k(0)|$, so that $\zeta^k(Tu) = T^{s_k}u$, and

$$D_k = \text{dens}\,\{n,\ u_n \neq u_{n+s_k}\},\ \text{ for } k \geq 1.$$

As a consequence of theorem 6.3, B. Host proved the following.

Lemma 6.7. *1) If ζ has discrete spectrum, then* $\lim_{k\to\infty} D_k = 0$.
2) If $\lim_{k\to\infty} D_k = 0$ *with a geometric rate of convergence, then ζ has discrete spectrum.*

We omit the proof, which is similar to the proof of the same fact in the constant-length case but using theorem 6.3 instead of theorem 6.2.

Recall the notation introduced in chapter 4 : for a non-empty word $\omega \in \mathscr{A}^*$ and $\alpha \in \mathscr{A}$, $L_\alpha(\omega)$ denotes the number of occurrences of the letter α in the word ω, and $L(\omega)$ is the composition vector of ω with $L_0(\omega)$ and $L_1(\omega)$ as components.

Definition 6.7. Two words A and B on $\mathscr{A}$ are called *equivalent* if $L(A) = L(B)$ and we shall speak of the equivalent pair $\begin{pmatrix} A \\ B \end{pmatrix}$, $\begin{pmatrix} A \\ A \end{pmatrix}$ being a *trivial equivalent pair.*

Note that for such a pair, the substituted words $\zeta(A)$ and $\zeta(B)$ are still equivalent.

Every equivalent pair may be decomposed into a finite sequence of smaller equivalent pairs; we are interested in *irreducible* pairs of nontrivially equivalent words, that is nontrivial equivalent pairs which do not split any more and we call *canonical splitting* the unique decomposition into irreducible pairs. For instance,

$$\begin{array}{l} \mathrm{A} = 01|\ 10|\ 011|\ 010|\ 001| \\ \mathrm{B} = 10|\ 10|\ 110|\ 010|\ 100| \end{array}$$

Pairs of infinite words on $\mathscr{A}$ also split in a unique way into irreducible pairs and this will be used to estimate D_k. We analyze some examples.

First Examples

1. Define ζ by $\zeta(0) = 001$, $\zeta(1) = 00$, and $u = \zeta^\infty(0) = 00100100001\cdots$ Fix $k \geq 1$. The canonical splitting of the pair $(u, T^{s_k}u)$ begins as follows.

$$\begin{array}{rl} u = & \zeta^k(0)|\ \zeta^k(0)\zeta^k(1)|\ \zeta^k(0)|\ \zeta^k(0)\zeta^k(1)|\ \zeta^k(0)\zeta^k(0)\cdots \\ T^{s_k}u = & \zeta^k(0)|\ \zeta^k(1)\zeta^k(0)|\ \zeta^k(0)|\ \zeta^k(1)\zeta^k(0)|\ \zeta^k(0)\cdots \end{array}$$

It is clear that every word 01 in u is followed by 0 and provides an irreducible equivalent pair $\begin{pmatrix} 01 \\ 10 \end{pmatrix}$ under the action of T. In order to enumerate the non-coincidences between u and $T^{s_k}u$, it is thus sufficient to enumerate the non-coincidences between $\zeta^k(01)$ and $\zeta^k(10)$. More precisely, let N_k be this latter number and $E_{N,k} = \{n \leq N,\ u_n \neq u_{n+s_k}\}$. Then, we have

$$\text{Card } E_{N,k} \sim N_k \cdot L_{\zeta^k(01)}(u_0 \cdots u_N)$$

and

$$\begin{aligned} D_k &= \lim_{N \to \infty} \frac{\text{Card } E_{N,k}}{N} = N_k\, \mu([\zeta^k(01)]) \\ &= \frac{N_k}{\theta^k}\, \mu([01]), \end{aligned}$$

where θ is the Perron-Frobenius eigenvalue of ζ.
We are left with the estimate of N_k, thus we compare N_{k+1} and N_k.

$$\begin{aligned} \zeta^{k+1}(01) &= \zeta^k(00)\ \zeta^k(100) \\ \zeta^{k+1}(10) &= \zeta^k(00)\ \zeta^k(001) \end{aligned}$$

so that $N_k = M_k$, the number of non-coincidences between $\zeta^k(100)$ and $\zeta^k(001)$. Now,

$$\begin{aligned} \zeta^{k+1}(100) &= \zeta^k(00)\ \zeta^k(001)\ \zeta^k(001) \\ \zeta^{k+1}(001) &= \zeta^k(00)\ \zeta^k(100)\ \zeta^k(100) \end{aligned}$$

and $M_{k+1} = 2M_k$. We deduce that $M_k = 2^{k+1}$, $N_k = 2^k$ and $D_k = (2/\theta)^k \mu([01])$. As $2 < \theta < 3$, D_k tends to zero geometrically and we conclude with lemma 6.7.

2. Suppose now that $\zeta(0) = 01100$, $\zeta(1) = 010$. Since

$$M(\zeta) = \begin{pmatrix} 3 & 2 \\ 2 & 1 \end{pmatrix}$$

the Perron-Frobenius eigenvalue $\theta = 2 + \sqrt{5}$. From

$$\begin{array}{rl} \text{u} = & 011|00|01|0|01|0|011|00|011|00\cdots \\ \text{Tu} = & 110|00|10|0|10|0|110|00|110|0\cdots \end{array}$$

we easily prove that the only number of non-coincidences between $\zeta^k(011)$ and $\zeta^k(110)$, also between $\zeta^k(01)$ and $\zeta^k(10)$, remains to be estimated. Denote by N_k and M_k respectively those two numbers. By splitting

$$\begin{array}{rl} \zeta(011) = & 01|100|0|10|01|0 \\ \zeta(110) = & 01|001|0|01|10|0 \end{array}$$

it appears a new nontrivial irreducible equivalent pair, $\begin{pmatrix}100\\001\end{pmatrix}$, which gives rise to P_k, the number of non-coincidences between $\zeta^k(100)$ and $\zeta^k(001)$. In the same way, we get the linear relations

$$\begin{aligned} N_{k+1} &= P_k + 2M_k \\ M_{k+1} &= P_k + M_k \\ P_{k+1} &= 2P_k + 2M_k \end{aligned}$$

the matrix of which, $M_b = \begin{pmatrix} 0 & 2 & 1 \\ 0 & 1 & 1 \\ 0 & 2 & 2 \end{pmatrix}$, admitting $\lambda = 3 + \sqrt{5} < \theta$ as dominant eigenvalue. Now,

$$\text{Card } E_{N,k} \sim \frac{N_k}{\theta^k}\mu([011]) + \frac{M_k}{\theta^k}\mu([01])$$

and $\lim_{N\to\infty} \text{Card } E_{N,k}/N \leq C(\lambda/\theta)^k$ for some constant C. Lemma 6.7 applies again and ζ must be discrete.

3. In [189] P. Michel observes that both ζ_1 defined by $\zeta_1(0) = 01$, $\zeta_1(1) = 1010$, and ζ_2 by $\zeta_2(0) = 01$, $\zeta_2(1) = 1001$, are discrete, while ζ_3 defined by $\zeta_3(0) = 01$, $\zeta_3(1) = 1100$ admits a mixed spectrum. Actually, he shows that the system (X_{ζ_3}, T) is metrically isomorphic to the following (X_η, T) where η is the constant-length substitution on the alphabet $A = \{a, b, c\}$ defined by $\eta(a) = abc$, $\eta(b) = bcb$, $\eta(c) = caa$ and admitting no coincidence.
With help of the previous method, we easily prove that ζ_1 and ζ_2 are discrete. On the other hand, for ζ_3, the same computations on coincidences lead to linear recurrent equations with the following matrix

$$M_b = \begin{pmatrix} 0 & 1 & 4 & 1 & 0 \\ 0 & 1 & 2 & 1 & 0 \\ 0 & 1 & 0 & 1 & 0 \\ 0 & 0 & 0 & 0 & 2 \\ 0 & 1 & 1 & 1 & 1 \end{pmatrix}$$

and dominant eigenvalue $\lambda = \theta = 3$. Thus, D_k cannot tend to zero.

Therefore, for a given substitution ζ with Perron-Frobenius eigenvalue θ, we are led to consider the *irreducible balanced pairs of* ζ, that is, nontrivial irreducible equivalent pairs occurring when splitting the pair of infinite words u and Tu.

Definition 6.8. We denote by $\widehat{A}$ the set of all nontrivial irreducible equivalent pairs of ζ.

If $\widehat{A}$ is finite, ζ induces a substitution on this new alphabet, the matrix of which being the so-called matrix M_b. We can state the following result : *If $\widehat{A}$ is a finite alphabet and if the dominant eigenvalue of the matrix M_b is smaller that θ, then ζ is discrete.*

Example 6.3. Let us return to the second example : $0 \to 01100$, $1 \to 010$. It is easily checked that $\widehat{A} = \left\{ \begin{pmatrix} 01 \\ 10 \end{pmatrix}, \begin{pmatrix} 100 \\ 001 \end{pmatrix}, \begin{pmatrix} 011 \\ 110 \end{pmatrix} \right\} =: \{\omega_1, \omega_2, \omega_3\}$ and that

$$\begin{aligned} \zeta(\omega_1) &= \omega_2\omega_1 \\ \zeta(\omega_2) &= \omega_2\omega_1\omega_2\omega_1 \\ \zeta(\omega_3) &= \omega_2\omega_1\omega_1 \end{aligned}$$

by identifying ω_1 and $\begin{pmatrix} 10 \\ 01 \end{pmatrix}$.

A remaining problem is to decide on ζ whether those two sufficient conditions are fulfilled.

Remark 6.6. An alternative to this algorithm consists in splitting the pair of infinite words $(\zeta^\infty(0), \zeta^\infty(1))$, as suggested in [189]; this will be efficient in the Pisot case.

6.3.2.2 Geometric Representation

One of the first famous geometric representations is undoubtedly the description of the Fibonacci dynamical system. The Perron-Frobenius eigenvalue of the Fibonacci substitution ζ, defined by $\zeta(0) = 01$ and $\zeta(1) = 0$, is the golden number $\theta =: 1 + \alpha$, $\alpha = (\sqrt{5} - 1)/2$. The Fibonacci dynamical system (X_ζ, T) is a coding of the rotation R_α on the interval $[0,1)$; more precisely, if $u = \zeta^\infty(0)$, it can be checked that $u_n = 0$ if n is such that $R^n\alpha \in [1-\alpha, 1)$ and $u_n = 1$ if $R^n\alpha \in [0, 1-\alpha)$ and this induces a semi-conjugacy between both systems (see [88]).

In [208] G. Rauzy considers the substitution defined on the three-letters alphabet $\{0,1,2\}$ by $\zeta(0) = 01$, $\zeta(1) = 02$, $\zeta(2) = 0$. The Perron-Frobenius eigenvalue θ satisfies the equation $X^3 = X^2 + X + 1$, so this substitution received the funny name: *Tribonacci*. Rauzy proved that ζ admits a geometric representation in the following form : the system (X_ζ, T) is semi-conjugate to the translation by the vector $\eta = (\theta^{-1}, \theta^{-2})$ on $\mathbf{R}^2/\mathbf{Z}^2$. More precisely, there exists a Borel partition of $\mathbf{R}^2/\mathbf{Z}^2$, $\Omega_0 \cup \Omega_1 \cup \Omega_2$, such that:

(i) the Ω_i are connected bounded open sets,
(ii) $\bigcup_i \Omega_i + \mathbf{Z}^2$ is dense in $\mathbf{R}^2$,
(iii) the $\Omega_i + \mathbf{Z}^2$ are disjoint.

If $u = \zeta^\infty(0)$, $u_n = i$ if and only if $n\eta \in \Omega_i + \mathbf{Z}^2$; this provides the semi-conjugacy between those two systems.

It is thus natural to ask about substitutions admitting such a geometric representation. Let us fix the usual notations : ζ is a primitive substitution on the alphabet $A = \{0, \ldots, s-1\}$, $u = \zeta^\infty(0)$ and $X = X_\zeta$. The problem can be formulated in the following way : *can one find a continuous map $\Phi : X \to \mathbf{R}^k$, $k < s$, such that:*

$$(a)\ \Phi \circ T = \tau \circ \Phi, \textit{ where } \tau \textit{ is a piecewise translation,}$$

$$(b)\ \Phi \circ \zeta = B \circ \Phi, \textit{ where } B \textit{ is a contraction.}$$

Here is an attempt of approach. Let $f_0, \ldots, f_{s-1} \in \mathbf{R}^k$ and let us consider Φ satisfying:

$$\Phi(u) = 0 \text{ and } \Phi(T^n u) = \Phi(T^{n-1}u) + f_{u_{n-1}} \quad n \geq 1.$$

Since u_{n-1} is the first component of $T^{n-1}u$, the translated $\tau(x)$ depends in fact on the first letter of x. If (b) holds, Φ is continuous and $\Phi(X)$, the closure of $(z_n)_{n\geq 0}$ with $z_n = \Phi(T^n u)$, is a compact set K. For $i = 0, \ldots, s-1$, consider $\Lambda_i = \{n,\ u_n = i\}$ and K_i the closure of $(z_n)_{n\in\Lambda_i}$. The ζ-system appears as a coding of a domain exchange if the K_i are disjoint in measure (or Φ one-to-one in measure) since $u_n = i$ if and only if $z_n \in K_i$.

Many problems raised in connection with geometric representations.

1. *Discrete spectrum*. Rauzy proved in addition that the dynamical system generated by the Tribonacci substitution is metrically isomorphic to an ergodic translation on a two-torus, and therefore, that it has pure point spectrum; which substitutions code the action of a toral translation ?
2. *Numeration system*. With the Tribonacci substitution, once more, one can associate a numeration system : every $n \in \mathbf{N}$ can be written in a unique way as

   $$n = \sum_{k\geq 0} \varepsilon_k(n)\, g_k, \quad \text{where } g_k = |\zeta^k(0)|$$

   and where the sequence $(\varepsilon_k(n))$ is well determined. What about general substitutions ? Beyond substitutions, dynamics of systems of numeration developed independently and a good survey with references can be found in [20].
3. *Cohomology*. The existence of a continuous map Φ, satisfying $\Phi(Tx) = \Phi(x) + f_{x_0}$ if $x = (x_n)_{n\geq 0} \in X$, is related to the existence of continuous generalized eigenfunctions.

Definition 6.9. The function $F \in L^2(X, \mu)$ is a *generalized eigenfunction* of the substitution dynamical system (X, μ, T) if there exists $\phi : A \to \mathbf{U}$ such that

$$F(Tx) = \phi(x_0)\, F(x), \quad x = (x_n)_{n\geq 0} \in X,$$

(so that F may be assumed unimodular). We precise the dependence by $F = F_\phi$.

As an example, if f is a classical eigenfunction corresponding to the eigenvalue $\lambda \in \mathbf{U}$, then $g = f \circ \zeta$ is a generalized eigenfunction since

$$g(Tx) = f(T^{|\zeta(x_0)|}\zeta(x)) = \lambda^{|\zeta(x_0)|}\, g(x).$$

Let D be the group of (continuous) eigenvalues of a primitive ζ, endowed with the discrete topology. The dual group $\widehat{D}$ of D is a compact group. If $\lambda \in D$, and if f_λ denotes the continuous associated eigenfunction normalized by $f_\lambda(u) = 1$, the group of (f_λ) may be identified with D. If we now define the function $\varepsilon : X \to \widehat{D}$ by $\varepsilon_x(\lambda) = f_\lambda(x)$, then

$$\varepsilon_{Tx}(\lambda) = \lambda\ \varepsilon_x(\lambda) \ \text{ for } x \in X, \lambda \in D.$$

When ζ is of constant length, $\mathbf{Z}$ is dense in $\widehat{D}$ and ε is easily seen to be onto. If, in addition, the family (f_λ) separates points on X, then ε is one-to-one. In this case, the systems (X_ζ, T) and $(\widehat{D}, R)$, where R is the rotation by $\gamma_0 : \lambda \mapsto \lambda$ on $\widehat{D}$, are metrically isomorphic, and ζ is discrete.
In the nonconstant-length case, ζ may happen to be weakly mixing so that D may be trivial (corollary 6.3). In order to precise the structure of the system (X_ζ, T), it is thus natural to consider G, the group of continuous generalized eigenfunctions: $\{F_\phi, \phi : A \to \mathbf{U}\}$, which obviously contains D. We now ask when $\varepsilon : X \to G$, defined by $\varepsilon_x(F_\phi) = F_\phi(x)$, provides a metric isomorphism.
Here is a brief analysis. For simplicity, suppose that the only coboundaries of ζ are the trivial ones, and let us return to the additive notation. If $M = M(\zeta)$, it transpose ${}^tM$ is the incidence matrix; we consider the subgroup of $\mathbf{R}^s$:

$$H = \{v \in \mathbf{R}^s,\ {}^tM^n v \text{ tends to } 0 \mod \mathbf{Z}^s\}.$$

The following can be proved : *F_ϕ is an additive and continuous generalized eigenfunction of ζ if and only if the s-dimensional vector $(\phi(\alpha))_{\alpha \in A}$ belongs to H.*
By using lemma 6.5, we decompose $H = H_1 \oplus H_2$, where H_1 is the contracting eigen-subspace, corresponding to the $\lambda, |\lambda| < 1$, and $H_2 = \{v \in \mathbf{R}^s,\ {}^tM^n v \in \mathbf{Z}^s \text{ ultimately}\}$. Note that if θ is a unitary PV-number ($\det M = \pm 1$), $H_2 = \{0\}$; in this case the isomorphism problem could be solved.

Example 6.4. 1. Consider ζ defined by $\zeta(0) = 001$ and $\zeta(1) = 00$. Then $M = \begin{pmatrix} 2 & 1 \\ 2 & 0 \end{pmatrix}$ and $\theta = 1 + \sqrt{3}$ is a PV-number; but θ is not unitary since $\det M = -2$.
It is easily checked that

$$H_1 = t \cdot \begin{pmatrix} -1 \\ 1+\sqrt{3} \end{pmatrix},\ t \in \mathbf{R},$$

and

$$H_2 = \{\begin{pmatrix} a \\ b \end{pmatrix}, \ a, b \text{ dyadic numbers}\}.$$

2. Consider ζ defined by $\zeta(0) = 00011$ and $\zeta(1) = 001$. Then $M = \begin{pmatrix} 3 & 2 \\ 2 & 1 \end{pmatrix}$ and $\theta = 2 + \sqrt{5}$ is a unitary PV-number. It is easily checked that

$$H_2 = \{0\} \ \text{ and } \ H_1 = t \cdot \begin{pmatrix} 4+2\sqrt{5} \\ 3+\sqrt{5} \end{pmatrix}, \ t \in \mathbf{R}.$$

In this case, the function F satisfying

$$\begin{aligned} F(Tx) &= F(x) + 4 + 2\sqrt{5} \ \text{ if } x_0 = 0 \\ &= F(x) + 3 + \sqrt{5} \ \ \text{ if } x_0 = 1 \end{aligned}$$

is a continuous additive generalized eigenfunction.

Many partial answers have been given to those problems and a good outline of the question can be found in chapter 7 of [88]. The particular case of Pisot substitutions is considered in the next item because of its specificity and of the important interest those substitutions raised in physics of quasi-crystals and tilings. An empirical conjecture states that all the Pisot substitution systems have pure point spectrum. This is true for a two-letters alphabet.

6.3.3 Pisot Substitutions

Recall that a *Pisot substitution* is automatically admissible [52]. We chose to begin with the unpublished–but frequently cited–result of B. Host [121].

Theorem 6.7 (B. Host). *Let ζ be a unimodular Pisot substitution on two symbols $\{0,1\}$, satisfying the coincidence condition (6.20) below. Then the associated system (X_ζ, μ, T) is conjugate to an irrational rotation; in particular, ζ is discrete.*

Proof. The *coincidence condition* will be stated later. We denote by $M = (M_{a,b})$ the composition matrix of ζ, by θ the Perron-Frobenius eigenvalue and by κ the Galois conjugate of θ $(0 < |\kappa| < 1)$; remind that, by assumption, $\theta\kappa = \pm 1$. The scheme of the proof goes as follows : we exhibit a geometric representation of the system, $F : X \to \mathbf{R}$, and the candidate $F \circ \pi$, where $\pi : \mathbf{R} \to \mathbf{R}/\mathbf{Z}$, is shown to be essentially one-to-one under an appropriate coincidence condition.

The following map has already been considered by M. Dekking [68].

Proposition 6.3. *There exist a continuous map $F : X_\zeta \to \mathbf{R}$, and t_0, $t_1 \in \mathbf{R}$ with $t_1 - t_0 = 1$, such that*

(i) For every $x \in X$, $F(Tx) - F(x) = t_{x_0}$ (where $x = x_0 x_1 \cdots$).

(ii) For every $x \in X$, $F(\zeta(x)) = \kappa F(x)$.

(iii) The restriction of F to both cylinders $[0]$ and $[1]$ is one-to-one, μ-almost everywhere.

Proof. (i)-(ii) We put $F(u) = 0$ where $u = \zeta^\infty(0)$ and for $x \in X := X_\zeta$,

$$F(Tx) = \begin{cases} F(x) + g_0 & \text{if } x_0 = 0 \\ F(x) + g_1 & \text{if } x_0 = 1 \end{cases}$$

where (g_0, g_1) is a left eigenvector corresponding to κ; observe that

$$g_0\mu([0]) + g_1\mu([1]) = 0. \tag{6.17}$$

By iterating this identity on x, we get that

$$\begin{aligned} F(T^n x) - F(x) &= g_{x_0} + \cdots + g_{x_{n-1}} \\ &= g_0 L_0(x_{[0,\,n-1]}) + g_1 L_1(x_{[0,\,n-1]}); \end{aligned}$$

in particular, using (5.16),

$$F(T^n u) = g_0 L_0(u_{[0,\,n-1]}) + g_1 L_1(u_{[0,\,n-1]}) = O(\kappa^n).$$

This proves that F can be extended to a continuous function on $X = \overline{O(u)}$, satisfying (ii) : indeed,

$$F(\zeta(x)) = \lim_{i\to\infty} F(\zeta(T^{n_i}u)) = \lim_{i\to\infty} F(T^{|\zeta(u_{[0,\,n_i-1]})|}u).$$

But, for every $n \geq 1$, by the choice of (g_0, g_1), one has

$$\begin{aligned} F(T^{|\zeta(u_{[0,\,n-1]})|}u) &= g_0 L_0(\zeta(u_{[0,\,n-1]})) + g_1 L_1(\zeta(u_{[0,\,n-1]})) \\ &= \kappa\left(g_0 L_0(u_{[0,\,n-1]}) + g_1 L_1(u_{[0,\,n-1]})\right) \\ &= \kappa\, F(T^n u) \end{aligned}$$

and finally, $F(\zeta x) = \kappa F(x)$ for every $x \in X$. Dividing by $g_0 - g_1$ if necessary, we get (i) and (ii).

(iii) Taking the projection by π in (i), we get

$$(\pi \circ F)(Tx) = (\pi \circ F)(x) + \alpha \tag{6.18}$$

where $\alpha := \pi(t_0) = \pi(t_1)$ is an irrational number. In fact, if t_0 were rational, g_0 and g_1 would be rationally dependent and, in view of (6.17), so would be $\mu[0]$ and $\mu[1]$

which is not the case with an irrational Perron-Frobenius eigenvalue. Now (6.18) implies that the sequence $((\pi \circ F)(T^n u))_{n\geq 0}$ is dense into $\mathbf{R}/\mathbf{Z}$ and $K = \pi(F(X))$ is nothing but the full circle. It follows that $m(F(X)) > 0$, where m denotes the Lebesgue measure on $\mathbf{R}$. If $a \in \{0,1\}$ and $k \geq 1$, we put

$$t_{a,k} = t_{\zeta(a)_0} + t_{\zeta(a)_1} + \cdots + t_{\zeta(a)_{k-1}}$$

so that, by (i),

$$F\Big(\bigcup_{\substack{a,k,\\ \zeta(a)_k=b}} T^k\zeta([a])\Big) = \bigcup_{\substack{a,k,\\ \zeta(a)_k=b}} F(\zeta([a])) + t_{a,k}, \quad \text{for every } b \in \{0,1\}.$$

Now, recall that, outside the countable set $\mathscr{D}$,

$$[b] = \bigcup_{\substack{a,k,\\ \zeta(a)_k=b}} T^k\zeta([a]) \quad \text{for every } b \in \{0,1\};$$

it follows that

$$\bigcup_{\substack{a,k,\\ \zeta(a)_k=b}} F(\zeta([a])) + t_{a,k} = F([b]) \tag{6.19}$$

up to a countable set; but the sets in the left member may fail to be disjoint (a.e.). Taking the Lebesgue measure of both sets, we still obtain

$$\begin{aligned} m(F([b])) &\leq \sum_a M_{a,b}\, m(F(\zeta([a]))) \\ &\leq \sum_a M_{a,b}|\kappa|\, m(F([a])) \ \text{ by } (ii), \ \text{ so that} \\ \theta m(F([b])) &\leq \sum_a M_{a,b}\, m(F([a])) \ \text{ since } \theta|\kappa| = 1. \end{aligned}$$

By Perron-Frobenius' theorem, all the inequalities become equalities and there exists a constant c such that $m(F([a])) = c\mu([a])$ for $a = 0,1$; in particular, both $F([0])$ and $F([1])$ must have positive measure. It follows that the union in (6.19) is disjoint m-a.e.. In the same way, for every $n \geq 1$, $F(\zeta^n[b])$ is the a.e. disjoint union of the sets

$$\bigcup_{\substack{a,k,\\ \zeta^n(a)_k=b}} F(\zeta^n([a])) + t_{a,k}^{(n)}$$

with the notation $t_{a,k}^{(n)} = t_{\zeta^n(a)_0} + t_{\zeta^n(a)_1} + \cdots + t_{\zeta^n(a)_{k-1}}$. Thus, for $b = 0,1$ and $n \geq 0$, F transforms the partition of $[b] : \mathscr{P}_n := (\{T^k\zeta^n[a]\} \cap [b]\})_{(a,k)}$ into a partition; this proves that $F_{|[b]}$ is μ-almost everywhere one-to-one since the family $(\mathscr{P}_n)$ spans the σ-algebra on $[b]$. □

This proposition, in fact, extends easily to any unimodular Pisot substitution (without any restriction on the alphabet) under the precise form given in subsection 6.3.2.2, but the following requires a two-letters alphabet.

Lemma 6.8. *If F is one-to-one almost everywhere on X, so is $\pi \circ F$.*

Proof. It is sufficient to establish that whenever $\pi \circ F(x) = \pi \circ F(y)$, one can find $n \geq 0$ such that $F(T^n x) = F(T^n y)$. So we fix $x, y \in X$ and we consider the set $\{n \geq 0,\ z_n := F(T^n x) - F(T^n y) = 0\}$. First of all, observe that $z_n \in \mathbf{Z}$ for every $n \geq 0$. Indeed, this is true for $n = 0$ and by (i), $z_{n+1} - z_n \in \{-1, 0, 1\}$. We are led to prove that the sequence z_n meets zero, in fact 0 is a recurrent point for this deterministic walk as we shall see. Let c and C be the bounds of F on X. By the minimality property of the system, both sets of integers : $A = \{n \geq 0,\ F(T^n x) > C - 1\}$ and $B = \{n \geq 0,\ F(T^n x) < c + 1\}$ have bounded gaps. If $n \in A$, $F(T^n y) \leq C < F(T^n x) + 1$ and $z_n \geq 0$; in the same way, if $n \in B$, $z_n \leq 0$. It follows that, given $m \leq n$, $m \in B$ and $n \in A$, one can find p such that $m \leq p \leq n$ and $z_p = 0$. □

Corollary 6.5. *If $\zeta(0)$ and $\zeta(1)$ begin (or end) with the same letter, then F is one-to-one almost everywhere on X and the system is discrete.*

Proof. If $\zeta(0)$ and $\zeta(1)$ begin with, say, 0, then $\zeta(X) \subset [0]$ and F is (a.e.) one-to-one on $\zeta(X)$ by proposition 6.3 (iii); the claim follows from the same proposition (ii). If $\zeta(0)$ and $\zeta(1)$ end with, say, 0, then $T^{-1}\zeta(X) \subset [0]$ and again, F is (a.e.) one-to-one on this set. But $F \circ T - F$ is constant on it and we come down to the previous case. We conclude with Lemma 6.8. □

We say that such a substitution is *proper*, and the theorem holds in this case.

We are thus left with the remaining nonproper case, where $\zeta(0)$ begins with 0 and $\zeta(1)$ begins with 1. Henceforth, we consider $u = \zeta^\infty(0)$ and $v = \zeta^\infty(1)$, and we define the following *coincidence condition* we shall make use of :

there exists $n \geq 1$ such that $u_{[0,\,n-1]}$ and $v_{[0,\,n-1]}$ are equivalent words and $u_n = v_n$ i.e.

$$L(u_{[0,\,n-1]}) = L(v_{[0,\,n-1]}) \text{ and } u_n = v_n. \tag{6.20}$$

Proposition 6.4. *F is one-to-one almost everywhere on X if and only if ζ satisfies the coincidence condition.*

Proof. ◁ Let $n \geq 1$ satisfying (6.20) and let $k \geq 0$ be such that both $|\zeta^k(0)| > n$ and $|\zeta^k(1)| > n$. Let now $x \neq y$ be such that $F(x) = F(y)$. By proposition 6.3 (iii), except for a negligible subset of such (x, y), one must have $x_0 \neq y_0$. Suppose for instance $x_0 = 0$ and $y_0 = 1$; thus $\zeta^k(x)_{[0,\,n]} = u_{[0,\,n]}$, $\zeta^k(y)_{[0,\,n]} = v_{[0,\,n]}$, so that, according to (6.20), $\zeta^k(x)_{[0,\,n-1]}$ and $\zeta^k(x)_{[0,\,n-1]}$ are equivalent words. It follows from proposition 6.3 (i) that

$$F(T^n\zeta^k(x)) - F(\zeta^k(x)) = F(T^n\zeta^k(y)) - F(\zeta^k(y));$$

but, $F(\zeta^k(x)) = F(\zeta^k(y))$ by (ii) of the same proposition, so that, finally,

$$F(T^n\zeta^k(x)) = F(T^n\zeta^k(y)).$$

Now, the second assumption $u_n = v_n$ gives $(T^n\zeta^k(x))_0 = (T^n\zeta^k(y))_0$, and, by proposition 6.3 (iii) again, one has $T^n\zeta^k(x) = T^n\zeta^k(y)$ whence $x = y$ (up to a negligible set). The condition is thus sufficient.

▷ Before proving the necessity, we give a combinatorial interpretation of the coincidence condition. Consider the canonical splitting of the pair $\begin{pmatrix} u \\ v \end{pmatrix}$ in irreducible **nontrivial as well as trivial** pairs. Let Λ be the set of indices of the bars in this splitting. For instance, if $\zeta(0) = 001$ and $\zeta(1) = 10$, the canonical splitting gives:

$$\begin{array}{r|c|c|c|c|c|c|c|c|c|l} \mathrm{u} = & 001 & 0 & 01 & 10 & 0 & 0 & 1 & 001 & 10 & 100\cdots \\ \mathrm{v} = & 100 & 0 & 10 & 01 & 0 & 0 & 1 & 100 & 01 & 001\cdots \end{array}$$

and $\Lambda = \{0,3,4,6,8,9,10,11,14,16,19,\ldots\}$. With the same arguments as in the proof of lemma 6.8 one can prove that :

Lemma 6.9. *Λ has bounded gaps.*

The coincidence condition exactly means that Λ contains two consecutive integers.

Now, suppose condition (6.20) is failing; then $u_n \neq v_n$ for every $n \in \Lambda$. On the other hand, since $L(u_{[0,\,n-1]}) = L(v_{[0,\,n-1]})$ and $F(u) = 0 = F(v)$, we have $F(T^n u) = F(T^n v)$ for every $n \in \Lambda$. Consider Z, the closure in $X \times X$ of the set $\{(T^n u, T^n v),\ n \in \Lambda\}$. For each $(x,y) \in Z$, there exists a sequence (n_i) of elements of Λ such that

$$x = \lim_{i\to\infty} T^{n_i}u \ \text{ and } \ y = \lim_{i\to\infty} T^{n_i}v.$$

Since $F(T^{n_i}u) = F(T^{n_i}v)$ for every i, then $F(x) = F(y)$. But $u_{n_i} \neq v_{n_i}$, in particular $x_0 \neq y_0$ and thus, $x \neq y$. If F were (a.e.) one-to-one, one would have $Z \subset (N \times X) \cup (X \times N)$ where N is a negligible subset of X and this provides a contradiction as we shall see. Indeed, if Σ is the closed orbit of (u,v) under $T \times T$,

$$\Sigma = \bigcup_{j\geq 0} (T \times T)^j Z$$

by lemma 6.9. And the next lemma, applied with Σ, would imply that μ is zero. Whence the proposition. □

Lemma 6.10. *Let p_1, p_2 be the canonical projections from $X \times X$ onto X and consider Σ some $T \times T$-invariant closed subset of $X \times X$. Then, for every Borel set $A \subset X$, one has*

$$\mu(p_1(p_2^{-1}(A) \cap \Sigma)) \geq \mu(A).$$

Proof. We give a quick proof of this technical lemma. Consider the measure $\nu = (\mu \times \mu)_{|\Sigma}$, a $T \times T$-invariant measure supported by Σ. Invoking the unique ergodicity of the substitution system, the T-invariant measure ν_2 defined on $(X, \mathscr{B})$ by

$$\nu_2(A) = \nu(p_2^{-1}(A)), \quad A \in \mathscr{B}$$

is nothing but μ. It follows, for every Borel set $A \subset X$, that

$$\begin{aligned}\mu(A) &= \nu(p_2^{-1}(A)) = (\mu \times \mu)(p_2^{-1}(A) \cap \Sigma) \\ &\leq \mu[p_1(p_2^{-1}(A) \cap \Sigma)]\end{aligned}$$

since μ is a probability measure. □

The proof of the theorem is complete. □

Remark 6.7. 1. The coincidence condition obviously holds for a proper substitution and this proposition appears as a generalization of corollary 6.5.
2. The coincidence condition, via lemma 6.9, implies that the set of distinct irreducible pairs occurring in the decomposition of the pair (u,v) is finite. This combinatorial approach of the spectrum problem is in fact equivalent to the previous one : If $\begin{pmatrix}0\\0\end{pmatrix}$ or $\begin{pmatrix}1\\1\end{pmatrix}$ occurs in some iterate of any irreducible pair and if the set of irreducible pairs in (u,v) is finite, then the coincidence condition is satisfied.

Actually, the *Pisot Conjecture* is true for any two-letters Pisot substitution. This has been proved by M. Hollander and B. Solomyak, by using a slightly different splitting method due to A. Livshits [117,169]. The concluding argument is given by the combinatorial result of Barge and Diamond [21].

Theorem 6.8 (Barge & Diamond). *Every Pisot substitution on two symbols satisfies the coincidence condition.*

Theorem 6.9 (Hollander & Solomyak). *Any two-letters Pisot substitution is discrete.*

What can be extended to the nonunimodular case, more generally to Pisot substitutions over $d > 2$ letters ? P. Arnoux, S. Ito and Y. Sano have studied the case of unimodular Pisot substitutions on $d > 2$ letters, while A. Siegel has obtained a geometric representation for general Pisot substitutions. The considerable amount of results, after Host and Rauzy, cannot be summarized in some lines. Complements on the Rauzy fractal, and recent improvements on explicit geometric realizations, numeration systems and tiling problems can be found with references in the already cited chapter in [88].

Chapter 7
Matrices of Measures

We now deal exclusively (unless specific mention of the contrary) with systems arising from **primitive and aperiodic** substitutions of **constant length**–more generally automata–and our purpose, till the end of chapter 11, is to give a constructive description of the spectral multiplicity and maximal spectral type of such systems. We shall prefer to refer to as the spectrum of the substitution.

In this chapter, we start with elementary observations on these spectral characteristics, by introducing a matrix of correlation measures that we denote by Σ. In the last two parts of the chapter, we consider more general matrices of measures and the action of characters of Δ, the Gelfand spectrum of $M(\mathbf{T})$, on those matrices.

7.1 Correlation Matrix

Let ζ be a primitive and aperiodic substitution of length q, hence, ζ is bilaterally recognizable (theorem 5.8). There exists a countable set $\mathscr{D}$ such that T is a bijection : $X\backslash\mathscr{D} \to X\backslash\mathscr{D}$ and we can confine ourselves to the system $(X\backslash\mathscr{D}, T, \mu)$ metrically isomorphic to the substitution bilateral subshift, for which corollary 5.11 holds. Thus below, by abuse of notation, we will write X instead of $X\backslash\mathscr{D}$, and the metric results of the two previous chapters may be applied.

We already know the discrete part of the spectrum of ζ but nothing about its continuous part, which is present as soon as ζ, or its pure base (item 6.3.1.2), admits no coincidence.

7.1.1 Spectral Multiplicity

Recall that the spectral multiplicity of a unitary operator U acting on H is the cardinality (may be infinite) of the maximal set $\mathscr{F}$ of U-orthogonal elements of H, with identical spectral measures. Here $U = U_T$ is a unitary operator as already observed (corollary 5.8). A first approximation of the multiplicity is given in the following.

M. Queffélec, *Substitution Dynamical Systems – Spectral Analysis: Second Edition*, Lecture Notes in Mathematics 1294, DOI 10.1007/978-3-642-11212-6_7,

© Springer-Verlag Berlin Heidelberg 2010

Proposition 7.1. *The spectral multiplicity m of ζ is bounded by s =Card A.*

Proof. This proposition is a straightforward consequence of the metric structure of the system (X,T) (subsection 5.6.3). We recall the notation $[U,f]$ for the closure of the linear span of the $\{U^n f,\ n \in \mathbf{Z}\}$, if $f \in L^2(X,\mu)$. Consider now, for every $n \geq 0$,

$$H_n = [U, (\mathbf{1}_{\zeta^n[\alpha]}, \alpha \in A)].$$

From the identity

$$\zeta^n[\alpha] = \bigcup_{\substack{k,\beta \\ \alpha=\zeta(\beta)_k}} T^{kq^n}(\zeta^{n+1}[\beta]) \qquad (\mu - a.e), \tag{7.1}$$

we derive the inclusion $H_n \subset H_{n+1}$ for every $n \geq 0$. The proposition 5.20 thus admits the following interpretation

$$\overline{\bigcup_{n\geq 0} H_n} = L^2(X,\mu). \tag{7.2}$$

Since the spectral multiplicity of U restricted to H_n is bounded by s for every $n \geq 0$, we conclude with help of proposition 2.12 □

Remark 7.1. 1. This proposition remains valid for nonconstant-length substitutions, since it relies on propositions 5.20 and 5.21, established in the general case.
2. We shall see later that the multiplicity of U on H_n is in fact independent of n in the constant-length case.
3. Another way of getting this estimate of spectral multiplicity, is to observe that a system arising from a primitive substitution is a special case of *finite rank* dynamical system; a precise definition of a dynamical system with rank $\leq r$, as well as a spectral analysis, can be found in [193]. Here is a rough idea of such a system $(X,\mathscr{B},\mu,T)$, T being invertible. Given a partition $\mathscr{P} = (P_j)_{1\leq j\leq k}$ of X, we define a *coding* of X in the usual way : to every $x \in X$ is associated the so-called *$\mathscr{P}$-name* $(\alpha_i)_{i\in\mathbf{Z}}$ of x, defined by

$$\alpha_i = j \quad \text{if} \quad T^i x \in P_j;$$

for every $n \geq 1$, the finite sequence $(\alpha_i)_{i=0,1,\ldots n-1}$ is called the *$\mathscr{P}-n$-name* of x. $(X,\mathscr{B},\mu,T)$ has rank $\leq r$ if, for some $\mathscr{P}$ and n large enough, "most" $\mathscr{P}-n$-names are almost juxtapositions of words which can be picked up from a finite set of r given words $\{B_1^n,\ldots,B_r^n\}$. As a consequence, the spectral multiplicity is majorized by the rank of the system.
Since every word in $\mathscr{L}_\zeta$ is the juxtaposition of $\zeta^k(\alpha)$, $k \geq 0$, $\alpha \in A$, the substitution system has finite rank $\leq s$, whence the proposition.

7.1.2 Maximal Spectral Type

We shall see that $\sigma_{\max}$, the maximal spectral type of ζ, can easily be obtained from the spectral measures $\sigma_{\mathbf{1}_{[\alpha]},\mathbf{1}_{[\beta]}}$, $\alpha,\beta \in A$.

Proposition 7.2. *The maximal spectral type $\sigma_{\max}$ of ζ is such that*

$$\sigma_{\max} \sim \sum_{n\geq 0,\ \alpha\in A} 2^{-n}\ \sigma_{\mathbf{1}_{\zeta^n[\alpha]}}.$$

Proof. Let us set $\sigma = \sum_{n\geq 0,\ \alpha\in A} 2^{-n}\ \sigma_{\mathbf{1}_{\zeta^n[\alpha]}}$; then, clearly, $\sigma \ll \sigma_{\max}$ by definition of $\sigma_{\max}$. Now, referring to subsection 5.6.3 for the notations, if $f \in L^2(X,\mu)$, $f_n = E^{\mathscr{B}_n}(f)$ is exactly the orthogonal projection of f onto H_n; it follows that $\sigma_{f_n} \ll \sigma$ since

$$\sigma_{\mathbf{1}_{\zeta^n[\alpha]},\mathbf{1}_{\zeta^p[\beta]}} \ll \sigma_{\mathbf{1}_{\zeta^n[\alpha]}} \quad \text{and} \quad \sigma_{\mathbf{1}_{\zeta^n[\alpha]},\mathbf{1}_{\zeta^p[\beta]}} \ll \sigma_{\mathbf{1}_{\zeta^p[\beta]}}.$$

As f_n converges to f in $L^2(X,\mu)$ (proposition 5.21), σ_{f_n} converges to σ_f by corollary 2.2 and $\sigma_{f_n} \ll \sigma_f$, which was to be proved. □

It is easy to build $\sigma_{\mathbf{1}_{\zeta^n[\alpha]}}$ from the measures $\sigma_{\mathbf{1}_{\zeta^{n+1}[\beta]}}$, $\beta \in A$, by making use again of identity (7.1). In the opposite direction, we have

Proposition 7.3. *For every $n \geq 0$ and $\alpha \in A$,*

$$\widehat{\sigma}_{\mathbf{1}_{\zeta^n[\alpha]}}(k) = \begin{cases} 0 & \text{if } q^n \text{ does not divide } k \\[2ex] \frac{1}{q^n}\widehat{\sigma}_{\mathbf{1}_{[\alpha]}}(m) & \text{if } k = mq^n. \end{cases} \tag{7.3}$$

Proof. Without loss of generality, we may assume that $n = 1$. Fix $\alpha \in A$. We may write

$$\begin{aligned} \widehat{\sigma}_{\mathbf{1}_{\zeta[\alpha]}}(k) &= < U^k\mathbf{1}_{\zeta[\alpha]}, \mathbf{1}_{\zeta[\alpha]} > \\ &= \int_X \mathbf{1}_{\zeta[\alpha]} \circ T^k \cdot \mathbf{1}_{\zeta[\alpha]}\, d\mu \\ &= \mu\left(T^{-k}\zeta[\alpha] \cap \zeta[\alpha]\right) \end{aligned}$$

If $x \in \zeta(X)$, $T^k x \in \zeta(X)$ if and only if q divides k (corollary 5.11), so that

$$\widehat{\sigma}_{\mathbf{1}_{\zeta[\alpha]}}(k) = 0 \quad \text{if } q \text{ does not divide } k.$$

If now $k = qm$, $T^{-qm}\zeta[\alpha] \cap \zeta[\alpha] = \zeta(T^{-m}\zeta[\alpha]) \cap \zeta[\alpha]$, so that, by the same corollary,

$$\widehat{\sigma}_{\mathbf{1}_{\zeta[\alpha]}}(k) = \mu\Big(\zeta(T^{-m}[\alpha]\cap[\alpha])\Big)$$

$$= \tfrac{1}{q}\mu(T^{-m}[\alpha]\cap[\alpha])$$

$$= \tfrac{1}{q}\widehat{\sigma}_{\mathbf{1}_{[\alpha]}}(m).$$

□

In order to give an interpretation of this last proposition, we need some notations : if $n \geq 1$ is fixed, D_n denotes the finite subgroup of q-adic rational numbers in $\mathbf{T}$, generated by $1/q^n$, $D = \cup_{n\geq 0} D_n$, ω_n is the Haar measure of D_n and S_q is the q-shift on $\mathbf{T}$ i.e. the transformation $x \mapsto qx \mod 2\pi$.

Definition 7.1. If ρ is a measure in $M(\mathbf{T})$, $\pi(\rho)$ denotes the measure defined by

$$\begin{cases} \widehat{\pi(\rho)}(k) &= 0 \text{ if } q \text{ does not divide } k \\ \widehat{\pi(\rho)}(qm) = \widehat{\rho}(m). \end{cases}$$

$\pi(\rho)$ will be called the *invariant-contracted* measure associated to ρ and can be constructed as follows :

Dividing the interval $[0, 2\pi[$ into $\bigcup_{j=0}^{q-1}[\frac{2\pi j}{q}, \frac{2\pi(j+1)}{q}[$, we contract ρ on the interval $[0, \frac{2\pi}{q}[$, and extend this measure to $[0, 2\pi[$ by reproducing the measure obtained in this way on each $[\frac{2\pi j}{q}, \frac{2\pi(j+1)}{q}[$.

Identity (7.3) now admits the following reformulation:

$$\sigma_{\mathbf{1}_{\zeta[\alpha]}} = \frac{1}{q}\pi(\sigma_{\mathbf{1}_{[\alpha]}}), \ \forall \alpha \in A.$$

More generally, for every $\rho \in M(\mathbf{T})$, observe that

$$(S_q \circ \pi)(\rho) = \rho$$

$$\text{and } (\pi \circ S_q)(\rho) = \rho * \omega_1 := \tfrac{1}{q}\textstyle\sum_{j=0}^{q-1} \rho * \delta_{\frac{2\pi j}{q}}.$$

This last measure is invariant under the translations of D_1 (D_1-invariant). In particular, if ρ is already *q-invariant* (that means $S_q(\rho) = \rho$) then $\pi(\rho)$ is exactly $\rho * \omega_1$. If now ρ is a spectral measure σ_f, where $f \in L^2(X, \mu)$, then

$$\pi(\sigma_f) = q\sigma_g$$

with

$$g(x) = \begin{cases} f \circ \zeta^{-1}(x) & \text{if } x \in \zeta(X) \\ 0 & \text{otherwise.} \end{cases} \tag{7.4}$$

Proposition 7.3 suggests the following comments.

Remark 7.2. 1. From this proposition follows the structure of the group of eigenvalues of ζ : if e^{it} is an eigenvalue of ζ, or equivalently, if $\sigma_f\{t\} \neq 0$ for some $f \in L^2(X,\mu)$, then $\sigma_g\{t/q\} \neq 0$ where g is defined by (7.4). Actually,

$$\begin{aligned}
\sigma_g\{t/q\} &= \lim_{N\to\infty} \tfrac{1}{N} \textstyle\sum_{n<N} \widehat{\sigma}_g(n) e^{-int/q} \\
&= \lim_{N\to\infty} \tfrac{1}{N} \textstyle\sum_{n<N} \tfrac{1}{q} \pi(\hat{\sigma}_f)(n) e^{-int/q} \\
&= \lim_{N\to\infty} \tfrac{1}{qN} \textstyle\sum_{m<N/q} \widehat{\sigma}_f(m) e^{-imt} \\
&= \tfrac{1}{q^2} \sigma_f\{t\}.
\end{aligned}$$

2. The maximal spectral type $\sigma_{\max}$ is equivalent to the measure $\sum_{n\geq 0} 2^{-n}\pi^n(\sigma)$ where σ is the sum $\sum_{\alpha\in A} \sigma_{\mathbf{1}_{[\alpha]}}$. Henceforth, we shall say that $\sigma_{\max}$ is *generated* by the measures $\sigma_{\mathbf{1}_{[\alpha]}}$, $\alpha \in A$, and we are thus led to determine those measures.
3. Recall the decomposition of $L^2(X,\mu) = H_d \oplus H_c$ relative to the spectrum of ζ, the spectrum being purely discrete on H_d and purely continuous on H_c. If $f_1,\ldots,f_k$ can be found in such a way that : the spectral measures σ_{f_j} are q-invariant and D-invariant, with, moreover, $H_c = [U, f_1, \ldots f_k]$, then the continuous part of $\sigma_{\max}$ is equivalent to $\sum_{j=1}^k \sigma_{f_j}$.

7.1.3 The Correlation Matrix Σ

From now on, we shall make use of the notation

$$\sigma_{\alpha\beta} := \sigma_{\mathbf{1}_{[\alpha]},\mathbf{1}_{[\beta]}}, \quad \sigma_\alpha = \sigma_{\alpha\alpha}, \quad \alpha,\beta \in A.$$

We denote by Σ the $s \times s$- matrix with entries $\sigma_{\alpha\beta}$ in $M(\mathbf{T})$

Proposition 7.4. *Let $u = (u_n)_{n\geq 0}$ be a fixed point of ζ and τ any map $A \to \mathbf{C}$. Then the infinite sequence $\tau(u) = (\tau(u_n))_n$ admits a unique correlation measure the expression of which is*

$$\sum_{\alpha,\beta\in A} \tau(\alpha)\overline{\tau(\beta)}\sigma_{\alpha\beta}.$$

Proof. We decompose τ into $\sum_{\alpha\in A} \tau(\alpha)\tau_\alpha$, where $\tau_\alpha(\beta) = \delta_{\alpha\beta}$ the Kronecker symbol, so that $\left(\tau_\alpha(u)\right)_n = 1$ if $u_n = \alpha$, 0 otherwise. Now

$$\widehat{\sigma}_{\alpha\beta}(k) = \int_X \mathbf{1}_{[\alpha]} \circ T^k \cdot \mathbf{1}_{[\beta]} \, d\mu$$

$$= \lim_{N\to\infty} \frac{1}{N} \sum_{n+k<N} (\mathbf{1}_{[\alpha]} \circ T^{n+k})(u) \cdot (\mathbf{1}_{[\beta]} \circ T^n)(u)$$

$$= \lim_{N\to\infty} \frac{1}{N} \text{ Card } \{n+k<N,\ u_{n+k} = \alpha,\ u_n = \beta\},$$

so that $\sigma_{\alpha\beta}$ is the mixed correlation measure of the sequences $\tau_\alpha(u)$ and $\tau_\beta(u)$; in particular, σ_α measures the correlation between the occurrences of α in u. The proposition follows from the properties of the spectral family. □

Remark 7.3. 1. From the primitivity assumption, recall that $X = \overline{O(x)}$ for every $x \in X$ so that

$$\widehat{\sigma}_{\alpha\beta}(k) = \lim_{N\to\infty} \frac{1}{q^N} \text{ Card } \{n+k<q^N,\ \zeta^N(\gamma)_{n+k} = \alpha,\ \zeta^N(\gamma)_n = \beta\} \tag{7.5}$$

for every $\gamma \in A$
2. If t is a q-automatic sequence (see section 5.1), t is the image of a substitutive sequence u by some $f : A \to \mathbf{C}$; if the underlying substitution is primitive and aperiodic, t also admits a unique correlation measure.

Turning back to the matrix Σ, we point out some of its elementary properties.

Proposition 7.5.

$$\begin{cases} (i) \quad \Sigma^* := {}^t\overline{\Sigma} = \Sigma \\ (ii) \quad \sum_{\alpha,\beta\in A} \sigma_{\alpha\beta} = \delta_0 \\ (iii) \quad \sigma_{\alpha\beta}\{0\} = \mu([\alpha])\mu([\beta]), \quad \alpha,\beta \in A. \end{cases}$$

Proof. (i) is obvious since $\widehat{\sigma}_{\alpha\beta}$ is real and $\widehat{\sigma}_{\alpha\beta}(k) = \widehat{\sigma}_{\alpha\beta}(-k)$ for every $k \in \mathbf{Z}$, by construction.

(ii) From the identity $\sum_{\alpha\in A} \mathbf{1}_{[\alpha]} = 1$, we get $\delta_0 = \sigma_{\sum_{\alpha\in A} \mathbf{1}_{[\alpha]}} = \sum_{\alpha,\beta\in A} \sigma_{\alpha\beta}$, and analogously, $\sum_{\alpha\in A} \sigma_{\alpha\beta} = \mu([\beta])\delta_0$.

(iii)

$$\sigma_{\alpha\beta}\{0\} = \lim_{N\to\infty} \frac{1}{N} \sum_{n<N} \widehat{\sigma}_{\alpha\beta}(n)$$

$$= \lim_{N\to\infty} \frac{1}{N} \sum_{n<N} \mu(T^{-n}[\alpha] \cap [\beta])$$

and by ergodicity of the system, the limit is $\mu([\alpha])\mu([\beta])$. □

A less elementary result is the following:

Proposition 7.6. *Let $\lambda \in \mathbf{U}$ be an eigenvalue of T; then $\sigma_{\alpha\beta}\{\lambda\} = \langle \mathbf{1}_{[\alpha]}, f_\lambda \rangle \langle f_\lambda, \mathbf{1}_{[\beta]} \rangle$ if f_λ is the normalized eigenfunction corresponding to λ.*

Proof. This is an immediate consequence of the uniquely ergodic Wiener-Wintner theorem (theorem 4.10); since the eigenvalues of the system all are simple and continuous (section 6.1),

$$\lim_{N\to\infty}\frac{1}{N}\sum_{n<N} f(T^n x)\lambda^{-n} = \begin{cases} 0 & \text{if } \lambda \text{ is not an eigenvalue;} \\ P_\lambda f & \text{if } \lambda \text{ is an eigenvalue,} \end{cases}$$

where $P_\lambda f = \langle f, f_\lambda\rangle f_\lambda$ is the projection of f to the one-dimensional eigenspace corresponding to λ, generated by the normalized eigenfunction f_λ. Taking $f = \mathbf{1}_{[\alpha]}$ we get as above

$$\begin{aligned} \sigma_{\alpha\beta}\{\lambda\} &= \lim_{N\to\infty}\int_{\mathbf{T}}\Big(\frac{1}{N}\sum_{n<N} e^{int}\lambda^{-n}\Big)\, d\sigma_{\alpha\beta}(t) \\ &= \lim_{N\to\infty}\frac{1}{N}\sum_{n<N}\lambda^{-n}\langle \mathbf{1}_{[\alpha]}\circ T^n, \mathbf{1}_{[\beta]}\rangle \\ &= \Big\langle \lim_{N\to\infty}\frac{1}{N}\sum_{n<N}\lambda^{-n}\mathbf{1}_{[\alpha]}\circ T^n, \mathbf{1}_{[\beta]}\Big\rangle. \end{aligned}$$

The result follows from theorem 4.10. □

Note that, if the height h of the constant-length ζ is one, the only possible discrete measures in Σ are the δ_x with $x \in \mathbf{Z}(q)$, but the two last propositions are valid for any primitive and aperiodic substitution.

7.2 Positive Definite Matrices of Measures

7.2.1 Definitions and Properties

More generally, we consider in this section $s \times s$-dimensional matrices whose entries are (complex) measures in $M(\mathbf{T})$. If $\mu \in M(\mathbf{T})$, recall that $\check{\mu}$ is the measure defined on Borel sets by

$$\check{\mu}(E) = \overline{\mu(-E)}.$$

Also, if $M = (\mu_{ij})_{1\le i,j\le s}$ is such a matrix of measures, M^* is the matrix with entries $\check{\mu}_{ij}$, $1 \le i, j \le s$. As expected, for every $n \in \mathbf{Z}$, $\widehat{M}(n)$ denotes the scalar matrix whose entries are $\widehat{\mu}_{ij}(n)$.

Definition 7.2. The matrix of measures M is said to be *hermitian* if $M = M^*$. In terms of Fourier transform, this means

$$\widehat{M}(n) = (\widehat{M}(-n))^* = {}^t\overline{\widehat{M}(-n)}$$

for every $n \in \mathbf{Z}$.

Proposition 7.7. *Let M be a hermitian matrix of measures. The following are equivalent.*

(i) For every Borel set A in $\mathbf{T}$, the scalar matrix $(\mu_{ij}(A))_{1\leq i,j\leq s}$ is positive definite.

(ii) For every s-dimensional vector $(f_1,\dots,f_s)$ of bounded measurable functions, then $\sum\limits_{1\leq i,j\leq s} \mu_{ij}(f_i\overline{f_j}) \geq 0$.

(iii) Let μ denote the positive measure $\sum_{1\leq i,j\leq s}|\mu_{ij}|$. For every positive definite matrix of functions of $L^\infty(\mathbf{T},\mu)$, (f_{ij}), then $\sum\limits_{1\leq i,j\leq s} \mu_{ij}(f_{ij}) \geq 0$.

(A definite positive matrix of L^∞-functions (f_{ij}) is such that, for μ-a.e. t, the scalar matrix $(f_{ij}(t))$ is positive definite.)

Proof. ⋆ Assume (iii) and choose $f_{ij} = \lambda_i \overline{\lambda_j} \mathbf{1}_A$, where A is any Borel set in $\mathbf{T}$ and the λ_i are complex numbers. Then we get (i).

⋆ If now (i) holds, let $f_1,\dots,f_s$ be bounded measurable functions. For every $\varepsilon > 0$, one can find simple functions $g_1,\dots,g_s$, of the form

$$g_i = \sum_k c_{ki} \mathbf{1}_{A_k},$$

such that $||g_i - f_i||_\infty \leq \varepsilon$ for every $1 \leq i \leq s$, where the partition (A_k) may be chosen independent of i (refining if necessary). By (i), each matrix $(\mu_{ij}(A_k))_{1\leq i,j\leq s}$ is positive definite, so that

$$\sum_{1\leq i,j\leq s} \mu_{ij}(g_i\overline{g_j}) = \sum_k \sum_{1\leq i,j\leq s} c_{ki}\overline{c_{kj}}\mu_{ij}(A_k) \geq 0,$$

and (ii) follows.

⋆ Finally, assume (ii) and consider $F = (f_{ij})$ a matrix of functions of $L^\infty(\mu)$ such that $F(t) = (f_{ij}(t)) \gg 0$ for μ-almost all t. Therefore, denoting by $(\lambda_k(t))$ the sequence of positive eigenvalues of $F(t)$, there exists a unitary matrix $U(t) = (u_{ij}(t))$ such that

$$f_{ij}(t) = \sum_k \lambda_k(t) u_{ki}(t)\overline{u_{kj}(t)} \quad \mu-a.e.$$

The functions λ_k and u_{ij} are bounded. In the case where the functions f_{ij} are simple, so are the λ_k and u_{ij} and (iii) is a direct consequence of (ii). Otherwise, given any $\varepsilon > 0$, every f_{ij} can be uniformly approximated up to ε on $\mathbf{T}$, by a bounded simple function ϕ_{ij} satisfying thus

$$\sum_{i,j} \lambda_i \overline{\lambda_j} \phi_{ij}(t) > \varepsilon \sum_i |\lambda_i|^2 \quad \mu-a.e.$$

The matrix of functions $(\phi_{ij}+\varepsilon\delta_{ij})$ is positive definite, and from the above first part,

$$\sum_{ij}\int_{\mathbf{T}}(\phi_{ij}+\varepsilon\delta_{ij})\,d\mu_{ij}\geq 0.$$

Letting ε to zero, we get

$$\sum_{ij}\int_{\mathbf{T}}\phi_{ij}\,d\mu_{ij}\geq 0,$$

and (iii) is proved. □

Definition 7.3. A matrix of measures satisfying one of these equivalent statements will be also called a *positive definite matrix* and one will simply denote this property by $M\gg 0$.

The following remarks are obvious, for a positive definite matrix $M=(\mu_{ij})$.
(a) The diagonal measures μ_{ii} are positive measures.

(b) Each μ_{ij} is absolutely continuous with respect to μ_{ii} and μ_{jj}, since, for every Borel set A in $\mathbf{T}$,

$$|\mu_{ij}(A)|^2\leq\mu_{ii}(A)\mu_{jj}(A).$$

Examples

1. If $\mu\in M(\mathbf{T})$ is a positive measure and $A=(a_{ij})$ is a positive definite scalar matrix, then $M=(\mu_{ij})$, with $\mu_{ij}=a_{ij}\mu$, is clearly $\gg 0$.
2. The correlation matrix Σ, defined above, is hermitian and $\gg 0$; indeed, if $(t_1,\ldots,t_s)\in\mathbf{C}^s$, the measure $\sum_{\alpha,\beta}t_\alpha\overline{t_\beta}\sigma_{\alpha\beta}$ is the spectral measure σ_f with $f=\sum_\alpha t_\alpha\mathbf{1}_{[\alpha]}$ and thus $\sigma_f(A)\geq 0$ for every Borel set A. This means that $(\sigma_{\alpha\beta}(A))$ is a positive definite scalar matrix and (i) is checked.
3. Recall that the Hadamard product (termwise) of two positive definite scalar matrices is again positive definite. Thus, it can be deduced from proposition 7.7 (iii) that the matrix of measures $(f_{ij}\cdot\mu_{ij})$ is again positive definite if (μ_{ij}) is a positive definite matrix of measures and (f_{ij}), a positive definite matrix of functions in $L^\infty(\mu)$.

We shall be needing the following remark.

Proposition 7.8. *Let M be a positive definite matrix of measures. The following are equivalent, for $v\in\mathbf{C}^s$, $v\neq 0$:*

(a) $Mv=0$
(b) $\widehat{M}(0)v=0$
(c) $\langle Mv,v\rangle=0$.

Proof. The first implication (a)$\Longrightarrow$(b) is evident. Consider now the measure $\mu:=\langle Mv,v\rangle=\sum_{ij}v_i\overline{v_j}\mu_{ij}$. By proposition 7.7 (i), μ is a positive measure and $\hat{\mu}(0)=0$ assuming (b); this establishes the second implication (b)$\Longrightarrow$(c). Finally, suppose that $\mu := \langle Mv, v\rangle = 0$, so that, for every Borel set A, $\mu(A) = \langle M(A)v, v\rangle = 0$. Since $M(A)$ is a positive definite scalar matrix, by Schwarz's inequality, $M(A)v = 0$; it follows that $Mv = 0$ and (a) holds. □

7.2.2 Decompositions of Matrices of Measures

As well as for scalar matrices, decomposition results for matrices of measures will be useful. We establish three independent such results.

Definition 7.4. We say that the matrix of measures, M, can be diagonalized, if there exist an invertible *scalar* matrix P and a diagonal matrix of measures D such that $P^{-1}MP = D$.

Proposition 7.9. *The matrix of measures, M, can be diagonalized if and only if the matrices $\widehat{M}(n)$, $n \in \mathbf{Z}$, can be diagonalized simultaneously.*

Proof. The necessity of the condition being evident, we suppose that the matrices $\widehat{M}(n)$ can be simultaneously reduced to diagonal form. There exist an invertible scalar matrix P and, for every $n \in \mathbf{Z}$, a diagonal scalar matrix D_n such that

$$P^{-1}\widehat{M}(n)P = (P^{-1}MP)\widehat{\ }(n) = D_n.$$

D_n is thus the nth Fourier coefficient of some matrix of measures, which, in turn, must be diagonal by Fourier unicity, and $P^{-1}MP = \Delta$ a diagonal matrix of measures. □

Remark 7.4. An hermitian matrix of measures may fail to be reducible to diagonal form in the above sense, as one can be convinced by the example

$$m = \begin{pmatrix} 0 & \mu \\ \check{\mu} & 0 \end{pmatrix}, \quad \text{with } \mu = \delta_0 + i\delta_\pi.$$

We now establish a polar decomposition for matrices of measures, which is an analogue of the scalar one.

Proposition 7.10. *Let $M = (\mu_{ij})$ be a matrix of measures and set, as above, $\mu = \sum_{ij} |\mu_{ij}|$. Then, there exist a positive definite matrix of measures, each of them being absolutely continuous with respect to μ, denoted by $|M|$, and a matrix U of functions in $L^\infty(\mu)$ such that,*

(i) $M = U\,|M|$
(ii) $|M| = U^*\,M$
(iii) U is a partial isometry.

Proof. For every (i,j), we can write $\mu_{ij} = f_{ij}\mu$ where $f_{ij} \in L^1(\mu)$ and we set $F = (f_{ij})$. For μ-almost all t, the scalar matrix $F(t)$ admits the decomposition

$$F(t) = U(t)\, P(t)$$

where $P(t) = \sqrt{F^*(t) \cdot F(t)}$ is a positive definite scalar matrix. If $P = (p_{ij})$, the functions $t \mapsto p_{ij}(t)$ are measurable and, using the identity $||P(t) \cdot x|| = ||F(t) \cdot x||$ for $x \in \mathbf{C}^n$, they are easily seen to satisfy

$$|p_{ij}(t)| \leq \big(\sum_{\ell} |p_{i\ell}(t)|^2\big)^{1/2} = \big(\sum_{\ell} |f_{i\ell}(t)|^2\big)^{1/2}.$$

We deduce that $|p_{ij}(t)| \leq \sum_{\ell} |f_{i\ell}(t)|$ which proves that each $p_{ij} \in L^1(\mu)$. Also, $|u_{ij}(t)| \leq 1$ for μ-almost all t by construction.

It remains to show that those u_{ij} can be chosen in a measurable way. Consider the map : $A \mapsto r(A)$ associating to the operator A its rank $r(A)$. Since this is an ℓ.s.c. map, and since $t \mapsto F(t)$ is a measurable map, so is $t \mapsto r(F(t))$. This provides a decomposition of $\mathbf{T} = \bigcup_{r=1}^n \Omega_r$ into disjoint Borel sets where

$$\Omega_r := \{t,\ r(F(t)) = r\} = \{t,\ r(P(t)) = r\}.$$

We restrict our attention to Ω_r for some fixed $r \leq n$, and thus to operators of constant rank. We denote by G_r the set of r-dimensional subspaces of $\mathbf{C}^n$, endowed with the Hausdorff metrics of their unit spheres; it is well-known that the maps

$$A \mapsto \operatorname{Im} A, \quad \text{and} \quad A \mapsto \operatorname{Im} A^{\perp}$$

are continuous ones from the set of all operators with rank r into G_r (resp. G_{n-r}). If $t \in \Omega_r$, we put $U(t)x = 0$ if $x \in \operatorname{Im} P(t)^{\perp}$, and $U(t)(P(t)y) = F(t)y$.

From the above remark, $t \in \Omega_r \mapsto U(t)$ is a measurable map and thereby, $t \mapsto U(t)$ is measurable on the whole of $\mathbf{T} = \bigcup_{r=1}^n \Omega_r$, as expected.

Notice that $U(t)$, so defined, is a partial isometry, and, since $F(t) = U(t)\, P(t)$, obviously $U(t)^*\, F(t) = U(t)^*\, U(t)\, P(t) = P(t)$. It follows that $M = U\ |M|$ with

$$|M| = P \cdot \mu$$

and one can check the decomposition to be intrinsic. □

We now refer to section 1.4. For an L-space L of $M(\mathbf{T})$, recall the decomposition $M(\mathbf{T}) = L \oplus L^{\perp}$, where $L^{\perp} = \{\mu \in M(\mathbf{T}),\ |\mu| \perp |\nu|$, for every $\nu \in L\}$. If M is a matrix of measures, M can be decomposed in a unique way

$$M = M_L + M_{L^{\perp}}$$

with obvious notations. Note the following useful result.

Proposition 7.11. *Let M be a matrix of measures admitting the decomposition $M = M_L + M_{L^\perp}$ relative to the L-space L of $M(\mathbf{T})$. Then*

$$M \gg 0 \text{ if and only if both } M_L \gg 0 \text{ and } M_{L^\perp} \gg 0.$$

Proof. Only the necessity needs a proof. If $M = (\mu_{ij})$, we consider the dominant positive measure $\mu = \sum_{ij} |\mu_{ij}|$ and its decomposition $\mu = \mu_L + \mu_{L^\perp}$. Since $\mu_L \perp \mu_{L^\perp}$, there exists a Borel set E in $\mathbf{T}$ such that $\mu_L(E^c) = 0$ and $\mu_{L^\perp}(E) = 0$. If now $\mu_{ij} = f_{ij} \cdot \mu$, then $(\mu_{ij})_L = f_{ij} \cdot \mu_L$ and $(\mu_{ij})_{L^\perp} = f_{ij} \cdot \mu_{L^\perp}$ by L-property and unicity of the decomposition.

Let A be any Borel set in $\mathbf{T}$. The scalar matrix

$$M_L(A) = M_L(A \cap E) = M(A \cap E)$$

and $M_L(A)$ is positive definite if M is supposed to be so. In the same way, $M_{L^\perp}(A) = M(A \cap E^c) \gg 0$ in turn. This proves that M_L and $M_{L^\perp}$ are positive definite matrices of measures. □

Examples

1. As a particularly important case, we shall encounter the decomposition $M = M_c + M_d$ into continuous and discrete parts. Then, $M \gg 0$ if and only if both M_c and M_d are $\gg 0$.
2. Consider $M = (\mu_{ij})$ and a positive measure $\mu \in M(\mathbf{T})$; for each pair (i, j), let $\mu_{ij} = \nu_{ij} + \rho_{ij}$ be the decomposition of μ_{ij} relative to μ : $|\nu_{ij}| \ll \mu$ and $\rho_{ij} \perp \mu$. The matrices (ν_{ij}) and (ρ_{ij}) are $\gg 0$ as soon as (μ_{ij}) is $\gg 0$.
3. The correlation matrix Σ admits the following decomposition relative to δ_0:

$$\Sigma = \Sigma\{0\} \cdot \delta_0 + \Sigma' \quad \text{where } \Sigma'\{0\} = 0.$$

The scalar matrix $\Sigma\{0\} = (\mu[\alpha]\mu[\beta])_{\alpha\beta}$ is clearly positive definite and so must be Σ'.

7.3 Characters on a Matrix of Measures

In this section, we study the behaviour of the characters $\chi \in \Delta := \Delta M(\mathbf{T})$, acting on a matrix of measures $M = (\mu_{ij})$. (The Gelfand spectrum Δ of $M(\mathbf{T})$ has been investigated in section 1.2).

If $\chi \in \Delta$ and $M = (\mu_{ij})$, χM denotes the new matrix of measures with entries $\chi_{\mu_{ij}} \cdot \mu_{ij}$, and whose Fourier transform is

$$(\chi M)\hat{}(\gamma) = \widehat{M}(\chi\gamma) = (\widehat{\mu}_{ij}(\chi\gamma))_{ij}$$

for every $\gamma \in \Gamma \simeq \mathbf{Z}$.

In case we are studying a fixed matrix $M = (\mu_{ij})$, we may restrict ourselves to $L^1(\mu)$, where μ, for instance, is the dominant positive measure $\sum_{ij} |\mu_{ij}|$. Then, $\chi_{\mu_{ij}} = \chi_\mu\ \mu_{ij}$-a.e, and the matrix χM is just the matrix deduced from M by pointwise multiplication by the function $\chi_\mu \in L^\infty(\mu)$.

We say that χ belongs to $\overline{\Gamma}(M)$ if there exists a sequence (γ_α) of elements of Γ such that

$$\lim_\alpha \widehat{\mu}_{ij}(\gamma_\alpha \gamma) = \widehat{\mu}_{ij}(\chi\gamma)$$

for every (i,j) and $\gamma \in \Gamma$. Here also, $\chi \in \overline{\Gamma}(\mu)$ for this measure μ.

We now intend to describe the characters χ which are constant on M in the following sense:

Definition 7.5. $\chi \in \Delta$ is said to be *constant* on M if there exists a scalar matrix $A := A(\chi)$ (or $B := B(\chi)$) such that

$$\chi M = MA \quad (\text{or } \chi M = BM), \tag{7.6}$$

which is equivalent, by Fourier transform, to saying that

$$\widehat{M}(\chi\gamma) = \widehat{M}(\gamma)\, A.$$

Remark 7.5. 1. Since $(\chi M)^* = \overline{\chi} M^*$, we see that $\chi M = BM$ is equivalent to $\overline{\chi} M^* = M^* B^*$. So, we may restrict ourselves to the first definition.

2. Let $M = M_c + M_d$ be the unique decomposition of M into discrete and continuous parts. Then, χM_c and χM_d are matrices of continuous and discrete measures respectively, so that

$$\chi M = MA \text{ if and only if } \chi M_c = M_c A \text{ and } \chi M_d = M_d A$$

by unicity of the decomposition of χM.

3. If $\chi M = MA$, by using proposition 7.10 we get

$$\chi|M| = \chi(U^* M) = U^* \chi M = U^* MA = |M| A.$$

In order to determine χ or A, we are thus allowed to assume M to be a positive-definite matrix.

4. Generally, the matrix A is not uniquely defined by (7.6). But, if $\widehat{M}(0)$ is an invertible matrix, A is clearly unique and equal to $(\widehat{M}(0))^{-1} \cdot \widehat{M}(\chi)$; in that case, χ is constant on M if and only if, putting $M_1 = (M(0))^{-1} \cdot M$, we have

$$\widehat{M}_1(\chi\gamma) = \widehat{M}_1(\gamma) \cdot \widehat{M}_1(\chi) \quad \text{for all } \gamma \in \Gamma. \tag{7.7}$$

(7.7) needs to be compared with (1.10).

If $\widehat{M}(0)$ is no longer invertible, we claim that there exists a choice of A satisfying both conditions:

$$\chi M = MA$$

and

$$Mv = 0 \Longleftrightarrow Av = 0. \tag{7.8}$$

Indeed, consider $K = \{v \in \mathbf{C}^s,\ Mv = 0\}$. By 3. we may assume M to be $\gg 0$, and K is exactly $\ker \widehat{M}(0)$ by proposition 7.8. Let now Q be any projection operator onto K; then $MAQ = 0$ since, by (7.6), A maps K into K. Therefore, letting $A' = A(I-Q)$, A' is different from A, A' satisfies (7.8) and $MA' = MA$, whence the claim.

We now are in a position to establish the following result.

Proposition 7.12. *Let M be a matrix of measures and $\chi \in \Delta$ be constant on M. Then*

(i) There exists a scalar matrix A satisfying : $\chi M = MA$, and each and every one of its eigenvalue has multiplicity one in the minimal polynomial of A.

(ii) If $\widehat{M}(0)$ is invertible, the unique A satisfying (7.6) can be reduced to diagonal form.

Proof. (i) Choose A satisfying both (7.6) and (7.8). Let λ be an eigenvalue of A and v be a vector in the Jordan subspace corresponding to λ. By definition of v, there exists $k \geq 1$ such that $(A - \lambda I)^k v = 0$ and then

$$M(A - \lambda I)^k v = 0. \tag{7.9}$$

Using (7.6), it can be checked, by induction on $j \geq 1$, that

$$M(A - \lambda I)^j = (\chi - \lambda)^j M$$

so that, from (7.9), $(\chi - \lambda)^k Mv = 0$. It follows that χ is constant on each component of the vector of measures Mv. This implies $M(A - \lambda I)v = (\chi - \lambda)Mv = 0$, or, equivalently

$$MAv = \lambda Mv.$$

If $w = Av - \lambda v$, this means $Mw = 0$, and since A satisfies (7.8), we must have $Aw = 0$ in turn. Combined with $(A - \lambda I)^{k-1} w = (A - \lambda I)^k v = 0$, we obtain that $\lambda^{k-1} w = 0$ and

$$w = 0 \quad \text{if } \lambda \neq 0.$$

But this exactly means that v is an eigenvector corresponding to λ and (i) is proved.

If $\lambda = 0$, we can only deduce that $A^2 v = Aw = 0$ and that the dimension of the Jordan subspace corresponding to $\lambda = 0$ is ≤ 2. In case $\widehat{M}(0)$ is invertible, $Mw = 0$ entails $\widehat{M}(0)w = 0$ and $w = 0$. Thus, the minimal polynomial of A has only simple roots and A can be diagonalized. □

This proposition says that a constant character χ on a matrix M with $\widehat{M}(0)$ invertible, is actually piecewise constant : more precisely, if $M_1 = P^{-1}MP$ where $P^{-1}AP = D$ is diagonal, then

$$\chi M_1 = M_1 D$$

or, if $D = [d_1, \ldots, d_s]$ and $M_1 = (\nu_{ij})$,

$$\chi \nu_{ij} = d_i \nu_{ij}.$$

The consequences below will be used later.

Corollary 7.1. *Let M be a matrix of measures such that $\widehat{M}(0) = I$, and $\chi_a M = M\,\widehat{M}(a)$ for every integer a, where χ_a is the limit in $\overline{\Gamma}(M)$ of the sequence $(\gamma_{aq^n})_n$. Then M can be diagonalized.*

Proposition 7.13. *Let M be a matrix of measures with $\widehat{M}(0)$ invertible and let χ and ϕ be two characters, both constant on M. Then, the associated scalar matrices are permutable.*

Proof. Assume that $\chi M = MA$ and $\phi M = MB$. The matrices of measures χM and ϕM are absolutely continuous, termwise, with respect to M. It follows from the obvious identity $\phi(\chi M) = \chi(\phi M)$ that

$$(MB)A = \phi MA = \chi MB = (MA)B$$

and $MAB = MBA$. Since $\widehat{M}(0)$ is invertible, we get $AB = BA$. □

Chapter 8
Matrix Riesz Products

Turning back to the correlation matrix $\Sigma = (\sigma_{\alpha\beta})$ associated, in the previous chapter, with the primitive and aperiodic substitution ζ of length q, we shall prove that Σ is the weak-star limit point of a product of matrices whose entries are trigonometric polynomials, in a way similar to the case of generalized Riesz products. This provides us with a constructive process to explicit Σ for special substitutions, such as commutative ones (Thue-Morse) but also for the Rudin-Shapiro substitution, and therefore, we will be able to deduce their maximal spectral type.

8.1 Σ as a Matrix Riesz Product

Throughout this chapter, R_j, for $j \in \{0,\dots,q-1\}$, will denote the matrix of the jth instruction ϕ_j of the associated automaton, with $0-1$ entries :

$$(R_j)_{\alpha\beta} = \begin{cases} 1 & \text{if and only if } \ \phi_j(\alpha) = \zeta(\alpha)_j = \beta, \\ 0 & \text{otherwise.} \end{cases}$$

Each R_j is stochastic by row, and $\sum_{j=0}^{q-1} R_j = {}^tM$, where M is the composition matrix of ζ. We then define the matrix polynomials

$$R(t) = R_0 + R_1 e^{it} + \cdots + R_{q-1} e^{i(q-1)t}$$

and

$$\Pi_n(t) = R(q^{n-1}t) \cdots R(t), \quad \text{for } n \geq 1.$$

Definition 8.1. Let (M_n) be a sequence of matrices whose entries are elements of $L^1(\mathbf{T})$, and let $M = (\mu_{\alpha\beta})$ be a matrix of measures of $M(\mathbf{T})$. We say that M is the *weak-star limit point* of the sequence (M_n), if, for every $\alpha, \beta \in A$ and $k \in \mathbf{Z}$,

M. Queffélec, *Substitution Dynamical Systems – Spectral Analysis: Second Edition*, Lecture Notes in Mathematics 1294, DOI 10.1007/978-3-642-11212-6_8,

© Springer-Verlag Berlin Heidelberg 2010

$$\widehat{\mu}_{\alpha\beta}(k) = \lim_{n\to\infty} \widehat{(M_n)}_{\alpha\beta}(k).$$

The following theorem will supply us with an algorithm for computing the correlation matrix Σ, which appears as a *matrix Riesz product.*

Theorem 8.1. *The matrix $s\Sigma$ is the weak-star limit point of the product $\frac{1}{q^n}\Pi_n^*\Pi_n$.*

Proof. First of all, observe that

$$\Pi_n(t) = \sum_{0\le k<q^n} Q_k e^{ikt} \tag{8.1}$$

where

$$Q_k = R_{k_{n-1}} R_{k_{n-2}} \cdots R_{k_0}$$

if the q-adic expansion of k is $k = k_0 + k_1 q + \cdots + k_{n-1}q^{n-1}$. Then,

$$\Pi_n^*(t)\Pi_n(t) = \Big(\sum_{0\le k<q^n} {}^tQ_k e^{-ikt}\Big)\Big(\sum_{0\le j<q^n} Q_j e^{ijt}\Big)$$

$$= \sum_{|r|<q^n} e^{-irt}\Big(\sum_{\substack{j+r<q^n\\ 0\le j<q^n}} {}^tQ_{j+r}Q_j\Big)$$

so that

$$\int_{\mathbf{T}} \big(\Pi_n^*(t)\Pi_n(t)\big)_{\alpha\beta}\, e^{irt}\,\frac{dt}{2\pi} = \sum_{\substack{j+r<q^n\\ 0\le j<q^n}} \big({}^tQ_{j+r}Q_j\big)_{\alpha\beta}$$

$$= \sum_{\substack{j+r<q^n\\ 0\le j<q^n}} \sum_{\gamma\in A} \big(Q_{j+r}\big)_{\gamma\alpha}\big(Q_j\big)_{\gamma\beta}.$$

But, expanding $j = j_0 + j_1 q + \cdots + j_{n-1}q^{n-1}$, and

$$\big(Q_j\big)_{\alpha\beta} = \sum_{\gamma_1,\ldots,\gamma_{n-1}} \big(R_{j_{n-1}}\big)_{\alpha\gamma_1}\big(R_{j_{n-2}}\big)_{\gamma_1\gamma_2}\cdots\big(R_{j_0}\big)_{\gamma_{n-1}\beta},$$

we see that $\big(Q_j\big)_{\alpha\beta} = 1$ if and only if there exist $\gamma_1,\ldots,\gamma_{n-1}$ in A such that

$$\phi_{j_{n-1}}(\alpha) = \gamma_1,\ \phi_{j_{n-2}}(\gamma_1) = \gamma_2,\ \ldots\ ,\ \phi_{j_0}(\gamma_{n-1}) = \beta,$$

or, equivalently, $\phi_{j_0}\circ\cdots\circ\phi_{j_{n-1}}(\alpha) = \beta$. Since $\zeta^n(\alpha)_j = \zeta^{n-1}\big(\zeta(\alpha)_{j_{n-1}}\big)_{j-j_{n-1}}$ (5.1), we have just proved that

$$\big(Q_j\big)_{\alpha\beta} = 1 \quad \text{if and only if} \quad \zeta^n(\alpha)_j = \beta.$$

Finally, recalling remark 7.5, we obtain for nonnegative r,

$$\lim_{n\to\infty}\frac{1}{q^n}\left(\Pi_n^*\Pi_n\right)_{\alpha\beta}\widehat{\ }(r) = \lim_{n\to\infty}\frac{1}{q^n}\sum_{\gamma\in A}\text{Card}\{j+r<q^n,\ \zeta^n(\gamma)_{j+r}=\alpha,\ \zeta^n(\gamma)_j=\beta\}$$

$$= \sum_{\gamma\in A}\widehat{\sigma}_{\alpha\beta}(r)$$

$$= s\,\widehat{\sigma}_{\alpha\beta}(r);$$

the matrix Σ being hermitian, the limit holds for every integer r and the theorem is proved. □

8.2 Examples of Maximal Spectral Types

We apply the previous theorem to compute the correlation measures $\sigma_{\alpha\beta}$ and thus, the maximal spectral type, in two examples.

8.2.1 Thue-Morse Sequence

The Thue-Morse sequence (m_n) on the alphabet $\{0,1\}$ is obtained as $\zeta^\infty(0)$ where $\zeta(0)=01$ and $\zeta(1)=10$, so that

$$R_0 = I \quad\text{and}\quad R_1 = \begin{pmatrix}0 & 1\\ 1 & 0\end{pmatrix}.$$

From the obvious identity :

$$(\Pi_{n+1}^*\Pi_{n+1})(t) = R^*(t)(\Pi_n^*\Pi_n)(2t)R(t),$$

we guess that the matrix $\Pi_n^*\Pi_n$ is of the form

$$\Pi_n^*\Pi_n = \begin{pmatrix}a_n & b_n\\ b_n & a_n\end{pmatrix}$$

where a_n and b_n are trigonometric polynomials satisfying the inductive relations

$$a_0 = 1,\quad b_0 = 0$$

$$a_{n+1}(t) = 2a_n(2t) + 2b_n(2t)\cos t$$
$$b_{n+1}(t) = 2b_n(2t) + 2a_n(2t)\cos t.$$

Combining these identities, we easily get

$$a_{n+1}(t)+b_{n+1}(t)=2(1+\cos t)(a_n(2t)+b_n(2t))$$

$$a_{n+1}(t)-b_{n+1}(t)=2(1-\cos t)(a_n(2t)-b_n(2t)).$$

Thereby we obtain

$$\frac{1}{2^n}(a_n+b_n)(t)=\prod_{j=0}^{n-1}(1+\cos(2^j t))$$

which converges weak-star to δ_0, and

$$\frac{1}{2^n}(a_n-b_n)(t)=\prod_{j=0}^{n-1}(1-\cos(2^j t))$$

which converges weak-star to ρ, the Riesz product constructed on the sequence (2^n) (subsection 4.3.3). Here is the expression of Σ

$$\Sigma=\frac{1}{4}\begin{pmatrix}\delta_0+\rho & \delta_0-\rho\\ \delta_0-\rho & \delta_0+\rho\end{pmatrix}$$

and we deduce from the invariance properties of ρ (section 3.3) the following consequence.

Corollary 8.1 (Keane). *Let f be in $L^2(X,\mu)$; then $f\in H_c$ in the decomposition relative to the Thue-Morse substitution if and only if $\sigma_f \ll \rho$, where* $\rho=\prod_{j=0}^{\infty}(1-\cos 2^j t)$.

8.2.2 Rudin-Shapiro Sequence

We introduced in subsection 4.2.3 the Rudin-Shapiro sequence, recursively defined by

$$\begin{cases} r_0=1\\ r_{2n}=r_n, \quad \text{and} \quad r_{2n+1}=(-1)^n r_n.\end{cases}$$

Actually this sequence has been constructed independently by Shapiro [226] and Rudin [217], in order to get uniform lower bounds for trigonometric sums of the form $\sum_{n<N}\varepsilon_n e^{int}$, where $\varepsilon_n=\pm1$. For almost all ±1-valued sequence (ε_n), it can be proved that $||\sum_{n<N}\varepsilon_n e^{int}||_\infty\le\sqrt{N\log N}$ by probabilistic methods, while obviously, in the opposite direction, $||\sum_{n<N}\varepsilon_n e^{int}||_\infty\ge\sqrt{N}$ by Parseval identity. The sequence (r_n) satisfies

$$||\sum_{n<N} r_n e^{int}||_\infty\le(2+\sqrt{2})\sqrt{N}$$

which is the sharper growth order (the constant can be improved [220]). Classes of real unimodular sequences (u_n), appearing as generalizations of the Rudin-Shapiro sequence [12, 13, 184, 211], have been proved to check the inequality

$$|| \sum_{n<N} u_n e^{int} ||_\infty = O(\sqrt{N}), \tag{8.2}$$

but certainly, an interesting result, if possible, would be a characterization of sequences (u_n) with $u_n = \pm 1$ and satisfying (8.2).

Besides, the sequence (r_n) turns out to be a 2-automatic sequence, obtained from a primitive substitution of length two on a four-letters alphabet [55]. For instance, consider

$$\zeta(0) = 0\,2$$
$$\zeta(1) = 3\,2$$
$$\zeta(2) = 0\,1$$
$$\zeta(3) = 3\,1$$

and $u = \zeta^\infty(0)$; then $r_n = \tau(u_n)$ where $\tau : A \to \{-1, 1\}$ projects $0, 2$ onto 1, and $1, 3$ onto -1. For such a choice of ζ,

$$R_0 = \begin{pmatrix} 1 & 0 & 0 & 0 \\ 0 & 0 & 0 & 1 \\ 1 & 0 & 0 & 0 \\ 0 & 0 & 0 & 1 \end{pmatrix} \quad \text{and} \quad R_1 = \begin{pmatrix} 0 & 0 & 1 & 0 \\ 0 & 0 & 1 & 0 \\ 0 & 1 & 0 & 0 \\ 0 & 1 & 0 & 0 \end{pmatrix}$$

so that, denoting as usual by γ the character $t \to e^{it}$, we get

$$R(t) = \begin{pmatrix} 1 & 0 & \gamma(t) & 0 \\ 0 & 0 & \gamma(t) & 1 \\ 1 & \gamma(t) & 0 & 0 \\ 0 & \gamma(t) & 0 & 1 \end{pmatrix}.$$

We deduce

$$(\Pi^*\Pi)(t) = (R^*R)(t) = \begin{pmatrix} 2 & \gamma(t) & \gamma(t) & 0 \\ \overline{\gamma}(t) & 2 & 0 & \overline{\gamma}(t) \\ \overline{\gamma}(t) & 0 & 2 & \overline{\gamma}(t) \\ 0 & \gamma(t) & \gamma(t) & 2 \end{pmatrix};$$

using the identity $(\Pi_{n+1}^*\Pi_{n+1})(t) = R^*(t)(\Pi_n^*\Pi_n)(2t)R(t)$ again, we can prove inductively that

$$\Pi_n^*\Pi_n = \begin{pmatrix} P_n & T_n & T_n & Q_n \\ \overline{T}_n & P_n & Q_n & \overline{T}_n \\ \overline{T}_n & Q_n & P_n & \overline{T}_n \\ Q_n & T_n & T_n & P_n \end{pmatrix}$$

where P_n, Q_n and T_n are trigonometric polynomials with Fourier spectrum within $[0,\ldots,2^n-1]$, satisfying

$$P_1 = 2,\ Q_1 = 0\ T_1 = \gamma,$$

and for $n \geq 1$,

$$P_{n+1}(t) = 2P_n(2t) + T_n(2t) + \overline{T}_n(2t) \tag{8.3}$$

$$Q_{n+1}(t) = 2Q_n(2t) + T_n(2t) + \overline{T}_n(2t) \tag{8.4}$$

$$T_{n+1}(t) = \big(2^n + Q_{n+1}(t)\big)\gamma(t). \tag{8.5}$$

From (8.3) and (8.4) we get

$$(P_{n+1} - Q_{n+1})(t) = 2(P_n - Q_n)(t)$$

whence

$$P_{n+1}(t) = 2^{n+1} + Q_{n+1}(t).$$

Setting then $U_n = 2^n + Q_{n+1}$, one may write

$$P_{n+1} = U_n + 2^n, \quad Q_{n+1} = U_n - 2^n, \quad T_{n+1} = \gamma U_n$$

which leads to

$$U_n(t) = U_{n-1}(2t)\big(2 + \gamma(2t) + \overline{\gamma}(2t)\big)$$

by combining (8.4) and (8.5). Since $U_0 = 1$, we deduce that

$$\frac{1}{2^{n+1}} U_n(t) = \frac{1}{2} \prod_{j=1}^{n} (1 + \cos 2^j t)$$

which converges weak-star to the discrete measure $\frac{1}{4}(\delta_0 + \delta_\pi) =: \frac{1}{2}\omega$. Finally, the correlation matrix Σ is of the following form

$$\Sigma = \frac{1}{8} \begin{pmatrix} \omega + m & \gamma\omega & \gamma\omega & \omega - m \\ \overline{\gamma}\omega & \omega + m & \omega - m & \overline{\gamma}\omega \\ \overline{\gamma}\omega & \omega - m & \omega + m & \overline{\gamma}\omega \\ \omega - m & \gamma\omega & \gamma\omega & \omega + m \end{pmatrix}$$

where m is the Lebesgue measure on $\mathbf{T}$.

Corollary 8.2. *Let f be in $L^2(X,\mu)$; then $f \in H_c$ in the decomposition relative to the Rudin-Shapiro substitution if and only if $\sigma_f \ll m$.*

The following corollary has been obtained by Kamae, by using purely arithmetical arguments [135].

Corollary 8.3. *The correlation measure of the sequence (r_n) is the Lebesgue measure on* $\mathbf{T}$.

Proof. Since $r_n = \tau(u_n)$ with the above notations, we can compute σ, the correlation measure of the sequence (r_n), with help of proposition 7.4.

$$\begin{aligned}\sigma &= \sigma_0 + \sigma_1 + \sigma_2 + \sigma_3 + \sigma_{02} + \sigma_{20} + \sigma_{13} + \sigma_{31} \\ &\quad -(\sigma_{01} + \sigma_{10} + \sigma_{12} + \sigma_{21} + \sigma_{03} + \sigma_{30} + \sigma_{32} + \sigma_{23}) \\ &= \frac{1}{2}(\omega + m) - \frac{1}{2}(\omega - m) \\ &= m\end{aligned}$$

□

Remark 8.1. Interesting extensions beyond these two exceptional sequences are gathered in the last section.

8.2.3 *Q-mirror Sequences*

Investigating, by coding, the behaviour of the iterates $f^{(n)}$ of some *unimodal* function $f : [0,1] \to [0,1]$, Allouche and Cosnard in [10] encounter 2-automatic sequences constructed as follows :

Let $Q \geq 2$ be an integer and $\omega = \omega_0 \cdots \omega_{Q-1} \in \{0,1\}^*$ be a word of length Q with $\omega_{Q-1} = 1$. We juxtapose to ω the word ω_1, obtained by exchanging in ω the letters 0 and 1, except for the last one, still equal to 1; we continue with ω_2, similarly constructed from the word $\omega\omega_1$, and so on. We get this way a $0-1$-sequence, called a *Q-mirror* sequence.

For example, starting with $\omega = 11001$, we obtain the following sequence

$$11001\ 00111\ 0011011001\ 00110110001100100111\ \cdots$$

The Thue-Morse sequence is a special case of those, corresponding to $\omega = 11$ and $Q = 2$. In order to describe the 2-automaton which generates those sequences, we introduce the following notations : if ω is the initial word, $\widetilde{\omega}$ is the word obtained after exchanging the letters 0 and 1, and $\overline{\omega}$, the word obtained after exchanging the letters 0 and 1 except the last 1. Therefore, the Q-mirror sequence issued from ω can be so coded:

$$\omega\ \overline{\omega}\ \widetilde{\omega}\ \omega\ \widetilde{\omega}\ \widetilde{\overline{\omega}}\ \omega\ \overline{\omega}\ \widetilde{\omega}\ \widetilde{\overline{\omega}}\ \omega\ \widetilde{\omega}\ \omega\ \overline{\omega}\ \widetilde{\omega}\ \omega\ \cdots$$

Now, by using the symbols $0,1,2,3$ in place of $\omega, \overline{\omega}, \widetilde{\omega}, \widetilde{\overline{\omega}}$, we see that all these sequences can be generated by the substitution η of length 2, defined on the alphabet $\{0,1,2,3\}$ by

$$\eta(0) = 0\ 1$$
$$\eta(1) = 2\ 0$$
$$\eta(2) = 2\ 3$$
$$\eta(3) = 0\ 2$$

Denoting by ζ the Thue-Morse substitution, we claim that the systems (X_η, T) and (X_ζ, T) are topologically isomorphic. Indeed, setting $u = \eta^\infty(0)$, and $m = \zeta^\infty(0)$, the map $\tau : \{0,1,2,3\} \to \{0,1\}$ defined by $\tau(0) = \tau(3) = 0$, $\tau(1) = \tau(2) = 1$ induces a map π from X_η onto X_ζ such that $\pi(u) = m$ and commuting with the shifts. Now π^{-1} defined on m by

$$\begin{aligned} \pi^{-1}(0) = 0 &\quad \text{if} \quad 0 \text{ is followed by } 1 \\ 3 &\quad \text{if} \quad 0 \text{ is followed by } 0 \end{aligned}$$

$$\begin{aligned} \pi^{-1}(1) = 1 &\quad \text{if} \quad 1 \text{ is followed by } 1 \\ 2 &\quad \text{if} \quad 1 \text{ is followed by } 0 \end{aligned}$$

leads to the isomorphism.

However, the correlation matrix associated to η is not so readily computable and we keep this example in mind.

8.3 Commutative Automata

In this section, we study automata whose instructions ϕ_j are permutable. Such an automaton will be called a *commutative automaton*. The examples arising from commutative automata turn out to be direct generalizations of the Thue-Morse automaton where the identity ϕ_0 commutes with ϕ_1! We shall see how to compute easily the matrix Σ in those cases and how to deduce once more the maximal spectral type of the system (X, T).

Theorem 8.2. *Let ζ be a primitive and aperiodic substitution of constant length q, such that the associated q-automaton is commutative. Then the maximal spectral type of the system (X, T) is generated by s generalized Riesz products $\lambda_1, \ldots, \lambda_s$ constructed on $\{q^n\}$. As a consequence, $f \in H_c$ if and only if $\sigma_f \ll \sum_{\lambda_j \text{ continuous}} \lambda_j$.*

Proof. We first observe the following.

Lemma 8.1. *If the associated q-automaton is commutative, the matrix Σ can be diagonalized with respect to an orthonormal basis.*

Proof of the lemma. The automaton $(A, (\phi_j)_{0 \le j \le q-1})$ being primitive, the semi-group Φ of mappings $A \to A$, generated by the instructions ϕ_j, is thus transitive. If, besides, Φ is a commutative semi-group, Φ must consist of permutations of letters of A : invoking the transitivity indeed, if $\phi \in \Phi$ and $\alpha \in A$, one can find $\psi \in \Phi$ such that $\psi(\phi(\alpha)) = \alpha$; since, in addition, Φ is commutative, $\phi(\psi(\alpha)) = \psi(\phi(\alpha)) = \alpha$ and ϕ is clearly surjective. It follows that the matrices R_j are orthogonal scalar matrices, and commute; therefore, they can be simultaneously diagonalized with respect to an orthonormal basis and there exists a unitary scalar matrix P such that, for every $t \in \mathbf{T}$,

$$Q(t) = P^*\, R(t)\, P$$

is a diagonal matrix. The matrix of measures $\Delta = P^*\, \Sigma\, P$ is diagonal too, in view of

$$s\Delta = \text{weak}^* - \lim_{n\to\infty} \frac{1}{q^n} Q^*(t)\cdots Q^*(q^{n-1}t)\, Q(q^{n-1}t)\cdots Q(t),$$

which was the claim. □

Lemma 8.2. *Under the same assumptions, the following identities hold for* Σ : *if* $a, b \in \mathbf{Z}$ *and* $|b| < q^p$, *then*

$$\widehat{\Sigma}(aq^p + b) = \widehat{\Sigma}_p(b)\, \widehat{\Sigma}(a) + \widehat{\Sigma}_p(b - q^p)\, \widehat{\Sigma}(a+1), \tag{8.6}$$

where Σ_p *denotes the product* $\frac{1}{q^p}\, \Pi_p^*\, \Pi_p$.

Proof of the lemma. To begin with, we forget the commutativity property and we establish identities for general automata. Since Σ is hermitian, we restrict ourselves to nonnegative a and b. Let $n \geq p$ be such that $aq^p + b < q^n$ and $a < q^{n-p}$. We decompose

$$\Sigma_n(t) = \frac{1}{q^p} \Pi_p^*(t)\, \Sigma_{n-p}(q^p t)\, \Pi_p(t).$$

The Fourier spectrum of Π_p lies within $\{-q^p+1, \ldots, -1, 0\}$ and we recall, for any matrix M whose entries are functions of $L^1(\mathbf{T})$ or measures in $M(\mathbf{T})$, the following relations

$$(M^*)\widehat{\ }(\ell) = (\widehat{M}(-\ell))^*,$$

also

$$(M(qt))\widehat{\ }(k) = \begin{cases} \widehat{M}(k/q) & \text{if } \; q \text{ divides } k \\ 0 & \text{otherwise.} \end{cases}$$

We deduce from these remarks that

$$q^p \widehat{\Sigma}_n(aq^p + b) = \sum_{\substack{i,j,k \\ i+j+k = aq^p+b \\ q^p \text{ divides } j \\ -q^p < k \leq 0 \leq i < q^p}} (\widehat{\Pi}_p(-i))^*\, \widehat{\Sigma}_{n-p}(j/q^p)\, \widehat{\Pi}_p(k)$$

so that, putting in this sum $j = q^p \ell$ with $\ell \in \{0, 1, \ldots, q^{n-p} - 1\}$ and $k = -m$, we get

$$q^p \widehat{\Sigma}_n(aq^p + b) = \sum_{\substack{i,\ell,m \\ i-m+\ell q^p = aq^p+b \\ 0 \leq i,m < q^p}} (\widehat{\Pi}_p(-i))^*\, \widehat{\Sigma}_{n-p}(\ell)\, \widehat{\Pi}_p(-m).$$

The only possible values for $i-m$ are b and $b-q^p$, and consequently, the only values for ℓ are a and $a+1$, which leads to

$$\begin{aligned}\widehat{\Sigma}_n(aq^p+b) = &\frac{1}{q^p}\sum_{m=0}^{q^p-b-1}(\widehat{\Pi}_p(-m-b))^*\,\widehat{\Sigma}_{n-p}(a)\,\widehat{\Pi}_p(-m)\\ &+\frac{1}{q^p}\sum_{m=q^p-b}^{q^p-1}(\widehat{\Pi}_p(-m-b+q^p))^*\,\widehat{\Sigma}_{n-p}(a+1)\,\widehat{\Pi}_p(-m)\end{aligned} \tag{8.7}$$

that we shall use later.

We now turn back to commutative automata. According to the identities

$$\frac{1}{q^p}\sum_{m=0}^{q^p-b-1}(\widehat{\Pi}_p(-m-b))^*\,\widehat{\Pi}_p(-m) = \widehat{\Sigma}_p(b)$$

and

$$\frac{1}{q^p}\sum_{m=q^p-b}^{q^p-1}(\widehat{\Pi}_p(-m-b+q^p))^*\,\widehat{\Pi}_p(-m) = \widehat{\Sigma}_p(b-q^p),$$

(8.7) can be simplified into

$$\widehat{\Sigma}_n(aq^p+b) = \widehat{\Sigma}_p(b)\,\widehat{\Sigma}_n(a)+\widehat{\Sigma}_p(b-q^p)\,\widehat{\Sigma}_n(a+1),$$

thanks to the commutativity. By taking the limit over n and using theorem 8.1, we get the expected (8.6). $\square$

Remark 8.2. Letting $b=0$ and $p=1$ in (8.6), we obtain

$$\widehat{\Sigma}(aq) = \widehat{\Sigma}(a) \quad \text{for every } a\in\mathbf{Z}$$

according to the obvious identity $(R^*R)\hat{}(0) = \sum_{j=0}^{q-1}{}^tR_j\,R_j = qI$; this means that Σ is *q-invariant*.

We now deduce theorem 8.2 from above. Combining those two lemmas, we obtain the analogue of (8.6) for the diagonal matrix $\Delta = P^*\Sigma P$,

$$\widehat{\Delta}(aq^p+b) = \widehat{\Delta}_p(b)\,\widehat{\Delta}(a)+\widehat{\Delta}_p(b-q^p)\,\widehat{\Delta}(a+1), \tag{8.8}$$

with $\Delta_p = P^*\Sigma_pP$. Let now $\lambda = \Delta_{\alpha\alpha}$ be a measure on the diagonal of Δ. We derive from (8.8) a relation for λ

$$\widehat{\lambda}(aq^p+b) = (\widehat{\Delta}_p(b))_{\alpha\alpha}\,\widehat{\lambda}(a)+(\widehat{\Delta}_p(b-q^p))_{\alpha\alpha}\,\widehat{\lambda}(a+1),$$

which can be written

$$\lambda = F_p\cdot(\lambda*\omega_p)$$

where F_p is the trigonometric polynomial $(\Delta_p)_{\alpha\alpha}$ and ω_p, as usual, the Haar measure of the subgroup generated by $2\pi/q^p$. As $\lim_{p\to\infty}(\widehat{F}_p(b-q^p)) = 0$, we deduce that λ is the weak-star limit point of the sequence $(F_p \cdot dt)$. It remains to find the form of F_p. From

$$\Delta_p(t) = \Delta_{p-1}(t)\Delta_1(q^{p-1}t)$$

we deduce that

$$F_p(t) = G(t) \cdots G(q^{p-1}t), \quad \text{with} \quad G := (\Delta_1)_{\alpha\alpha}.$$

Notice that $\Delta_1 = \frac{1}{q}P^*(R^*R)P$ is a positive definite matrix of trigonometric polynomials so that, each of its diagonal terms is nonnegative and so is G. We have thus proved that every measure of Δ is a generalized Riesz product, obtained as the weak*-limit of $\prod_{n<N} G(q^n t)\, dt$, with a nonnegative trigonometric polynomial G. Recall that these measures are q-invariant and D-quasi-invariant, hence, by the purity law, they are either purely discrete or purely continuous, whence the description of H_c. □

Example 8.1. Consider again the Thue-Morse automaton.

Choose $P = P^* = \frac{1}{\sqrt{2}}\begin{pmatrix} 1 & 1 \\ 1 & -1 \end{pmatrix}$, then P^*R_1P is diagonal and

$$\Delta_1(t) = \frac{1}{2}P^*(R^*R)(t)P = \begin{pmatrix} 1+\cos t & 0 \\ 0 & 1-\cos t \end{pmatrix} = \begin{pmatrix} G_0 & 0 \\ 0 & G_1 \end{pmatrix}.$$

Applying the previous theorem, we deduce that Σ is similar to $\begin{pmatrix} \delta_0 & 0 \\ 0 & \rho \end{pmatrix}$, where

$$\delta_0 = \text{weak}^* - \lim_{N \to \infty} \prod_{n<N} G_0(2^n t), \qquad \rho = \text{weak}^* - \lim_{N \to \infty} \prod_{n<N} G_1(2^n t).$$

Remark 8.3. 1. The next chapter is devoted to automata with one-to-one instructions, commutative or not, with a rather different approach. The diagonal measures of Δ will appear as correlation measures of explicit sequences associated with the automaton, which can be easily described, specifically in the commutative (bijective) case. So we give no other example in this section.

2. It follows from (8.6) that the matrix $M = s\Sigma$ is strongly mixing with respect to the q-shift S_q since

$$\lim_{p \to \infty} \widehat{M}(aq^p + b) = \widehat{M}(b)\, \widehat{M}(a).$$

If χ_a denotes the character of $\overline{\Gamma}(M)$ obtained as the weak*-limit point on M of the sequence (γ_{aq^p}), this last property is equivalent to

$$\chi_a M = M\, A \quad \text{with} \quad A = \widehat{M}(a).$$

As $\widehat{M}(0) = I$, we may apply the results of chapter 7 : $\widehat{M}(a)$ can be diagonalized, and since the $(\widehat{M}(a))_{a \in \mathbf{Z}}$ are permutable, the same holds for M. Thus, lemma 8.1 turns out to be a consequence of lemma 8.2.

The question arising now is whether such a property of strong mixing can be expected in the general case. Actually, this seems to be false in the non-commutative case, but later, we will be led to consider a bigger matrix of correlation measures, always enjoying this property.

8.4 Automatic Sums

Many contributions are concerned with the behaviour of sums related to q-automatic sequences, more specifically the behaviour of such trigonometric sums, whose estimate turns out to be relevant for uniform distribution and pointwise ergodic theorem along subsequences.

Let ζ be a constant-length substitution on the alphabet A with a fixed point $u = (u_n)$ and let $\tau : A \to \mathbf{C}$; we consider

$$S_N(t) = \sum_{n<N} \tau(u_n)e^{int}$$

and ask for punctual or uniform estimate of S_N.

The most famous result involves the Rudin-Shapiro sequence as previously quoted : $||\sum_{n<N} r_n e^{int}||_\infty \leq C\sqrt{N}$, and the following result, related to the Thue-Morse sequence, is noteworthy but rather specific.

Proposition 8.1. *Let* (ε_n) *be the Thue-Morse sequence on* $\{\pm 1\}$, *that is* $\varepsilon_n = (-1)^{m_n}$ *if* $(m_n) = \zeta^\infty(0)$ *and* ζ *the Thue-Morse substitution on* $\{0,1\}$. *Then*

$$\Big\| \sum_{n<N} \varepsilon_n e^{int} \Big\|_\infty \leq 3N^{1-\delta}, \quad \text{with } \delta = \frac{1}{4}\log_2(27/16) > 0.$$

Proof. We first pay attention to sums over dyadic blocks, in view to make use of the mirror property of the sequence. By that, we mean the following symmetry of the blocks : since $\zeta^n(0) = \zeta^{n-1}(0)\zeta^{n-1}(1)$, clearly, for every N,

$$\varepsilon_{[0,\, 2^{N-1}-1]} = -\varepsilon_{[2^{N-1},\, 2^N-1]};$$

also

$$S_{2^N}(t) = S_{2^{N-1}}(t) + \sum_{n=2^{N-1}}^{2^N-1} \varepsilon_n e^{int} = (1 - e^{i2^{N-1}t})S_{2^{N-1}}(t)$$

and

$$S_{2^N}(t) = \prod_{n=0}^{N-1} (1 - e^{i2^n t}).$$

Gathering the terms by pair and using the inequality

$$\sup_{x\in\mathbf{R}}|\sin x\cdot\sin 2x|\leq\frac{4}{3\sqrt{3}}=:c<1,$$

lead to

$$|S_{2^N}(t)|=2^N\prod_{0}^{N-1}|\sin 2^{n-1}t|\leq 2^N c^{N/2}=:2^{N\alpha}.$$

The interpolation argument is easy : if $N=2^{N_1}+\cdots+2^{N_k}=m+2^{N_k}$, with $N_1<N_2<\cdots<N_k$ and $m<2^{N_k}$, we observe that

$$S_{m+2^{N_k}}(t)=S_{2^{N_k}}(t)+e^{i2^{N_k}t}S_m(t)$$

so that, finally,

$$\begin{aligned}|S_N(t)|&\leq|S_{2^{N_1}}(t)|+\cdots+|S_{2^{N_k}}(t)|\leq 2^{N_1\alpha}+\cdots+2^{N_k\alpha}\\&\leq\frac{2^{N_k\alpha}}{1-2^{-\alpha}}\leq 3N^{\alpha}\end{aligned}$$

and the estimate follows. □

This estimate, combined with theorem 4.11, furnishes a direct proof of the following result, also consequence of theorem 4.12 on return times [38, 164, 165].

Theorem 8.3 (E. Lesigne). *The pointwise ergodic theorem holds along the Thue-Morse integral sequence $\Lambda=\{k_1<k_2<\cdots\}$ where $m_{k_n}=1$.*

Both Thue-Morse and Rudin-Shapiro sequences enjoy remarkable arithmetic properties and the question whether some similar estimates can be obtained for general automatic sequences has been raised afterwards. A partial answer already appears in [69]. Let us denote by $M:=M(\zeta)$ the composition matrix of some substitution ζ of length q, and let $|q_1|$ be the second eigenvalue in modulus of M, that is : $q>|q_1|\geq|\lambda|$ for any other eigenvalue λ of M. Recall that the right normalized eigenvector corresponding to Perron-Frobenius eigenvalue q is nothing but $(\mu[\beta])_{\beta\in A}$, where μ is the unique T-invariant probability measure on X_ζ.

Proposition 8.2. *If $\tau:A\to\mathbf{C}$ is such that $\sum_{\beta\in A}\tau(\beta)\mu[\beta]=0$ and $u=\zeta(u)$, then*

$$\sum_{n<N}\tau(u_n)=O(N^{\delta}),\quad\text{where }\ \delta=\frac{\log|q_1|}{\log q}.$$

Proof. We begin with a formulation of the trigonometric sum :

$$T_{n,\alpha}(t)=\sum_{k<q^n}\tau(v_k)e^{ikt}$$

where $v_k = \zeta^n(\alpha)_k$ is the kth letter of the word $\zeta^n(\alpha)$. If e_α is the s-dimensional vector $(0,\ldots,0,\underbrace{1}_{\alpha^{\text{th}}},0,\ldots,0)$ and if τ also denotes the vector $(\tau(0),\ldots,\tau(s-1))$ in $\mathbf{C}^s$, then, using the notations of the first section,

$$\begin{aligned}
T_{n,\alpha}(t) &= \sum_{\beta\in A}\Big(\sum_{\substack{0\le k<q^n\\ \zeta^n(\alpha)_k=\beta}} e^{ikt}\Big)\tau(\beta)\\
&= \sum_{\beta\in A}\sum_{0\le k<q^n} (Q_k)_{\alpha\beta}\, e^{ikt}\ \tau(\beta)\\
&= \sum_{\beta\in A} (\Pi_n(t))_{\alpha\beta}\ \tau(\beta)\\
&= \langle e_\alpha, \Pi_n(t)\cdot\tau\rangle.
\end{aligned}$$

In particular, $T_{n,\alpha} := T_{n,\alpha}(0) = \langle e_\alpha, {}^tM^n\cdot\tau\rangle$ since $\Pi_n(0) = \big(\sum_{j=0}^{q-1} R_j\big)^n$. We now decompose $M = qP + W$, where P is the projection onto the eigenspace corresponding to the Perron-Frobenius eigenvalue q, parallel to the subspace corresponding to the $\lambda \neq q$, and W satisfies $PW = WP = 0$. Thus,

$$T_{n,\alpha} = \langle M^n e_\alpha, \tau\rangle = q^n\langle Pe_\alpha,\tau\rangle + \langle W^n e_\alpha,\tau\rangle.$$

Now,

$$\begin{aligned}
\langle Pe_\alpha,\tau\rangle &= \lim_{p\to\infty}\sum_{\beta\in A} q^{-p} L_\beta(\zeta^p(\alpha))\ \tau(\beta)\\
&= \sum_{\beta\in A}\tau(\beta)\mu[\beta] \quad \text{according to proposition 5.8}\\
&= 0 \quad \text{from our assumption.}
\end{aligned}$$

The spectral radius of W being $|q_1| < q$, we get the asymptotic estimates

$$T_{n,\alpha} = O(||W||_{sp}^n) = O(|q_1|^n)$$

and

$$T_{n,\alpha} = O(q^{n\delta}) \quad \text{where } \delta = \frac{\log|q_1|}{\log q}. \tag{8.9}$$

To achieve the estimate of $S_N = \sum_{n<N}\tau(u_n)$ with $u = \zeta^\infty(0)$, we proceed classically as follows : writing $N = \sum_{j\le n}\varepsilon_j q^j$ with $\in \{0,1,\ldots,q-1\}$, and $\varepsilon_n \neq 0$, we have

$$S_N = \sum_{j=0}^{n}\sum_{m=1}^{\varepsilon_j q^j}\tau(u_{m+\varepsilon_{j+1}q^{j+1}+\cdots+\varepsilon_n q^n}).$$

Now,

$$\begin{aligned} u_{m+q^{j+1}r} &= \left(\zeta^{n+1}(0)\right)_{m+q^{j+1}r} \\ &= \left(\zeta^{j+1}(\alpha_j)\right)_m \quad \text{where } \alpha_j = \left(\zeta^{n-j}(0)\right)_r \end{aligned}$$

as observed in (5.1); we thus are led to estimate sums like $\sum_0^{aq^j-1} \tau(\zeta^{j+1}(\alpha)_m)$ and simpler, $\sum_{rq^j}^{(r+1)q^j-1} \tau(\zeta^{j+1}(\alpha)_m), 0 \le r < a$; by putting $m = \ell + rq^j$ and $\alpha_r = \zeta(\alpha)_r$, this latter sum can finally be reduced to $\sum_{r=0}^{a-1}\sum_0^{q^j-1} \tau(\zeta^j(\alpha_r)_\ell)$. We deduce from (8.9) that

$$\sum_0^{aq^j-1} \tau(\zeta^{j+1}(\alpha)_m) = O(a \cdot q^{j\delta}),$$

and

$$S_N = \sum_{j=0}^{n} O(\varepsilon_j \cdot q^{j\delta}) = O(N^\delta) \quad \text{since } q^n \le N < q^{n+1}.$$

□

Remark 8.4. We wish to indicate various directions of generalization.

1. In [13] J.P. Allouche and M. Mendes France consider sums related to the Rudin-Shapiro sequence and extend the extremal property it enjoys to 2-multiplicative sequences :

   Theorem 8.4. *If $(f(n))$ is a unimodular, 2-multiplicative sequence, then*

   $$\Big|\sum_{n<N} f(n) e^{2i\pi t u(n)}\Big| \le C(t) N^{\alpha(t)} \quad \text{for } t \in \mathbf{R}/\mathbf{Z}$$

   where $\alpha(t) = (1 + \log_2(1 + |\cos \pi t|))/2$, $C(t) \le 2 + \sqrt{2}$ and $u(n)$ counts the number of occurrences of 11 *in the binary expansion of n.*

   Recall that $r_n = (-1)^{u(n)}$ is the nth-component of the Rudin-Shapiro sequence. See also [12, 40, 45, 220] for other patterns.
2. In [58, 73, 194], the authors focus on the ± 1-Thue-Morse sequence (t_n) and they study trigonometric sums in restriction to arithmetical progressions. For instance J.M. Dumont proved in [73] that

   $$\sum_{\substack{n \equiv j \pmod r \\ n<N}} t_n = O(N^{\alpha(r)})$$

   for every odd integer r, where the exponent $\alpha(r)$ is best possible; when $r = 3$, $\alpha(r) = \log 3/\log 4$ and in this case, J. Coquet has proved a summation formula

   $$\sum_{n<N} t_{3n} = N^{\log 3/\log 4} F(N)$$

involving a continuous nowhere differentiable function F, formula which has been extended to a class of substitutive sequences in [74]. With a different point of view, see also [233].
By studying the Thue-Morse sequence in restriction to prime numbers, Mauduit and Rivat [181] proved that, asymptotically, half of the primes, as expected, have odd digit sum to base 2. One of their results is the following :

Theorem 8.5. *Let $\mathscr{P} = 2,3,5,\ldots$ be the sequence of prime numbers and let $\pi(x)$ be the number of primes less than x. Then*

$$\sum_{\substack{p\leq x \\ p\in\mathscr{P}}} t_p = o(\pi(x))$$

3. The link between trigonometric sums, distribution modulo 2π of real sequences and mean ergodic theorem has been sketched in sections 2.6 and 4.3. If now $\Lambda := \{k_1 < k_2 < \cdots < k_n < \cdots\}$ is a sequence of integers, it is well-known that the real sequence $(k_n x)$ is uniformly distributed modulo 2π for almost every x. The set of such x is called the *normal set* associated with Λ and its algebraic description is an interesting problem.
After Cobham [56], we say that the sequence $\Lambda = \{k_1 < k_2 < \cdots\}$ is *recognizable* if the sequence $\mathbf{1}_\Lambda$ is q-automatic for some $q \geq 2$. As for example, we already met the integral sequence $\{1 < 2 < 4 < 7 < 8 < 11 < 13 < 14 < 16 < \cdots\}$ arising from the Thue-Morse sequence. What is the normal set for a recognizable sequence ? In [180], C. Mauduit deals with sums like

$$\sum_{\substack{n<N \\ u_n=\alpha}} e^{2\pi i n\xi}, \quad \xi \in \mathbf{R}$$

where u is any automatic sequence in $A^{\mathbf{N}}$ and $\alpha \in A$. A characterization of (non necessarily primitive) automata for which the normal set consists exactly of irrational numbers is obtained.
As mentioned in [38], recognizable sequences are candidates for pointwise ergodic theorem, and not only for mean ergodic theorem.

Chapter 9
Bijective Automata

In this chapter, we study a special case of automata, called bijective automata, whose instructions are permutations of the letters of A, including the case of additionally commutative automata in the previous sense. In order to describe the measures generating the maximal spectral type, we investigate the structure of the associated system with a new viewpoint. In the commutative bijective case, we give a complete description of those measures, and we touch on the spectral multiplicity problem which will be discussed later. In the non-commutative case, we only succeed in exhibiting strongly mixing generating measures. Note that the sequences arising from bijective substitutions are related to generalized Morse sequences studied in [156, 175, 176].

Definition 9.1. The q-automaton $(A, (\phi_j)_{0 \leq j \leq q-1})$ is said to be a *bijective automaton*, if the instructions ϕ_j are one-to-one maps from A onto A, or equivalently, if the matrices R_j (defined in section 8.1) are nonsingular. The substitution of constant length q is called, in turn, a bijective substitution if the associated automaton is bijective.

Note that in this case, the substitution must be pure ($h(\zeta) = 1$).

Example 9.1. Such are the two substitutions

$$\begin{array}{lcl} 0 \to 0\,2\,1 & & 0 \to 0\,1\,3 \\ 0 \to 1\,0\,0 & \text{and} & 1 \to 1\,0\,2 \\ 0 \to 2\,1\,2 & & 2 \to 2\,3\,1 \\ & & 3 \to 3\,2\,0 \end{array}$$

the first one being non-commutative.

As a first result, by using identity (8.7) of the previous chapter, we prove that the positive measure $\mathrm{Tr}(\Sigma)$ is q-invariant (Tr denoting the trace of the matrix). Letting $p = 1$ and $b = 0$ in (8.7), we get

$$\mathrm{Tr}(\Sigma)\widehat{\ }(aq) = \frac{1}{q} \sum_{\gamma, \delta \in A} \widehat{\sigma}_{\gamma\delta}(a) \Big(\sum_{\alpha \in A} \sum_{j=0}^{q-1} (R_j)_{\gamma\alpha} (R_j)_{\delta\alpha} \Big).$$

M. Queffélec, *Substitution Dynamical Systems – Spectral Analysis: Second Edition*, Lecture Notes in Mathematics 1294, DOI 10.1007/978-3-642-11212-6_9,

© Springer-Verlag Berlin Heidelberg 2010

The sum inside brackets is nothing but the coincidence number between the two words $\zeta(\gamma)$ and $\zeta(\delta)$. This number is q if $\gamma = \delta$, 0 otherwise, since the instructions are one-to-one. We deduce from this remark that

$$\mathrm{Tr}(\Sigma)\hat{}(aq) = \mathrm{Tr}(\Sigma)\hat{}(a) \quad \text{for } a \in \mathbf{Z}.$$

Also note that, thanks to Dekking's theorem 6.6, the spectrum is never purely discrete.

In view to obtain further results we need a new interpretation of the system (X_ζ, T), in terms of group extension of the q-odometer.

9.1 Structure of Bijective Substitution Systems

9.1.1 Extension of the q-Odometer

We recall a few notations about the *q-odometer* that we already met in subsection 3.6.2. $\mathbf{Z}_q = \varprojlim(\mathbf{Z}/q^n\mathbf{Z})$ denotes the set of q-adic integers, usually identified with $\{0,1,\dots,q-1\}^{\mathbf{N}}$ endowed with the addition mod q with carry over. If, in addition, $\{0,1,\dots,q-1\}^{\mathbf{N}}$ is endowed with the product topology, $\mathbf{Z}_q$ becomes a compact additive group containing $\mathbf{Z}$ as a dense subgroup. It is convenient to represent the element $\kappa = (k_j)_{j\geq 0}$ of $\mathbf{Z}_q$ by the formal infinite expansion

$$\kappa = k_0 + k_1 q + \cdots + k_n q^n + \cdots \quad \text{with } 0 \leq k_i \leq q-1,$$

or, equivalently, by a sequence of integers $(j_n)_{n\geq 1}$ satisfying

$$0 \leq j_1 \leq q-1 \quad j_{n+1} \equiv j_n \mod q^n,$$

keeping in mind the relationship between both notations given by

$$j_{n+1} = k_0 + k_1 q + \cdots + k_n q^n.$$

We now define the transformation τ on $\mathbf{Z}_q$ as the addition of 1 (with carry) on $\{0,1,\dots,q-1\}^{\mathbf{N}}$.

Definition 9.2. The *q-odometer* is the topological dynamical system $(\mathbf{Z}_q, \tau)$ considered as a symbolic system.

This system is minimal and uniquely ergodic with the Haar measure as unique invariant measure. The Haar measure λ_q on $\mathbf{Z}_q$ identified with $\{0,1,\dots,q-1\}^{\mathbf{N}}$ is the product measure $\otimes(\frac{1}{q}\sum_{0\leq j\leq q-1}\delta_j)$.

We start with an important fact, valid more generally for a primitive and aperiodic substitution of constant length q, which asserts that the q-odometer is a factor of the system (X_ζ, μ, T); this is once more a consequence of the recognizability property.

Proposition 9.1. *If ζ is a primitive and aperiodic substitution of constant length q, there exists a factor map from (X_ζ, T) to the q-odometer.*

Definition 9.3. Let ζ be a primitive and aperiodic substitution of constant length q and $x \in X_\zeta$. We call a *code* for x, any q-adic integer κ such that, for every $n \geq 1$,

$$x \in T^{j_n} \zeta^n (X_\zeta).$$

Proof. We shall prove that μ-almost every $x \in X_\zeta$ admits a unique code $\kappa := \kappa(x)$. Referring to section 5.6, we consider $Y = X_\zeta \backslash \mathscr{D}$ the set of $y \in X_\zeta$ with a unique decomposition for every $k \geq 0$ into $y = T^k x$ with $x \in X_\zeta$. We showed that $\mu(X_\zeta \backslash Y) = 0$ and also that $(\mathscr{P}_n)$ with

$$\mathscr{P}_n = \{T^j \zeta^n [\alpha], \;\; \alpha \in A, \;\; 0 \leq j \leq q^n - 1\},$$

is a nested sequence of partitions of Y.

If now $x \in Y$, there exist, for every $n \geq 1$, a unique letter α_n and a unique integer $j_n \leq q^n - 1$ such that $x \in T^{j_n} \zeta^n [\alpha_n]$. Obviously, $j_{n+1} \equiv j_n \mod q^n$, and, if $j_{n+1} = j_n + k_n q^n$,

$$T^{j_n} \zeta^n \left(T^{k_n} \zeta[\alpha_{n+1}]\right) = T^{j_{n+1}} \zeta^{n+1} [\alpha_{n+1}] \subset T^{j_n} \zeta^n [\alpha_n]$$

which implies that

$$\zeta(\alpha_{n+1})_{k_n} = \alpha_n.$$

We thus may write, for every $n \geq 1$, $x = T^{j_n} x^{(n)}$ with $x^{(n)} = T^{-j_n} x \in \zeta^n(X_\zeta)$. The q-adic integer $\kappa(x) = (j_n)$ is a unique code for x, and this holds for every $x \in Y$. The so-defined map $x \mapsto \kappa(x)$ is clearly onto and the proposition is proved. □

Now, suppose that ζ, in addition, is a bijective substitution. The system (X_ζ, T) turns out to be semi-conjugate to a skew-product over the q-odometer, actually to a *s-points extension of the q-odometer*.

Theorem 9.1. *Let ζ be a bijective primitive substitution of constant length q on A with Card A=s. Then, the systems (X_ζ, μ, T) and $(\mathbf{Z}_q \times A, \lambda_q \times \lambda_A, S)$ are metrically isomorphic, where*

$$S(\kappa, \alpha) = (\kappa + 1, \psi_\kappa(\alpha))$$

and

$$\psi_\kappa = \phi_{j_n+1} \circ \phi_{j_n}^{-1}$$

if n is the first index for which $j_n < q^n - 1$. Note that $\phi_{j_n} = \phi_{k_0} \circ \cdots \circ \phi_{k_n}$ and the probability measure λ_A is equidistributed on A.

Proof. Retaining the notations of the previous proposition, we naturally consider $f : Y \to \mathbf{Z}_q \times A$ defined by

$$f(x) = (\kappa, x_0)$$

where $\kappa := \kappa(x)$ is the unique code of $x \in Y$ and x_0 the first letter of x. We aim to prove that f is a bijection up to negligible sets and then, that f exchanges the transformations T and S. The systems being uniquely ergodic, the theorem will follow.

$\star$ For λ_q- almost all $\kappa = (j_n) \in \mathbf{Z}_q$ and all $\alpha \in A$, we shall construct a unique $x \in X_\zeta$ such that $f(x) = (\kappa, \alpha)$. Since ζ is a bijective substitution, we may consider for every n the unique letter $\alpha_n \in A$ such that $\zeta^n(\alpha_n)_{j_n} = \alpha$. Putting then $\alpha_1 = \phi_{k_0}^{-1}(\alpha)$ and $\alpha_{n+1} = \phi_{k_n}^{-1}(\alpha_n)$, it can be proved by induction on $n \geq 1$ that

$$\zeta^{n+1}(\alpha_{n+1})_{j_{n+1}} = \zeta^n(\phi_{k_n}(\alpha_{n+1}))_{j_n} = \zeta^n(\alpha_n)_{j_n} = \alpha. \tag{9.1}$$

We restrict ourselves to the κ, infinitely many k_i of which being different from $q-1$. Then $j_n < q^n - 1$ for n large enough. Consider the finite word

$$\omega_n = T^{j_n}\zeta^n(\alpha_n) = \zeta^n(\alpha_n)_{[j_n+1,\, q^n[};$$

by (9.1), we have

$$(\omega_{n+1})_{j_{n+1}+\ell} = (\omega_n)_{j_n+\ell} \text{ as long as } j_n + \ell \leq q^n - 1,$$

and the sequence of finite words (ω_n) converges to a limit $x \in X_\zeta$ since $q^n - j_n \to \infty$, by our assumption on κ; but this occurs for λ_q- almost all κ and we have proved that f is one-to-one from Y to $\mathbf{Z}_q$ and "almost" onto.

$\star$ It remains to establish that for $x \in Y$, $f(Tx) = S(f(x)) = (\kappa + 1, x_1)$ if $x = x_0x_1\cdots$. Clearly this identity reduces to $\psi_\kappa(x_0) = x_1$. But remembering that $\kappa = \kappa(x)$ is the unique code for x, there exists $x' \in X_\zeta$, $x' = x_0'x_1'\cdots$ such that

$$x = T^{k_0}\zeta(x'), \quad \text{and} \quad x_0 = \phi_{k_0}(x_0').$$

If $k_0 < q-1$, $x_1 = \phi_{k_0+1}(x_0')$ and

$$\psi_\kappa(x_0) = (\phi_{k_0+1} \circ \phi_{k_0}^{-1})(x_0) = \phi_{k_0+1}(x_0') = x_1.$$

If $k_0 = q-1$, we must write $x = T^{k_1q+q-1}\zeta^2(x'')$ for some $x'' \in X_\zeta$, and this time, $x_0 = (\phi_{q-1} \circ \phi_{k_1})(x_0'')$. In the same way,

$$\psi_\kappa(x_0) = (\phi_{j_2+1} \circ \phi_{j_2}^{-1})(x_0) = \phi_{j_2+1}(x_0'') = x_1$$

if $k_1 + 1 < q-1$ and so on $\cdots$ The proof of the theorem is complete. □

Remark 9.1. In the non-bijective case, the system (X_ζ, T) is an extension of the q-odometer $(\mathbf{Z}_q, \tau)$ with $C(\zeta)$ (instead of s) fibers, where the column number $C(\zeta)$ has been defined in section 6.3.2 (see [68, 175]), but the factor map is not so easily explicit.

Examples

1. *Thue-Morse automaton.* In this case, $A = \{0,1\}$, $q = 2$, ϕ_0 is the identity on A and ϕ_1 is the permutation $(01) \to (10)$. If $\kappa = (j_n) \in \mathbf{Z}_2$, we compute easily ψ_κ to be ϕ_1^n where n is the first index such that $j_n < 2^n - 1$; indeed, this index n is characterized by $k_0 = k_1 = \cdots = k_{n-2} = 1$ and $k_{n-1} = 0$ if, as above, $j_n = k_0 + k_1 q + \cdots + k_{n-1} q^{n-1}$; thus $j_n = 2^{n-1} - 1$ and $\psi_\kappa = \phi_{2^{n-1}} \circ \phi_{2^{n-1}-1} = \phi_1 \circ \phi_1^{n-1}$, whence the claim.
   We now check that $S^k(0,0) = (k, m_k)$ where (m_n) is the Thue-Morse sequence on A. This is clear for the first components according to the iteration
   $$(0,0) \xrightarrow{S} (1,1) \xrightarrow{S} (01,1) \xrightarrow{S} (11,0) \xrightarrow{S} (001,1) \xrightarrow{S} (101,0) \cdots$$
   If now we set $S^k(0,0) = (k, u_k)$, observe that $u_k = u_{k-1}$ if the first zero binary digit of k occurs with an odd index, $u_k = 1 - u_{k-1}$ otherwise. But this property characterizes the Thue-Morse sequence too and $u_k = m_k$.
   The Thue-Morse dynamical system (X, T) is thus metrically isomorphic to the following skew product over the two-odometer : $(\mathbf{Z}_2 \times \{\pm 1\}, \tau_\varphi)$ where
   $$\tau_\varphi(z, \varepsilon) = (z+1, \varepsilon \cdot \varphi(z))$$
   with
   $$\varphi(z) = (-1)^{\sum_{n=0}^{\infty}((z+1)_n - z_n)}. \tag{9.2}$$
   One easily checks that
   $$\varphi(z) = (-1)^{k+1}$$
   where k is the index of first occurrence of zero in z, and that
   $$(-1)^{m_n} = \varphi^{(n)}(0) := \varphi(0) \cdots \varphi(n-1),\ n \geq 1.$$
   This Thue-Morse cocycle (9.2) satisfies the equation $\varphi(z) = \varphi(2z)\varphi(2z+1)$ and is entirely defined by $\varphi([0]) = -1$. This can fruitfully be compared with Ledrappier's result at the end of chapter 3. This representation of the Thue-Morse system has successfully been exploited in [137].
2. *Rudin-Shapiro automaton.* This automaton is not bijective and the column number $C(\zeta) = 2$. However, if κ codes $x \in X_\zeta$, clearly k_0 determines x_0 and X_ζ can be identified with $\{(\kappa, \alpha) \in \mathbf{Z}_2 \times A$ where $\alpha \in A_{k_0}\}$, with $A_0 = \{0,3\}$ and $A_1 = \{1,2\}$.
   The dynamical system generated by the Rudin-Shapiro sequence (r_n) in turn is metrically isomorphic to a similar skew product over the two-odometer $(\mathbf{Z}_2 \times \{\pm 1\}, \tau_\psi)$, with ψ satisfying $\psi(z) = \psi(2z)\psi(2z+1)$, $\psi([00]) = 1$ and

$\psi([01]) = -1$. This can be deduced from the arithmetical description of the sequence : $r_n = (-1)^{f(n)}$ where $f(n)$ counts the occurrences of the pattern 11 in the binary expansion of n; and conversely,

$$r_n = \psi^{(n)}(0) := \psi(0)\cdots\psi(n-1),\ n \geq 1.$$

Remark 9.2. It follows from proposition 3.15 that each correlation measure in both previous examples must be a pure measure.

9.1.2 Group Automaton

We shall associate to the bijective substitution ζ a substitution on a finite group, whose spectral study will appear to be easier. If we denote by G_A the group of permutations of A, each instruction ϕ_j acts on G_A in a natural way, setting for $\phi \in G_A$, $\phi_j(\phi) = \phi_j \circ \phi$. Now ζ induces a new substitution θ on the alphabet G_A by the formula

$$\theta(\phi) = \phi_0(\phi)\phi_1(\phi)\cdots\phi_{q-1}(\phi).$$

Notice that θ inherits some of the properties of ζ but might fail to be primitive. Whence the useful following remarks.

Remark 9.3. 1. Replacing, if necessary, ζ by some suitable ζ^j, we may assume that $\phi_0 = id$, the identity of the group G_A; it suffices, for that, to choose j as the order of the permutation ϕ_0. Thus, $\theta(\phi)$ begins with ϕ for every $\phi \in G_A$.

2. We now restrict the alphabet to the subgroup G of G_A generated by the permutations ϕ_j, $0 \leq j \leq q-1$, in other words, we preserve only the "letters" occurring in $\theta^\infty(id)$ since

$$\theta^\infty(id)_k = \phi_{k_0} \circ \cdots \circ \phi_{k_{n-1}} \quad \text{if} \quad k = k_0 + k_1 q + \cdots + k_{n-1}q^{n-1} < q^n.$$

The substitution θ is clearly irreducible on G : actually, for every pair (ϕ, ψ) of elements of G, $\psi\phi^{-1}$ occurs in $\theta^n(id)$ for n large enough, so that, ψ occurs in $\theta^n(id) \circ \phi = \theta^n(\phi)$. Combined with the hypothesis $\phi_0 = id$, this implies that θ is primitive.

From now on, for sake of simplicity, we assume $\phi_0 = id$ and we write ζ instead of ζ^j (which generates the same system). Note that for every α, $\zeta^\infty(\alpha)$ is a fixed point of ζ.

Definition 9.4. We denote by (X_G, T_G) the minimal and unique ergodic system associated to the primitive and bijective substitution θ on G.

Recall that X_G may be viewed as the orbit closure of $\theta^\infty(id)$ under the shift T_G on $G^{\mathbf{N}}$ (restricted to X_G). Observe first the following.

Lemma 9.1. *The unique T_G- invariant probability measure ν on X_G is uniformly distributed on the cylinders $[\phi]$, $\phi \in G$.*

Proof. As θ is primitive,

$$\nu([\phi]) = \lim_{N\to\infty} \frac{1}{q^N} \text{ Card } \{k < q^N,\ \theta^N(\psi)_k = \phi\}$$

for every $\psi, \phi \in G$. But $\theta^N(\psi)_k = \phi$ means $\theta^N(\psi\phi^{-1})_k = id$ and $\nu([\phi]) = \nu([id]) = 1/\text{Card } G$. □

The system (X_ζ, T) is clearly a *factor* of the system (X_G, T_G). Actually, fix $\alpha \in A$ and consider the projection map $\pi = X_G \to X_\zeta$, defined by

$$\pi(g) = g(\alpha), \quad \text{if } g = (g_n)_n \in G;$$

then π is continuous, onto and commutes with the shifts : $\pi \circ T_G = T \circ \pi$; from unique ergodicity of both systems, it follows that $\pi(\nu) = \mu$.

Proposition 9.2. *When the group G is abelian, the systems (X_ζ, T, μ) and (X_G, T_G, ν) are metrically isomorphic.*

Proof. We have just to show that the projection map π is one-to-one. Let g, g' be two elements of X_G satisfying $g(\alpha) = g'(\alpha)$; by primitivity of ζ, for every $\beta \in A$, there exists $\psi \in G$ such that $\psi(\alpha) = \beta$. Now, using the assumption on G, we have for every $n \geq 0$,

$$\begin{aligned} g_n(\beta) &= g_n(\psi(\alpha)) = \psi(g_n(\alpha)) = \psi(g'_n(\alpha)) = g'_n(\psi(\alpha)) \\ &= g'_n(\beta), \quad \text{for every } \beta \in A, \end{aligned}$$

and $g = g'$. □

In the non-abelian case, we can only prove that μ-almost $x \in X_\zeta$ admits a unique lifting $g \in X_G$, provided the first letter $g_0 \in G$ is given; indeed, if $x \in Y$, x admits a unique code κ and any $g \in G$ such that $\pi(g) = x$ admits the same code. But g is entirely determined by (κ, g_0) (theorem 9.1) and in turn, so is x.

$$\begin{array}{ccc} g \in X_G & \longrightarrow & (\kappa, g_0) \in \mathbf{Z}_q \times G \\ \pi \downarrow & & \pi \downarrow \\ x \in X_\zeta & \longrightarrow & (\kappa, x_0) \in \mathbf{Z}_q \times A \end{array}$$

Example 9.2. On $A = \{0, 1, 2\}$ consider the non-commutative automaton whose instructions are the permutations

$$\phi_0 = id, \quad \phi_1 = (012), \quad \phi_2 = (120).$$

Then $\pi : g \to g(0)$ maps both $\theta^\infty(\phi_0)$ and $\theta^\infty(\phi_1)$ onto $u = \zeta^\infty(0)$ and is not one-to-one.

The projection map π provides an isometric imbedding of $L^2(X_\zeta, \mu)$ into $L^2(X_G, \nu)$, defined by $\widetilde{f}(g) = f(\pi(g))$ on $L^2(X, \mu)$. If P is the orthogonal projection from $L^2(X_G, \nu)$ onto $L^2(X_\zeta, \mu)$, clearly $\sigma_{Ph} \ll \sigma_h$ for every $h \in L^2(X_G, \nu)$. Whence the already quoted remark in the general case :

Proposition 9.3. *The maximal spectral type of the system (X_G, T_G) dominates the one of the initial system (X_ζ, T).*

9.2 Spectral Study of Bijective Substitutions

We now wish to make the spectral analysis of the system (X_G, T_G) associated as above with the system (X_ζ, T), where ζ is a bijective and primitive substitution of length q. According to the foregoing, we shall get, in this way, a complete description of the spectrum of ζ in the abelian case, but only an approach of it for non-abelian G's.

Recall the notation $\sigma_{\alpha\beta}$ for the mixed correlation measure of both sequences $(\tau_\alpha(u_n))_n$ and $(\tau_\beta(u_n))_n$, $u = (u_n)_n$ being a fixed point of ζ and $\tau_\alpha : A \to \mathbf{C}$ defined by $\tau_\alpha(\beta) = \delta_{\alpha\beta}$ (the Kronecker symbol). In the following, if $(\phi_n)_n = \theta^\infty(id)$ and τ and τ' are two maps from G into $\mathbf{C}$, $\sigma_{\tau,\tau'}$ will denote the mixed correlation measure of both sequences $(\tau(\phi_n))_n$ and $(\tau'(\phi_n))_n$, and $\sigma_\tau := \sigma_{\tau,\tau}$. Remember that for nonnegative k,

$$\widehat{\sigma}_{\tau,\tau'}(k) = \lim_{N\to\infty} \frac{1}{q^N} \sum_{n+k<q^N} \tau(\phi_{n+k})\overline{\tau'(\phi_n)}. \tag{9.3}$$

The mixed correlation measure of the occurrences of $\phi, \psi \in G$ in the sequence $(\phi_n)_n$, denoted by $\sigma_{\phi,\psi}$, is characterized by

$$\widehat{\sigma}_{\phi,\psi}(k) = \lim_{N\to\infty} \frac{1}{q^N} \text{Card}\,\{n+k < q^N,\ \theta^N(id)_{n+k} = \phi,\ \theta^N(id)_n = \psi\}$$

for nonnegative k. But, we clearly have

$$\widehat{\sigma}_{\phi,\psi}(k) = \lim_{N\to\infty} \frac{1}{q^N} \text{Card}\,\{n+k < q^N,\ \theta^N(\psi^{-1})_{n+k} = \phi\psi^{-1},\ \theta^N(\psi^{-1})_n = id\}$$

so that, by the primitivity property of θ, the following identity holds for every ϕ, $\psi \in G$:

$$\sigma_{\phi,\psi} = \sigma_{\phi\psi^{-1},id}. \tag{9.4}$$

9.2.1 Abelian Case

We start with bijective and commutative automata (or constant-length substitutions). In this case, we observed that Σ can be diagonalized (chapter 8) and we are now in a position to exhibit a basis of functions with respect to which, σ has a diagonal form. Notice that the cardinality of G, denoted from now on by $|G|$, must be equal to s when G is abelian.

Proposition 9.4. *The maximal spectral type of the system (X_ζ, T) is generated by the s correlation measures σ_γ, where γ runs over $\widehat{G}$, the dual group of G.*

Proof. We already know that the systems (X_ζ, T) and (X_G, T_G) have identical spectrum (proposition 9.2). Thus, we may restrict our attention to the matrix Σ associated with the substitution θ. Every map $\tau : G \to \mathbf{C}$ can be decomposed into a finite Fourier sum

$$\tau = \frac{1}{|G|} \sum_{\gamma \in \widehat{G}} \widehat{\tau}(\gamma)\, \gamma;$$

therefore, σ_τ defined by (9.3), is a linear combination of correlation measures of the form $\sigma_{\gamma,\xi}$, with $\gamma, \xi \in \widehat{G}$ (also defined by (9.3)). On the other hand, we claim that, for arbitrary $\gamma, \xi \in \widehat{G}$,

$$\sigma_{\gamma,\xi} = \begin{cases} |G| \; \sum_{\phi \in G} \gamma(\phi)\, \sigma_{\phi, id} & \text{if} \quad \gamma = \xi \\ \\ 0 \qquad \text{otherwise} \end{cases} \tag{9.5}$$

Actually, for $k \geq 0$,

$$\begin{aligned} \widehat{\sigma}_{\gamma,\xi}(k) &= \lim_{N \to \infty} \frac{1}{q^N} \sum_{n+k<q^N} \gamma(\phi_{n+k}) \overline{\xi(\phi_n)} \\ &= \lim_{N \to \infty} \frac{1}{q^N} \sum_{n+k<q^N} \sum_{\substack{\phi_{n+k}=\phi \\ \phi_n=\psi}} \gamma(\phi) \overline{\xi(\psi)} \\ &= \sum_{\phi,\psi \in G} \gamma(\phi) \overline{\xi(\psi)}\; \widehat{\sigma}_{\phi,\psi}(k). \end{aligned}$$

Applying identity (9.4), we deduce that

$$\begin{aligned} \sigma_{\gamma,\xi} &= \sum_{\phi,\psi \in G} \gamma(\phi\psi^{-1}) \gamma(\psi) \overline{\xi}(\psi)\; \sigma_{\phi\psi^{-1}, id} \\ &= \sum_{\phi \in G} \Big(\sum_{\psi \in G} \gamma(\psi) \overline{\xi}(\psi) \Big) \gamma(\phi) \sigma_{\phi, id} \end{aligned}$$

and the orthogonality property of the character group leads to (9.5) □

This means that the correlation matrix Σ, associated with θ, can be diagonalized with respect to the basis consisting of the s-dimensional vectors $\{\gamma = (\gamma(\phi))_{\phi\in G}, \gamma \in \widehat{G}\}$.

We are now able to give the precise expression of those measures σ_γ.

Proposition 9.5. *For each $\gamma \in \widehat{G}$, σ_γ is the generalized Riesz product*

$$\sigma_\gamma = \prod_{n\geq 0} P_\gamma(q^n t)$$

where

$$P_\gamma(t) = \frac{1}{q}|\gamma(\phi_0) + \gamma(\phi_1)e^{it} + \cdots + \gamma(\phi_{q-1})e^{i(q-1)t}|^2.$$

Proof. The generalized Riesz products have been extensively studied in chapter 1. Since $\phi_0 = 1$ and

$$\phi_n = \phi_{n_0} \circ \phi_{n_1} \circ \cdots \circ \phi_{n_k}, \text{ if } n = n_0 + n_1 q + \cdots + n_k q^k$$

we have, by putting $v_n = \gamma(\phi_n)$,

$$v_{aq^p+b} = \gamma(\phi_{aq^p+b}) = \gamma(\phi_b \phi_a)$$

$$= v_b v_a \quad \text{if} \quad b < q^p.$$

This proves that the sequence $(v_n)_n$ is strongly q-multiplicative and we already identified the correlation measure of such a sequence as $\prod_{n\geq 0} P(q^n t)$ with $P(t) = \frac{1}{q}\Big|\sum_{j=0}^{q-1} v_j e^{ijt}\Big|^2$ in subsection 4.3.3. □

Recall that these measures obey a dichotomy law, so that, for $\gamma \neq \gamma' \in \widehat{G}$, σ_γ and $\sigma_{\gamma'}$ are either identical or mutually singular. Since $\sigma_{\gamma,\gamma'} = 0$ by (9.5), an identity $\sigma_\gamma = \sigma_{\gamma'}$ for some pair $\gamma \neq \gamma'$ would give rise to multiplicity ≥ 2 for the system (X_ζ, T). We shall discuss this problem in the last chapter, but we may already have an outline through the following examples.

Examples Consider the commutative automata $(A, (\phi_j)_j)$.

1. $A = \{0,1,2\}$, $\phi_0 = id$ and $\phi_1 = (201)$. Then $G = \{\phi_0, \phi_1, \phi_2\}$ where $\phi_2 = \phi_1^2 = (120)$. The mutually singular generalized Riesz products

$$\delta_0, \ \prod_{n\geq 0} \frac{1}{2}|1 + e^{2i\pi/3}e^{i2^n t}|^2 \ \text{and} \ \prod_{n\geq 0} \frac{1}{2}|1 + e^{-2i\pi/3}e^{i2^n t}|^2$$

generate the maximal spectral type (and the spectrum is simple).

2. $A = \{0,1\}$, $\phi_0 = \phi_1 = \cdots = \phi_{q-2} = id$ and $\phi_{q-1} = (10)$. Here $G = \mathbf{Z}/2\mathbf{Z}$ and the two generating measures are

$$\delta_0 \text{ and } \prod_{n \geq 0} \frac{1}{q} |1 + e^{iq^n t} + \cdots + e^{i(q-2)q^n t} - e^{i(q-1)q^n t}|^2$$

(the spectrum is simple again).

3. $A = \{0,1,2,3\}$, $\phi_0 = id$, $\phi_1 = (1032)$ and $\phi_2 = (3210)$. In this case, $G = \mathbf{Z}/2\mathbf{Z} \times \mathbf{Z}/2\mathbf{Z}$ and the mutually singular generating measures are

$$\delta_0, \prod_{n \geq 0} \frac{1}{3} |1 + e^{i3^n t} - e^{2i3^n t}|^2, \prod_{n \geq 0} \frac{1}{3} |1 - e^{i3^n t} + e^{2i3^n t}|^2$$

and

$$\prod_{n \geq 0} \frac{1}{3} |1 - e^{i3^n t} - e^{2i3^n t}|^2.$$

The mutual singularity of these measures can be deduced from corollary 1.1 and the simplicity of the spectrum follows.

4. $A = \{0,1,2\}$, $\phi_0 = id$, $\phi_1 = (120)$ and $\phi_2 = id$. In that case, $G = \mathbf{Z}/3\mathbf{Z}$ and the measures σ_γ are now δ_0, $\prod_{n \geq 0} P_\gamma(2^n t)$ and $\prod_{n \geq 0} P_{\gamma'}(2^n t)$ with

$$P_\gamma(t) = \frac{1}{3} |1 + e^{2i\pi/3} e^{it} + e^{2it}|^2,$$

and

$$P_{\gamma'}(t) = \frac{1}{3} |1 + e^{-2i\pi/3} e^{it} + e^{2it}|^2.$$

One easily checks that $P_\gamma = P_{\gamma'}$, hence $\sigma_\gamma = \sigma_{\gamma'}$, though $\sigma_{\gamma,\gamma'} = 0$ still holds. The spectral multiplicity is ≥ 2 in this case, in fact exactly equal to 2 [101].

It is not difficult to produce many such examples in which symmetry gives rise to spectral multiplicity 2. The question now is whether a multiplicity greater than 2 might happen with commutative automata ?

9.2.2 Non-Abelian Case

Now, suppose that G is a non-abelian group. The spectral study of the system (X_G, T_G) can be carried out in a similar way as we shall see, however, the description of the generating measures still has to be completed.

We denote by $\mathscr{R}$ the set of non-isomorphic unitary irreducible representations of G. Recall that $\rho = (\rho_{i,j})_{1 \leq ij \leq r} \in \mathscr{R}$ is an homomorphism from G into the set of unitary operators on some finite-dimensional vector space. The *trace* of $\rho \in \mathscr{R}$, sometimes denoted by χ_ρ and defined by

$$\chi_\rho(g) = \mathrm{Tr}(\rho(g)) \quad \text{for} \quad g \in G,$$

is called the *character of the representation*, and we shall see that these characters play the same role as the elements of $\widehat{G}$ in the abelian case. We refer to [225] for other classical notations and facts we shall make use of.

Theorem 9.2. *The maximal spectral type of the system (X_G, T_G) is generated by the correlation measures $\sigma_{\mathrm{Tr}(\rho)}$, where ρ runs over $\mathscr{R}$, the set of non-isomorphic unitary irreducible representations of G.*

Proof. If $\rho \in \mathscr{R}$, the entries ρ_{ij}, considered as functions on G, separate the points on G, and thus, span the space of mappings from G into $\mathbf{C}$. It follows that every spectral measure $\sigma_{f,h}$, where f and h are complex functions on G, is a linear combination of correlation measures of the form

$$\sigma_{\rho_{ij},\rho'_{k\ell}}, \quad \rho, \rho' \in \mathscr{R}.$$

We claim that

$$\sigma_{\rho_{ij},\rho'_{k\ell}} = \begin{cases} 0 & \text{if } \rho \neq \rho' \\ 0 & \text{if } \rho = \rho' \text{ and } j \neq \ell \\ \dfrac{|G|}{r} \displaystyle\sum_{\phi \in G} \rho_{ik}(\phi)\, \sigma_{\phi, id} & \text{if } \rho = \rho',\ j = \ell. \end{cases} \tag{9.6}$$

(r is the dimension of ρ).
Indeed, following the scheme of the abelian case, we get readily

$$\begin{aligned} \sigma_{\rho_{ij},\rho'_{k\ell}} &= \sum_{\phi,\psi \in G} \rho_{ij}(\phi)\overline{\rho'_{k\ell}(\psi)}\, \sigma_{\phi,\psi} \\ &= \sum_{\phi,\psi \in G} \Big(\sum_p \rho_{ip}(\phi\psi^{-1})\rho_{pj}(\psi)\Big)\overline{\rho'_{k\ell}(\psi)}\, \sigma_{\phi\psi^{-1}, id} \end{aligned}$$

by (9.4). But, ρ' being unitary, $\overline{\rho'_{k\ell}(\psi)} = \rho'_{\ell k}(\psi^{-1})$ and

$$\sigma_{\rho_{ij},\rho'_{k\ell}} = \sum_p \Big(\sum_{\phi \in G} \rho_{ip}(\phi)\sigma_{\phi, id}\Big)\Big(\sum_{\psi \in G} \rho_{pj}(\psi)\rho'_{\ell k}(\psi^{-1})\Big).$$

From Schur's lemma [225], we deduce

$$\sum_{\psi \in G} \rho_{pj}(\psi)\rho'_{\ell k}(\psi^{-1}) = \begin{cases} 0 & \text{if} \quad \rho \neq \rho' \\ 0 & \text{if} \quad \rho = \rho', \quad \text{either} \quad p \neq k \text{ or } j \neq \ell \\ \dfrac{|G|}{r} & \text{if} \quad \rho = \rho', \quad p = k \text{ and } j = \ell. \end{cases}$$

and our claim (9.6) follows. Observe that the measure $\sigma_{\rho_{ij},\rho_{kj}}$ does not depend on j. In particular $\sigma_{\rho_{ij}} = \sigma_{\rho_{ii}}$ for every i, j and the maximal spectral type of (X_G, T_G) is generated by the $(\sigma_{\rho_{ii}})$'s. Besides,

$$\begin{aligned} \sigma_{\mathrm{Tr}(\rho)} &= \sigma_{\sum_i \rho_{ii}} = \textstyle\sum_{i,j} \sigma_{\rho_{ii},\rho_{jj}} \\ &= \textstyle\sum_i \sigma_{\rho_{ii}} \quad \text{according to } (9.6) \end{aligned}$$

and the proof is complete. □

We now establish some properties of the generating measures $\sigma_{\mathrm{Tr}(\rho)}$, $\rho \in \mathscr{R}$.

First of all, it follows from (9.6) that

$$\sigma_{\mathrm{Tr}(\rho),\mathrm{Tr}(\rho')} = 0 \quad \text{if} \quad \rho \neq \rho'$$

and

$$\sigma_{\mathrm{Tr}(\rho)} = \frac{|G|}{r} \sum_{\phi \in G} \mathrm{Tr}(\rho(\phi))\sigma_{\phi, id}. \tag{9.7}$$

In particular, *the characters $Tr(\rho)$, $\rho \in \mathscr{R}$, are U_G-orthogonal functions* (U_G the unitary operator associated to T_G). Moreover,

Theorem 9.3. *If $\rho \in \mathscr{R}$, then $\sigma_{\mathrm{Tr}(\rho)}$ is a probability measure, strongly mixing with respect to the q-shift S_q (q-strongly mixing in short).*

Proof. Fix $\rho \in \mathscr{R}$ and put $r = \dim \rho$. By (9.7),

$$\begin{aligned} \widehat{\sigma}_{\mathrm{Tr}(\rho)}(0) &= \frac{|G|}{r} \sum_{\phi \in G} \mathrm{Tr}(\rho(\phi))\widehat{\sigma}_{\phi, id}(0) \\ &= \frac{|G|}{r} \mathrm{Tr}(\rho(\mathrm{id}))\frac{1}{|G|} \quad \text{with lemma 9.1} \\ &= 1. \end{aligned}$$

Now, consider the sequence of $r \times r$ matrices $M_m := \rho(\phi_m)$, $m \geq 0$, where ϕ_m, as above, is defined by $\phi_{m_0+m_1q+\cdots+m_kq^k} = \phi_{m_0}\phi_{m_1}\cdots\phi_{m_k}$. We easily check that

$$M_{aq^p+b} = M_b\, M_a \quad \text{for} \quad a \geq 0,\ b < q^p. \tag{9.8}$$

The hermitian matrix of measures σ_M defined (if possible) by

$$\widehat{\sigma_M}(k) = \lim_{N\to\infty} \sum_{n+k<q^N} M_{n+k}\, M_n^*, \quad k \geq 0,$$

is a positive definite matrix of measures (chapter 7), related to the correlation measures $\sigma_{\rho_{ij},\rho_{k\ell}}$, $\rho \in \mathscr{R}$, in the following way :

$$\widehat{(\sigma_M)}_{ij}(k) = \lim_{N\to\infty} \frac{1}{q^N} \sum_{n+k<q^N} \sum_p \rho_{ip}(\phi_{n+k})\overline{\rho_{jp}(\phi_n)}$$

$$= \sum_p \widehat{\sigma}_{\rho_{ip},\rho_{jp}}(k).$$

Applying (9.6) again, we deduce that

$$(\sigma_M)_{ij} = |G| \sum_{\phi\in G} \rho_{ij}(\phi)\, \sigma_{\phi,id}, \tag{9.9}$$

which, combined with (9.7), gives us in particular $\mathrm{Tr}(\sigma_M) = n\sigma_{\mathrm{Tr}(\rho)}$. The multiplicativity property (9.8) of the sequence (M_m) gives rise to the following

$$\lim_{p\to\infty} \mathrm{Tr}(\sigma_M)\hat{}(aq^p+b) = \mathrm{Tr}(\widehat{\sigma}_M(b)\, \widehat{\sigma}_M(a)), \quad a,b \geq 0$$

which does not lead to the claim. It is more advisable to write, using (9.9),

$$(\sigma_M)_{ij}\hat{}(aq^p+b) =$$

$$\lim_{N\to\infty} \frac{|G|}{q^N} \sum_{\phi\in G} \rho_{ij}(\phi)\ \mathrm{Card}\,\{n+aq^p+b<q^N,\ \phi_{n+aq^p+b}=\phi,\ \phi_n=id\}.$$

Denoting the above Card by C_N and setting $n = mq^p + \ell$, we get the following estimate :

$$C_N \sim \mathrm{Card}\,\{\ell<q^p-b, m+a<q^{N-p},\ \phi_{\ell+b}\phi_{m+a}=\phi,\ \phi_\ell\phi_m=id\}+$$

$$\mathrm{Card}\,\{q^p-b\leq\ell<q^p, m+a+1<q^{N-p},\ \phi_{\ell+b-q^p}\phi_{m+a+1}=\phi,\ \phi_\ell\phi_m=id\}. \tag{9.10}$$

The first term in (9.10) is also

$$\sum_{\psi,\chi\in G} \mathrm{Card}\,\{m+a<q^{N-p},\ \phi_{m+a}=\psi,\ \phi_m=\chi\}\times$$

$$\mathrm{Card}\,\{\ell<q^p-b,\ \phi_{\ell+b}\psi=\phi,\ \phi_\ell\chi=id\},$$

so that, by dividing by q^N and taking the limit on N, this term reduces to

$$\sum_{\psi,\chi\in G} \widehat{\sigma}_{\psi,\chi}(a)\,\frac{1}{q^p}\mathrm{Card}\,\{\ell < q^p - b,\ \phi_{\ell+b} = \phi\psi^{-1},\ \phi_\ell = \chi^{-1}\}.$$

Proceeding in the same way with the second term, we get a quantity which vanishes as p goes to infinity; finally, turning back to σ_M, we obtain that

$$L := \lim_{p\to\infty} (\sigma_M)_{ij}\widehat{\ }(aq^p + b) = |G| \sum_{\phi,\chi,\psi\in G} \rho_{ij}(\phi)\widehat{\sigma}_{\psi,\chi}(a)\widehat{\sigma}_{\phi\psi^{-1},\chi^{-1}}(b);$$

setting $\phi\psi^{-1}\chi = g$ and using once more (9.4), the right-hand side is easily seen to factorize as follows

$$L = |G|\textstyle\sum_{g,\chi,\psi\in G}\left(\sum_k \rho_{ik}(g)\rho_{kj}(\chi^{-1}\psi)\right)\widehat{\sigma}_{\psi\chi^{-1},id}(a)\widehat{\sigma}_{g,id}(b)$$

$$= \textstyle\sum_k (\sigma_M)_{ik}\widehat{\ }(b)\cdot\sum_{\chi,\psi\in G}\rho_{kj}(\chi^{-1}\psi)\widehat{\sigma}_{\psi\chi^{-1},id}(a)$$

according to (9.9). It remains to compare the last sum with $\widehat{\sigma}_{\mathrm{Tr}(\rho)}(a)$. So we put $\psi = h\chi$ in this sum which therefore becomes

$$\sum_{\chi,h\in G} \rho_{kj}(\chi^{-1}h\chi)\widehat{\sigma}_{h,id}(a) = \sum_h \widehat{\sigma}_{h,id}(a)\Big(\sum_{\ell,m}\rho_{\ell m}(h)\sum_{\chi\in G}\rho_{k\ell}(\chi^{-1})\rho_{mj}(\chi)\Big);$$

thanks to Schur's lemma, this expression inside brackets reduces to

$$\begin{cases} 0 & \text{if} \quad k \neq j \\ \dfrac{|G|}{r}\displaystyle\sum_\ell \rho_{\ell\ell}(h) & \text{if} \quad k = j \end{cases}$$

We finally get for L the value

$$\frac{|G|}{r}\sum_h \widehat{\sigma}_{h,id}(a)\sum_\ell \rho_{\ell\ell}(h)$$

which is nothing but $\widehat{\sigma}_{\mathrm{Tr}(\rho)}(a)$ in view of (9.9). We summarize all those computations by the next formula :

$$\lim_{p\to\infty} (\sigma_M)_{ij}\widehat{\ }(aq^p + b) = (\sigma_M)_{ij}\widehat{\ }(b)\ \widehat{\sigma}_{\mathrm{Tr}(\rho)}(a). \tag{9.11}$$

This can be written in the more compact form

$$\chi_a\sigma_M = \sigma_M\,A$$

where $\chi_a = \lim_{p\to\infty} \gamma_{aq^p}$ in $\overline{\Gamma}(M)$ and A is the scalar matrix λI with $\lambda = \widehat{\sigma}_{\mathrm{Tr}(\rho)}(a)$. In particular,

$$\chi_a Tr(\sigma_M) = \widehat{\sigma}_{\mathrm{Tr}(\rho)}(a)\ Tr(\sigma_M)$$

and

$$\chi_a \sigma_{\mathrm{Tr}(\rho)} = \widehat{\sigma}_{\mathrm{Tr}(\rho)}(a)\sigma_{\mathrm{Tr}(\rho)}$$

for every $\rho \in \mathscr{R}$. The theorem is proved. □

Let c be the number of conjugacy classes of G; c is the cardinality of the set $\mathscr{R}$ and the c spectral measures $\sigma_{\mathrm{Tr}(\rho)}$ are pairwise either equal or mutually singular. The maximal spectral type of the system (X_G, T_G), henceforth that one of the initial system (proposition 9.3), is generated by those measures. To conclude,

Corollary 9.1. *The maximal spectral type of the system arising from a bijective substitution is generated by* at most *c q-strongly mixing probability measures of the form* $\sigma_{\mathrm{Tr}(\rho)}$, *ρ unitary irreducible representations of G.*

The study of the correlation measures $\sigma_{\rho_{ij},\rho_{kj}}$ given by (9.6), shows that the system (X_G, ν, T_G) never happens to have simple spectrum. More precisely,

Proposition 9.6. *The spectral multiplicity of the system* (X_G, ν, T_G) *is at least* $\sup_{\rho\in\mathscr{R}}(\dim\rho) \geq 2$.

Proof. Fix $\rho \in \mathscr{R}$ and $i \leq r = \dim\rho$. From (9.6), $\sigma_{\rho_{ij},\rho_{i\ell}} = 0$ if $j \neq \ell$ but with

$$\sigma_{\rho_{ij}} = \sigma_{\rho_{i\ell}} = \frac{|G|}{r} \sum_{\phi\in G} \rho_{ii}(\phi)\ \sigma_{\phi,id} = \sigma_{\rho_{ii}}$$

as already observed. This means that the functions $\rho_{i\ell}$, $1 \leq \ell \leq r$, are U_G-orthogonal, with identical spectral measures and the spectral multiplicity of the system is not less than $r = \dim\rho$ (corollary 2.10), whence the proposition. □

Remark 9.4. This estimate of the spectral multiplicity m_G can be improved by investigating more precisely the spectral measures $\sigma_{\rho_{ii}}$ since $\sigma_{\mathrm{Tr}(\rho)} = \sum_i \sigma_{\rho_{ii}}$. Some of them may happen to be identical. In this case, fixing $j \leq r$, for every $i \neq k$ such that $\sigma_{\rho_{ii}} = \sigma_{\rho_{kk}} = \sigma$, and for every $\ell \neq j$, we have together

$$\sigma_{\rho_{i\ell},\rho_{kj}} = 0 \quad \text{and} \quad \sigma_{\rho_{i\ell}} = \sigma_{\rho_{kj}} = \sigma.$$

This implies that $m_G \geq \sup_{\rho\in\mathscr{R}}(p(\dim\rho - 1))$ if the same measure σ appears $p \leq r$ times. In addition, distinct representations may happen to give rise to the same correlation measure $\sigma_{\mathrm{Tr}(\rho)}$, so that, exact estimate of m_G seems out of reach at the moment.

Example 9.3. Let $A = \{0,1,2\}$, $\phi_0 = id$, $\phi_1 = (021)$, $\phi_2 = (120)$. In this case $G = S_3$ the symmetric group, and we enumerate $G = \{id, c_0, c_1, c_2, r, r^{-1}\}$ where c_i is the permutation fixing i and r is the cyclic permutation (120).

There are three conjugacy classes in G, $C_0 = \{id\}$, $C_1 = \{c_0, c_1, c_2\}$ and $C_2 = \{r, r^{-1}\}$, and thus, three irreducible unitary representations ρ_0, ρ_1, ρ_2 with dimension $1, 1, 2$ respectively; they are defined by

$$\rho_0(g) = 1 \quad \text{for all } \ g \in G,$$

$$\rho_1(g) = \varepsilon(g) \ \text{the signature of } \ g$$

and

$$\rho_2(r) = \begin{pmatrix} \omega & 0 \\ 0 & \overline{\omega} \end{pmatrix}, \quad \rho_2(c_0) = \begin{pmatrix} 0 & 1 \\ 1 & 0 \end{pmatrix}, \quad \rho_2(c_1) = \begin{pmatrix} 0 & \overline{\omega} \\ \omega & 0 \end{pmatrix},$$

$$\rho_2(c_2) = \begin{pmatrix} 0 & \omega \\ \overline{\omega} & 0 \end{pmatrix} \quad \text{where} \quad \omega = e^{2i\pi/3}.$$

We denote by $\chi_i := \mathrm{Tr}(\rho_i)$, the character of the representation ρ_i. Then

$$\chi_0(g) = 1 \quad \text{for all } \ g \in G,$$

$$\chi_1(id) = 1 = \chi_1(r) \qquad \text{and} \quad \chi_1(c_i) = -1, \ \ i = 0, 1, 2,$$

$$\chi_2(id) = 2, \ \chi_2(r) = -1 \quad \text{and} \quad \chi_2(c_i) = 0, \ \ i = 0, 1, 2$$

The substitution θ, associated with ζ, is defined on G by

$$\theta(\phi) = \phi \ c_0(\phi) \ r(\phi)$$

with our notations. Coding the permutations by

$$c_0 = 0, \ c_1 = 1, \ c_2 = 2, \ r = 3, \ r^{-1} = 4, \ \text{and id} = 5$$

and using the group law of S_3, we get a symbolic representation for θ of the form

$$\begin{array}{|c|} \hline 0 \\ 1 \\ 2 \\ \hline \end{array} \longrightarrow \begin{array}{|c|} \hline 0 \\ 1 \\ 2 \\ \hline \end{array} \begin{array}{c} 5 \\ 3 \\ 4 \end{array} \begin{array}{|c|} \hline 2 \\ 0 \\ 1 \\ \hline \end{array}$$

$$\begin{array}{|c|} \hline 3 \\ 4 \\ 5 \\ \hline \end{array} \longrightarrow \begin{array}{c} 3 \\ 4 \\ 5 \end{array} \begin{array}{|c|} \hline 1 \\ 2 \\ 0 \\ \hline \end{array} \begin{array}{c} 4 \\ 5 \\ 3 \end{array}$$

The maximal spectral type of the system (X_G, ν, T_G) is generated by the correlation measures σ_{χ_i} of the sequences $(\chi_i(\phi_n))_n$, for $i = 0, 1, 2$. Clearly $\sigma_{\chi_0} = \delta_0$ and we claim that $\sigma_{\chi_1} = \delta_\pi$. Actually, noticing that $\chi_1 = -1$ on the class $C_0 = \{0,1,2\}$ and 1 elsewhere, we deduce from the rule of the game for θ that $\chi_1(\phi_n) = (-1)^n$, whence $\sigma_{\chi_1} = \delta_\pi$.

If the last measure σ_{χ_2} were discrete, the spectrum of the initial system in turn would be purely discrete; but ζ is pure and admits no coincidence. Thus σ_{χ_2} must be continuous. Letting $u_n = \chi_2(\phi_n)$, it follows from the above diagram that $u_{2n+1} = 0$ and that the first components are

$$2\,0-1\,0\,2\,0-1\,0-1\,0\,2\,0\,2\,0-1\,0-1\,0-1\,0-1\,0-1\,0-1\,0\,2\,0\,2\,0\,2\,0-1\,0-1\,0\,2\,0-1\,0\,2\,0-1\,0-1\,0-1\cdots$$

Also, σ_{χ_2} is 3-strongly mixing and satisfies $\widehat{\sigma}_{\chi_2}(2n+1) = 0$ for every n. We shall identify this measure in the last chapter.

The sequence $((-1)^n)_n$ does not belong to X_ζ whose height is one, and thus, the maximal spectral type of the initial system (X_ζ, T) is generated by σ_{χ_0} and σ_{χ_2}.

Chapter 10
Maximal Spectral Type of General Automata

In this chapter, we prove the main theorem relative to the spectral study of primitive and aperiodic substitutions of length q or q-automata. Since only the continuous part of the spectrum has to be described, we may restrict our attention to pure substitutions without loss of generality (according to item 6.3.1.2); we shall get the following : the maximal spectral type is generated by $k \leq s$ probability measures which are strongly mixing with respect to the q-shift S_q (in case the height is one). To get this result, we associate to the substitution ζ, a new substitution defined on $A \times A$, the alphabet consisting in pairs of letters of A, and whose correlation matrix always enjoys wonderful properties.

10.1 The Coincidence Matrix C

Throughout this chapter, ζ will denote a primitive and aperiodic substitution of length q. In chapter 8, we obtained an identity satisfied by the correlation matrix Σ in the most general case (8.7), a special form of which being

$$\begin{aligned}\widehat{\Sigma}_n(aq+b) &= \frac{1}{q}\sum_{m=0}^{q-b-1} (\widehat{R}(-m-b))^* \, \widehat{\Sigma}_{n-1}(a) \, \widehat{R}(-m) \\ &+ \frac{1}{q}\sum_{m=q-b}^{q-1} (\widehat{R}(-m-b+q))^* \, \widehat{\Sigma}_{n-1}(a+1) \, \widehat{R}(-m),\end{aligned}$$

by specifying $p = 1$; now, letting n go to infinity, we get the relation

$$\widehat{\Sigma}(aq+b) = \frac{1}{q}\sum_{m=0}^{q-b-1} {}^t R_{m+b} \, \widehat{\Sigma}(a) \, R_m + \frac{1}{q}\sum_{m=q-b}^{q-1} {}^t R_{m+b-q} \widehat{\Sigma}(a+1) \, R_m, \tag{10.1}$$

by definition of R. The matrix Σ is no longer q-invariant, and more precisely, taking $b = 0$ in (10.1), we see that

M. Queffélec, *Substitution Dynamical Systems – Spectral Analysis: Second Edition*, Lecture Notes in Mathematics 1294, DOI 10.1007/978-3-642-11212-6_10,
© Springer-Verlag Berlin Heidelberg 2010

$$\widehat{\sigma}_{\alpha\beta}(aq) = \frac{1}{q} \sum_{\gamma,\delta \in A} C^{\gamma\delta}_{\alpha\beta}\, \widehat{\sigma}_{\gamma\delta}(a) \tag{10.2}$$

where

$$C^{\gamma\delta}_{\alpha\beta} = \sum_{j=0}^{q-1} (R_j)_{\gamma\alpha}\, (R_j)_{\delta\beta}. \tag{10.3}$$

Recalling that $(R_j)_{\alpha\beta} = 1$ if $\phi_j(\alpha) = \beta$ and 0 otherwise, we see that

$$(R_j)_{\gamma\alpha}\, (R_j)_{\delta\beta} = 1 \text{ if } \phi_j(\gamma) = \alpha \text{ and } \phi_j(\delta) = \beta,$$

$$= 0 \text{ otherwise}$$

so that, $C^{\gamma\delta}_{\alpha\beta}$ is the number of occurrences of α in $\zeta(\gamma)$ and β in $\zeta(\delta)$ *on the same place*.

Definition 10.1. We call the *coincidence matrix* of ζ and denote by $C := C(\zeta)$, the $s^2 \times s^2$ matrix whose entries are $C^{\gamma\delta}_{\alpha\beta}$, with (α,β) and (γ,δ) running on $A \times A$ (the notations come from the tensor calculus).

With help of these notations, (10.2) results into

$$S_q(\Sigma) = \frac{1}{q} C\, \Sigma, \tag{10.4}$$

$S_q(\Sigma)$ denoting the pullback of Σ under the q-shift.

10.1.1 Properties of C

In the sequel, we identify the alphabet A with $\{0,\dots,s-1\}$.

1. *C is the composition matrix of some substitution.* Consider in fact the substitution η defined on $A \times A$ by

$$\eta\binom{\alpha}{\beta} := \binom{\zeta(\alpha)}{\zeta(\beta)} = \binom{\zeta(\alpha)_0}{\zeta(\beta)_0}\binom{\zeta(\alpha)_1}{\zeta(\beta)_1}\cdots\binom{\zeta(\alpha)_{q-1}}{\zeta(\beta)_{q-1}}.$$

η is a substitution of length q on $A \times A$ and if $M := M(\eta)$, $M^{\gamma\delta}_{\alpha\beta}$ is the number of occurrences of $\binom{\alpha}{\beta}$ in the word $\eta\binom{\gamma}{\delta}$, that is to say $M^{\gamma\delta}_{\alpha\beta} = C^{\gamma\delta}_{\alpha\beta}$. (One allows oneself to write linearly $\eta((\alpha,\beta))$ instead of $\eta\binom{\alpha}{\beta}$ if no confusion with the two-words in $\mathcal{L}_\zeta$ could arise).

The dominant eigenvalue of C is thus $\theta = q$ and

$$\sum_{\alpha,\beta \in A} C^{\gamma\delta}_{\alpha\beta} = q \quad \text{for all } (\gamma,\delta) \in A \times A.$$

But C may even fail to be irreducible, and the multiplicity of the eigenvalue q is ≥ 1.

2. *C or η possesses k ergodic classes with $1 \leq k \leq s$.*
We recall the notion of ergodic class in this context. If $(\alpha,\beta) \in A \times A$, $O((\alpha,\beta))$ denotes the orbit-set of (α,β) under η, that is the set of all pairs of letters occurring in some $\eta^n((\alpha,\beta)), n \geq 1$. Such minimal sets with respect to inclusion are disjoint sets since, obviously, $O((\alpha,\beta))$ contains $O((\gamma,\delta))$ for every (γ,δ) in $O((\alpha,\beta))$. These minimal orbit-sets are called *ergodic classes of C* and denoted by $E_0, E_1, \dots, E_{k-1}$. E_0, the orbit-set of $(0,0)$ (or any (β,β)), is the class of all $(\alpha,\alpha), \alpha \in A$ and always appears as an ergodic class of C. The class of all remaining pairs will be denoted by T (T for transient); now, if $(\alpha,\beta)) \in T$, $O((\alpha,\beta))$ need intersect one of the ergodic classes $E_j, j \geq 1$, ; we are thus led to decompose T into

$$T = T_0 \cup T_1 \cup \dots \cup T_{k-1} \cup T'$$

where T_j, for every $0 \leq j \leq k-1$, consists in pairs of T ultimately absorbed by E_j, while remaining pairs of T' are visiting T infinitely often under η.

It remains to prove that $k \leq s$.
From the primitivity hypothesis on ζ, every orbit must have at least s elements since all the letters necessarily appear in each component of the sequence $\eta^n((\alpha,\beta))$, $n \geq 0$. From the identity

$$A \times A = \left(\bigcup_{j=0}^{k-1} E_j\right) \cup T$$

we derive $s^2 \geq ks$, whence $k \leq s$.

It would be interesting to have a more precise upper bound for k, in particular in the bijective case. A related question is to decide in which case the E_j are symmetric, (that is $(\alpha,\beta) \in E_j$ if and only if $(\beta,\alpha) \in E_j$).

3. *By changing ζ into a suitable ζ^j if necessary, C restricted to any ergodic class is primitive* . When restricted to any ergodic class, C is irreducible. Actually, let E be an ergodic class. If $(\alpha,\beta) \in E$, then $O((\alpha,\beta)) = E$ by the minimality of E, so that, for every $(\gamma,\delta) \in E$, $(\gamma,\delta) \in \eta^n(\alpha,\beta)$ for some $n \geq 1$ already. Now, by changing ζ into a suitable ζ^j if necessary, we can find in any ergodic class, an element (α,β) such that $\eta((\alpha,\beta))$ begins with (α,β). This gives the claim (see chapter 4). Note that C restricted to E_0 is always a primitive matrix since $\eta((0,0))$ begins with $(0,0)$. Besides, $C_{|E_0} = M(\zeta)$.

From now on, we write ζ instead of ζ^j, that is C is assumed to be primitive in restriction to any ergodic class.

4. C gives a few information about the spectrum of the system (X_ζ, T).

Proposition 10.1. *The following are equivalent :*

(a) ζ admits a discrete spectrum.

(b) q is a simple eigenvalue of C.

(c) E_0 is the only ergodic class of C.

Lemma 10.1. *The multiplicity of the Perron-Frobenius eigenvalue q of C is exactly k, the number of ergodic classes of C.*

Proof of the lemma. This is a consequence of the above property 3. Denote by C_j the matrix C restricted to the ergodic class E_j, for $0 \leq j \leq k-1$, and by C_T, the restriction of C to the transient pairs T. There exists a permutation matrix P such that

$$P^{-1}CP = \begin{pmatrix} C_0 & & & & & N_0 \\ & C_1 & & 0 & & \cdot \\ & & \cdot & & & \cdot \\ & & & \cdot & & \cdot \\ & 0 & & & \cdot & \cdot \\ & & & & C_{k-1} & N_{k-1} \\ 0 & \cdot & \cdot & \cdot & \cdot & C_T \end{pmatrix}$$

where the matrices N_j are not all together equal to zero if C_T is not zero. C/q is a row-stochastic matrix (or C is q-stochastic) and thus, using Hadamard's formula for instance, every eigenvalue is easily seen to have modulus $< q$. Also, each matrix C_j being q-row-stochastic and primitive must admit q as a simple eigenvalue, whence the lemma. □

Lemma 10.2. *Let $B = (b_{ij})$ be an $n \times n$ q-row-stochastic matrix, with nonnegative integral entries such that $b_{1j} \geq 1$ for $j = 1, \ldots, n$. Then B admits q as a simple eigenvalue.*

Proof of the lemma. Let $x = (x_1, \ldots, x_n)$ a left eigenvector of B corresponding to the eigenvalue q. For $j = 1, \ldots, n$,

$$\sum_{i=2}^{n}(x_i - x_1)b_{ij} = \sum_{i=1}^{n} x_i b_{ij} - x_1 \sum_{i=1}^{n} b_{ij}$$

$$= q(x_j - x_1),$$

and the vector with components $(x_j - x_1)_{2 \leq j \leq n}$ is an eigenvector of the truncated matrix $\widetilde{B} = (b_{ij})_{2 \leq i,j \leq n}$ corresponding to the eigenvalue q. But

$$\sum_{i=2}^{n} b_{ij} = q - b_{1j} \leq q - 1 \quad \text{for every } j$$

and q cannot be an eigenvalue of $\widetilde{B}$ unless $x_j = x_1$ for $j = 2, \ldots, n$. This means that q has multiplicity one. □

Proof of the proposition. We shall make use of Dekking's criterion for a constant-length substitution to be discrete (theorem 6.6). Recall that, assuming the height to be one, ζ has a discrete spectrum if and only if ζ admits a coincidence, which means that there exist n and $k < q^n$ such that $\zeta^n(\alpha)_k = \zeta^n(\beta)_k$ for every $\alpha, \beta \in A$.
Now assume (a) and n as above. Since $C(\zeta^n) = C^n(\zeta)$, (a) is equivalent to

$$(C^n)^{\alpha\beta}_{\gamma\gamma} \geq 1 \text{ for some } \gamma \in A \text{ and for every } \alpha, \beta \in A.$$

Then (b) follows from lemma 10.2.

The implication (b)$\Longrightarrow$(c) is an immediate consequence of lemma 10.1.

Finally, suppose that E_0 is the only ergodic class of C. C is similar to some matrix

$$\begin{pmatrix} C_0 & N_0 \\ 0 & C_T \end{pmatrix}$$

with $N_0 \neq 0$. Now C_0 is nothing but $M(\zeta)$ as already observed so that C^n/q^n converges to a projection operator P whose $s^2 \times s^2$ matrix is

$$\begin{pmatrix} \mu([0]) & \cdot & \cdot & \cdot & \cdot & \cdot & \mu([0]) \\ \cdot & & & & & & \cdot \\ \cdot & & & & & & \cdot \\ \cdot & & & & & & \cdot \\ \mu([s-1]) & \cdot & \cdot & \cdot & \cdot & \cdot & \mu([s-1]) \\ 0 & \cdot & \cdot & \cdot & \cdot & \cdot & 0 \\ \cdot & & & & & & \cdot \\ \cdot & & & & & & \cdot \\ \cdot & & & & & & \cdot \\ 0 & \cdot & \cdot & \cdot & \cdot & \cdot & 0 \end{pmatrix}$$

since $(\mu([\alpha]))_{\alpha \in A}$ is the positive normalized eigenvector of M corresponding to the eigenvalue q. For n large enough, the entries in the first s lines of C^n must be positive integers and ζ admits a coincidence, whence (a). □

10.1.2 Examples

1. Observe that E_0 is the unique ergodic class of $C(\zeta)$ where ζ is defined by $\zeta(0) = 012$, $\zeta(1) = 020$, $\zeta(2) = 201$.

2. If ζ is a *bijective* substitution, C is both q-row-stochastic and q-line-stochastic; since each C_j enjoys this property, T must be empty.

For ζ defined by $\zeta(0) = 01$, $\zeta(1) = 12$, $\zeta(2) = 20$, C admits three ergodic classes : E_0, $E_1 = \{(0,1)(1,2)(2,0)\}$ and $E_2 = \{(1,0)(2,1)(0,2)\}$.
To the substitution ζ defined by $\zeta(0) = 001$, $\zeta(1) = 122$, $\zeta(2) = 210$, studied at the end of chapter 9, correspond two ergodic classes only, E_0 and $E_1 = \{(\alpha,\beta),\ \alpha \neq \beta\}$.

3. *Rudin-Shapiro automaton.* In this example, C possesses two ergodic classes, E_0 and $E_1 = \{(0,3)(2,1)(1,2)(3,0)\}$; the blocks $C_0 = C_1 = M(\zeta)$ while $C_T = 0$.

4. *Q-mirror automaton.* This is the automaton associated to the Q-mirror sequences and ζ defined by $\zeta(0) = 01$, $\zeta(1) = 20$, $\zeta(2) = 23$, $\zeta(3) = 02$. In this case too, C has two ergodic classes, E_0 and $E_1 = \{(0,2)(3,1)(1,3)(2,0)\}$ with $C_0 = C_1 = M(\zeta)$. But the class of transient pairs, T, is divided into $T_0 = \{(0,3)(2,1)(1,2)(3,0)\}$ and $T_1 = \{(0,1)(3,2)(2,3)(1,0)\}$. More precisely, C is similar to the following matrix

$$\begin{pmatrix} C_0 & 0 & N_0 & 0 \\ 0 & C_1 & 0 & N_1 \\ 0 & 0 & C_{T_0} & 0 \\ 0 & 0 & 0 & C_{T_1} \end{pmatrix}$$

10.2 The Projection Operator P

From now on, we assume the system (X_ζ, T) to have mixed spectrum; therefore, the coincidence matrix C admits $k \geq 2$ ergodic classes. We denote by F the subspace of left eigenvectors of C corresponding to the eigenvalue q; in other words, $v = (v(\alpha,\beta)) \in \mathbf{C}^{s^2}$ belongs to F if

$$\sum_{\alpha,\beta \in A} v(\alpha,\beta) C_{\alpha\beta}^{\gamma\delta} = q\, v(\gamma,\delta)$$

or, in short,

$$vC = qv \quad \text{with tensor multiplication.}$$

Then, $\dim F = k$, in particular, F contains the vector $\mathbf{1}$ whose components are $\mathbf{1}(\alpha,\beta) = 1$ for every $\alpha, \beta \in A$.

Definition 10.2. The *projection operator* onto F is denoted by $P := \lim_{n\to\infty} C^n/q^n$.

We wish to describe P and F. The following proposition is a version of corollary 5.1 in this framework.

Proposition 10.2. *If* $P = (P_{\alpha\beta}^{\gamma\delta})$, *then* $P_{\alpha\beta}^{\gamma\delta} = \sum_{j=0}^{k-1} p_j(\alpha,\beta) q_j(\gamma,\delta)$ *with* p_j, q_j *nonnegative vectors. Moreover,*

$$p_j(\alpha,\beta) = 0 \quad \text{if} \quad (\alpha,\beta) \notin E_j$$

and $\sum_{(\alpha,\beta)\in E_j} p_j(\alpha,\beta) = 1$,

$$q_j(\gamma,\delta) = \begin{cases} 1 & \text{if } (\gamma,\delta) \in E_j \\ 0 & \text{if } (\gamma,\delta) \in E_i,\ i \neq j \\ \in [0,1] & \text{if } (\gamma,\delta) \in T \end{cases}$$

and $\sum_j q_j(\gamma,\delta) = 1$.

Proof. If the matrix C is similar to the following

$$\begin{pmatrix} C_0 & & & & N_0 \\ & \cdot & 0 & & \cdot \\ & & \cdot & & \cdot \\ & 0 & \cdot & & \cdot \\ & & & C_{k-1} & N_{k-1} \\ 0 & \cdot & \cdot \quad \cdot & 0 & C_T \end{pmatrix}$$

then P is similar to the matrix

$$\begin{pmatrix} P_0 & & & & Q_0 \\ & \cdot & 0 & & \cdot \\ & & \cdot & & \cdot \\ & 0 & \cdot & & \cdot \\ & & & P_{k-1} & Q_{k-1} \\ 0 & \cdot & \cdot \quad \cdot & 0 & 0 \end{pmatrix}$$

where $P_j = \lim_{n\to\infty} C_j^n/q^n$ for $0 \leq j \leq k-1$. Denote by $(p_j(\alpha,\beta))_{\alpha,\beta}$ the positive eigenvector of the primitive and q-stochastic matrix C_j, corresponding to the eigenvalue q, extended by 0 out of E_j, and normalized by $\sum_{(\alpha,\beta)\in E_j} p_j(\alpha,\beta) = 1$. Since P_j has rank one, $(P_j)_{\alpha\beta}^{\gamma\delta} = p_j(\alpha,\beta)$ for every (γ,δ). Now, the relation $PC = qP$ entails

$$P_j N_j = Q_j(qI - C_T) \quad \text{for } 1 \leq j \leq k-1.$$

The spectral radius of C_T being $< q$ by definition of C_T, the matrix $(qI - C_T)$ is invertible and

$$Q_j = P_j N_j (qI - C_T)^{-1} =: P_j W_j.$$

Therefore, if $(\gamma,\delta) \in T$, the vector $q_j(\gamma,\delta) = \sum_{\alpha,\beta} (W_j)_{\alpha\beta}^{\gamma\delta}$ must be a nonnegative one. Also, from the q-stochasticity of C, we have

$$\sum_{\alpha,\beta} (N_0 + \cdots + N_{k-1})_{\alpha\beta}^{\gamma\delta} = \sum_{\alpha,\beta} (qI - C_T)_{\alpha\beta}^{\gamma\delta},$$

so that, the identity $\Sigma_j W_j = (\Sigma_j N_j)(qI - C_T)^{-1}$ implies

$$\Sigma_j q_j(\gamma,\delta) = \Sigma_{\alpha,\beta}\left(\Sigma_j (W_j)_{\alpha\beta}^{\gamma\delta}\right)$$

$$= 1 \quad \text{for every} \quad (\gamma,\delta),$$

and the description of P follows easily. □

Remark 10.1. Every stochastic matrix may be viewed as the transition matrix of some stationary Markov process, with a finite number of states. Then $p_j(\alpha,\beta)$ is the probability for (α,β) to belong to the class E_j and $q_j(\gamma,\delta)$, the probability for (γ,δ) to be eventually absorbed by E_j.

Here is a description of F.

Corollary 10.1. *The s^2-dimensional vector $v = (v(\alpha,\beta))$ belongs to F if and only if*

$$v(\alpha,\beta) = \sum_{j=0}^{k-1} v_j q_j(\alpha,\beta)$$

where the v_j are complex numbers; to be short, we shall denote such a vector v by $(v_0, v_1, \ldots, v_{k-1})$.

Proof. Since $vC = qv$ or, equivalently, $vP = v$, we get from the previous proposition,

$$v(\alpha,\beta) = \Sigma_{\gamma,\delta}\, v(\gamma,\delta)(P)_{\alpha\beta}^{\gamma\delta}$$

$$= \Sigma_j q_j(\alpha,\beta)\left(\Sigma_{\gamma,\delta}\, v(\gamma,\delta) p_j(\gamma,\delta)\right).$$

Thus, if $(\alpha,\beta) \in E_j$, $v(\alpha,\beta) = v_j := \Sigma_{\gamma,\delta}\, v(\gamma,\delta) p_j(\gamma,\delta)$, since $q_j(\alpha,\beta) = 1$ and $q_i(\alpha,\beta) = 0$ for $i \neq j$. This proves that every vector of F is constant on each ergodic class and

$$v(\alpha,\beta) = \sum_{j=0}^{k-1} q_j(\alpha,\beta) v_j \quad \text{for every} \quad (\alpha,\beta).$$

Conversely, it is easy to check that F contains the k independent vectors q_j, $0 \leq j \leq k-1$, defined in the proposition and that (q_j) is a basis of F. □

10.2.1 A Property of P

The projection operator P satisfies a special positivity property which will appear to be fundamental in the sequel. If we identify the s^2-dimensional vectors with linear operators on $\mathbf{C}^s$, we now prefer to consider P acting on $\mathscr{L}(\mathbf{C}^s)$ by the same formula :

if $v \in \mathscr{L}(\mathbf{C}^s)$ and $v = (v_{\alpha\beta})$ in the canonical basis of $\mathbf{C}^s$, then $w = Pv$ is the linear operator on $\mathbf{C}^s$ with entries

$$w_{\alpha\beta} = \sum_{\gamma,\delta} P^{\gamma\delta}_{\alpha\beta} v_{\gamma\delta};$$

in a similar way, $w = vP$ is defined by

$$w_{\gamma\delta} = \sum_{\alpha,\beta} v_{\alpha\beta} P^{\gamma\delta}_{\alpha\beta}.$$

Proposition 10.3. *The operator P maps the set of positive definite operators of $\mathscr{L}(\mathbf{C}^s)$ into itself, or, in short, if $v \in \mathscr{L}(\mathbf{C}^s)$ and $v \gg 0$, so are Pv and vP.*

Proof. It is sufficient to check that $\sum_{\alpha,\beta} v_\alpha \overline{v_\beta} P^{\gamma\delta}_{\alpha\beta}$ is a positive definite operator for every s-dimensional vector $v = (v_\alpha) \in \mathbf{C}^s$. So, let $(t_\gamma) \in \mathbf{C}^s$ and consider

$$\sum_{\gamma,\delta} \Big(\sum_{\alpha,\beta} v_\alpha \overline{v_\beta} P^{\gamma\delta}_{\alpha\beta} \Big) t_\gamma \overline{t_\delta} = \lim_{n\to\infty} \frac{1}{q^n} \sum_{\gamma,\delta} \Big(\sum_{\alpha,\beta} v_\alpha \overline{v_\beta} (C^n)^{\gamma\delta}_{\alpha\beta} \Big) t_\gamma \overline{t_\delta}. \tag{10.5}$$

For any fixed n, we now define the vector $V_{n,\alpha,\gamma}(t) \in \mathbf{C}^{q^n}$ as follows : in the word $\zeta^n(\gamma)$ of length q^n, we replace each letter α by the complex number t_γ and the letters $\beta \neq \alpha$ by 0. We thus get q^n-dimensional vectors satisfying

$$\langle V_{n,\alpha,\gamma}(t), V_{n,\beta,\delta}(t) \rangle = \sum_{\gamma,\delta} (C^n)^{\gamma\delta}_{\alpha\beta}\, t_\gamma \overline{t_\delta};$$

with these notations,

$$\sum_{\gamma,\delta} \Big(\sum_{\alpha,\beta} v_\alpha \overline{v_\beta} (C^n)^{\gamma\delta}_{\alpha\beta} \Big) t_\gamma \overline{t_\delta} = \Big| \sum_{\alpha} v_\alpha V_{n,\alpha,\gamma}(t) \Big|^2 \geq 0,$$

and so is

$$\sum_{\gamma,\delta} \Big(\sum_{\alpha,\beta} v_\alpha \overline{v_\beta} P^{\gamma\delta}_{\alpha\beta} \Big) t_\gamma \overline{t_\delta}$$

by (10.5). The property is established. □

Notation : From now on, we say that the vector $v = (v(\alpha,\beta))$ is *strongly positive* and write $v \gg 0$ if *the associated operator v of $\mathscr{L}(\mathbf{C}^s)$ is a positive definite operator* (not to be confused with a nonnegative or positive vector).

The previous property, shared by the operator P acting on $\mathscr{L}(\mathbf{C}^s)$, will be denoted by

$$P \underset{\text{op}}{\gg} 0.$$

The vectors (q_j), spanning the space F, generally are not strongly positive (not yet symmetric operators if the ergodic classes are not), but F contains such strongly positive vectors, for instance vP, for every already strongly positive vectors v, according to proposition 10.3; these vectors will play an important role in the study of Σ, in the next chapter, whence the following definition.

Definition 10.3. We denote by F_+ the *convex cone* consisting of strongly positive vectors of F.

10.2.2 Examples

We turn back to our favourite examples.

1. *Rudin-Shapiro Automaton.* Recall, in this case, that C has two ergodic classes E_0 and $E_1 = \{(0,3)(2,1)(1,2)(3,0)\}$. It is easy to compute

$$q_j(\alpha,\beta) = 1/2 \quad \text{for } (\alpha,\beta) \in T,\ j = 0,1,$$

so that every $v \in F$ is of the following form :

$$(\underbrace{v_0 \cdots v_0}_{E_0}\ \underbrace{v_1 \cdots v_1}_{E_1}\ \underbrace{\frac{v_0+v_1}{2} \cdots \frac{v_0+v_1}{2}}_{T}).$$

We intend to describe the convex cone F_+ of those $v = (v_0, v_1)$. For that, we put the quadratic form g defined on $t \in \mathbf{C}^4$ by

$$g(t) = \sum_{\alpha,\beta} t_\alpha \overline{t_\beta} v(\alpha,\beta)$$

into canonical form. Assuming $v_0 \neq 0$, we get

$$\frac{1}{v_0} g(t) = \left|t_0 + \frac{v_1}{v_0} t_3 + \frac{1}{2}\left(1 + \frac{v_1}{v_0}\right)(t_1 + t_2)\right|^2 + \left(1 - \frac{v_1^2}{v_0^2}\right)\left|t_3 + \frac{t_1+t_2}{2}\right|^2 + \frac{1}{2}\left(1 - \frac{v_1}{v_0}\right)^2 |t_1 - t_2|^2$$

and we deduce, for the Rudin-Shapiro automaton, that

$$F_+ = \{v = (v_0, v_1), |v_1| \leq v_0\}.$$

2. *Q-mirror Automaton.* Recall that C has two ergodic classes in this case, E_0 and $E_1 = \{(0,2)(3,1)(1,3)(2,0)\}$ while $T = T_0 \cup T_1$ with $T_0 = \{(0,3)(2,1)(1,2)(3,0)\}$ and $T_1 = \{(0,1)(3,2)(2,3)(1,0)\}$. From the canonical form of the matrix C, we derive

$$q_0(\alpha,\beta) = 1 \quad \text{if } (\alpha,\beta) \in E_0 \cup T_0, \; 0 \text{ otherwise}$$

$$q_1(\alpha,\beta) = 1 \quad \text{if } (\alpha,\beta) \in E_1 \cup T_1, \; 0 \text{ otherwise}$$

and then, $v \in F$ if and only if

$$v = (\underbrace{v_0 \cdots v_0}_{E_0} \; \underbrace{v_1 \cdots v_1}_{E_1} \; \underbrace{v_0 \cdots v_0}_{T_0} \; \underbrace{v_1 \cdots v_1}_{T_1}).$$

It is rather simple to compute for $v_0 \neq 0$

$$g(t) = \Sigma_{\alpha,\beta} t_\alpha \overline{t_\beta} v(\alpha,\beta)$$

$$= v_0 \left(\left| a + b\frac{v_1}{v_0} \right|^2 + |b|^2 \left(1 - \frac{v_1^2}{v_0^2} \right) \right)$$

where $a = (t_0 + t_3)$ and $b = (t_1 + t_2)$. For this automaton,

$$F_+ = \{v = (v_0, v_1), |v_1| \leq v_0\}.$$

3. *Bijective Automata.* We saw that $T = \emptyset$ since C/q and thus P are bi-stochastic matrices. Moreover,

$$P_{\alpha\beta}^{\gamma\delta} = p_j(\alpha,\beta) \quad \text{if } (\gamma,\delta) \in E_j$$

and the constraint $\Sigma_{\gamma,\delta} P_{\alpha\beta}^{\gamma\delta} = 1$ infers $p_j(\alpha,\beta) = 1/|E_j|$ if $(\alpha,\beta) \in E_j$.
When C has two ergodic classes only (for instance, this holds for the Thue-Morse automaton but also for the example at the end of subsection 8.2.2) then $E_1 = \{(\alpha,\beta), \alpha \neq \beta\}$, and

$$g(t) = v_0 \left(\Sigma_\alpha |t_\alpha|^2 \right) + v_1 \Sigma_{\alpha \neq \beta} t_\alpha \overline{t_\beta}$$

$$= (v_0 - v_1) \left(\Sigma_\alpha |t_\alpha|^2 \right) + v_1 |\Sigma_\alpha t_\alpha|^2.$$

We thus obtain that

$$F_+ = \{v = (v_0, v_1), -\frac{v_0}{s-1} \leq v_1 \leq v_0\}.$$

Remark 10.2. In the emblematic examples of Thue-Morse and Rudin-Shapiro automata, the maximal spectral type of the system is generated by those measures, associated with vectors arising from F_+ as it can easily be checked. Actually, in the Thue-Morse example,

$$\delta_0 = \Sigma_{\alpha,\beta} v(\alpha,\beta) \sigma_{\alpha\beta} \quad \text{with} \quad v = (1,1)$$
$$\text{and} \quad \rho = \Sigma_{\alpha,\beta} v(\alpha,\beta) \sigma_{\alpha\beta} \quad \text{with} \quad v = (1,-1);$$

likewise, in the Rudin-Shapiro example,

$$\begin{aligned} \delta_0 &= \Sigma_{\alpha,\beta}\,\nu(\alpha,\beta)\sigma_{\alpha\beta} \quad \text{with} \quad \nu=(1,1) \\ \text{and} \quad m &= \Sigma_{\alpha,\beta}\,\nu(\alpha,\beta)\sigma_{\alpha\beta} \quad \text{with} \quad \nu=(1,-1); \end{aligned}$$

We shall explain this fact in the last chapter.

10.3 The Bi-Correlation Matrix Z

We study in this section the matrix Z of correlation measures associated with the substitution η on the alphabet $A \times A$, although this substitution is not primitive; we shall establish properties for Z and get, as a consequence, that this matrix can almost be diagonalized, providing in this way the maximal spectral type of the initial substitution ζ.

For every $\alpha, \beta \in A$, $\sigma_{\alpha\beta}$ measures the correlation between the occurrences of both letters α and β in the fixed point u, and, thanks to the primitivity assumption on ζ, we already observed that

$$\lim_{N\to\infty} \frac{1}{q^N} \text{Card}\,\{n+k<q^N,\ \zeta^N(\gamma)_{n+k}=\alpha,\ \zeta^N(\gamma)_n=\beta\} = \widehat{\sigma}_{\alpha\beta}(k)$$

does not depend on $\gamma \in A$ (7.5). Actually, the existence of Σ may appear as a consequence of the Perron-Frobenius theorem, thereby, the existence of the projection operator $\lim_{n\to\infty} M^n/q^n$ where $M := M(\zeta)$.

Here, even if η fails to be primitive, we shall deduce, from the existence of $P = \lim_{n\to\infty} C^n/q^n$ where $C := M(\eta)$, that the following limits exist :

$$\lim_{N\to\infty} \frac{1}{q^N} \text{Card}\left\{n+k<q^N,\ (\eta^N(\tbinom{\gamma}{\delta}))_{n+k} = (\tbinom{\alpha}{\alpha'}),\ (\eta^N(\tbinom{\gamma}{\delta}))_n = (\tbinom{\beta}{\beta'})\right\} \tag{10.6}$$

for every $(\binom{\alpha}{\alpha'}), (\binom{\beta}{\beta'}), (\binom{\gamma}{\delta})$ in $A \times A$.

Proposition 10.4. *For every $k \in \mathbf{N}$, $\alpha,\beta,\gamma,\delta \in A$,*

$$\lim_{N\to\infty} \frac{1}{q^N} \text{Card}\left\{n+k<q^N,\ \zeta^N(\gamma)_{n+k}=\alpha,\ \zeta^N(\delta)_n=\beta\right\} \tag{10.7}$$

exists and is the Fourier transform in k of some measure of $M(\mathbf{T})$, that we shall denote by $\sigma_{\alpha\beta}^{\gamma\delta}$.

Proof. Clearly, Card $\left\{n+k<q^N,\ \zeta^N(\gamma)_{n+k}=\alpha,\ \zeta^N(\delta)_n=\beta\right\}$

$$=\sum_{\alpha',\beta'}\text{Card}\left\{n+k<q^N,\ (\eta^N(\begin{smallmatrix}\gamma\\ \delta\end{smallmatrix}))_{n+k}=(\begin{smallmatrix}\alpha\\ \alpha'\end{smallmatrix}),\ \eta^N((\begin{smallmatrix}\gamma\\ \delta\end{smallmatrix}))_n=(\begin{smallmatrix}\beta\\ \beta'\end{smallmatrix})\right\}$$

so that, we are reduced to prove that the limit exists in (10.6). This could be derived from a quite general version of the Perron Frobenius theorem but we give the proof in our context.

Fix B, a word on the alphabet $A\times A$, and denote by h_N the s^2-dimensional vector with components

$$h_N(\gamma,\delta)=\frac{1}{q^N}L_B\left(\eta^N(\begin{smallmatrix}\gamma\\ \delta\end{smallmatrix})\right),$$

(recalling the notation $L_B(C)$ for the occurrence number of B in the word C). If $\lim_{N\to\infty}h_N$ exists, B occurs in $\eta^N(\begin{smallmatrix}\gamma\\ \delta\end{smallmatrix})$ with a possibly zero frequency, depending on (γ,δ). But this will be sufficient to entail (10.6). First observe that, on the one hand,

$$L_B\left(\eta^N(\begin{smallmatrix}\gamma\\ \delta\end{smallmatrix})\right)\geq\Sigma_{\alpha,\beta}L_B\left(\eta^{N-1}(\begin{smallmatrix}\alpha\\ \beta\end{smallmatrix})\right)C_{\alpha\beta}^{\gamma\delta}$$

and

$$\leq\Sigma_{\alpha,\beta}L_B\left(\eta^{N-1}(\begin{smallmatrix}\alpha\\ \beta\end{smallmatrix})\right)C_{\alpha\beta}^{\gamma\delta}+|B|q$$

on the other hand; hence, using the identity $CP=qP$, we get

$$h_NP\geq h_{N-1}P \tag{10.8}$$

and

$$\frac{1}{q^p}\sum_{\alpha,\beta}h_N(\alpha,\beta)(C^p)_{\alpha\beta}^{\gamma\delta}\leq h_{N+p}(\gamma,\delta)\leq\frac{1}{q^p}\sum_{\alpha,\beta}h_N(\alpha,\beta)(C^p)_{\alpha\beta}^{\gamma\delta}+O(\frac{1}{q^N}). \tag{10.9}$$

Since the sequence (h_N) is uniformly bounded, it follows from (10.8) that h_NP converges to some limit ℓ. Now, letting $\underline{h}=\liminf h_N$ and $\bar{h}=\limsup h_N$, we deduce, by taking the limit on p in (10.9), that

$$(h_NP)(\gamma,\delta)\leq\underline{h}(\gamma,\delta)\leq\bar{h}(\gamma,\delta)\leq(h_NP)(\gamma,\delta)+O(\frac{1}{q^N})$$

and, therefore $\underline{h}=\bar{h}=\ell$, which proves the limit exists in (10.6) and (10.7).
Now, fix $k\in\mathbf{N}$ and denote by $\Gamma_{\alpha\beta}^{\gamma\delta}(k)$ the limit in (10.7). If we consider the so-defined q^N-dimensional vector $V_N(\begin{smallmatrix}\gamma\\ \alpha\end{smallmatrix})$, obtained by replacing in the word $\zeta^N(\gamma)$, α by 0 and any $\beta\neq\alpha$ by 1, we have

$$\Gamma_{\alpha\beta}^{\gamma\delta}(k)=\lim_{N\to\infty}\frac{1}{q^N}\langle T^kV_N(\begin{smallmatrix}\gamma\\ \alpha\end{smallmatrix}),V_N(\begin{smallmatrix}\delta\\ \beta\end{smallmatrix})\rangle.$$

Finally, we extend Γ to negative integers by putting

$$\Gamma_{\alpha\beta}^{\gamma\delta}(-k) = \overline{\Gamma_{\alpha\beta}^{\gamma\delta}(k)}, \quad k \geq 1.$$

$\Gamma_{\alpha\beta}^{\gamma\delta}$ is easily checked to be the Fourier transform of some measure called $\sigma_{\alpha\beta}^{\gamma\delta}$.

Note that, in case $\zeta^N(\gamma)$ and $\zeta^N(\delta)$ converge to some infinite words u and v respectively, $\sigma_{\alpha\beta}^{\gamma\delta}$ is just the mixed correlation measure of the sequences $\tau_\alpha(u)$ and $\tau_\beta(v)$. □

Remark 10.3. Since C is assumed to be primitive when restricted to an ergodic class, $\sigma_{\alpha\beta}^{\gamma\delta}$ only depends on the class E_j when $(\gamma,\delta) \in E_j$. In particular, $\sigma_{\alpha\beta}^{\gamma\gamma} = \sigma_{\alpha\beta}$ (ζ is primitive).

Definition 10.4. We denote by Z the $s^2 \times s^2$ matrix $(\sigma_{\alpha\beta}^{\gamma\delta})$ of measures in $M(\mathbf{T})$, indexed on pairs (α,β), (γ,δ) in $A \times A$, that one has better sometimes to consider as indexed on pairs $\binom{\gamma}{\alpha}$ and $\binom{\delta}{\beta}$.

Properties of Z

From the above remark, all the measures of Σ do appear in Z, actually in the first column of Z when indexed on (α,β), (γ,δ); but, in fact, there is no measure in Z singular with respect to Σ (or σ_{max}). This will be a consequence of the following important property of Z.

Proposition 10.5. *The matrix Z transform the positive definite operators of $\mathscr{L}(\mathbf{C}^s)$ into positive definite $s \times s$ matrices of measures in the following sense :*

If $v \in \mathscr{L}(\mathbf{C}^s)$ is $\gg 0$, the matrices of measures

$$(\mu_{\alpha\beta}) = \Big(\sum_{\gamma,\delta} \sigma_{\alpha\beta}^{\gamma\delta} v(\gamma,\delta)\Big)$$

and

$$(v_{\gamma\delta}) = \Big(\sum_{\alpha,\beta} v(\alpha,\beta)\sigma_{\alpha\beta}^{\gamma\delta}\Big)$$

are both positive definite matrices (and denoted by Zv and vZ respectively).

We refer to section 7.2 for the definitions. This proposition turns out to be a direct consequence of the following lemma.

Lemma 10.3. *The matrix Z, considered as* $\left(\sigma_{\binom{\gamma}{\alpha}\binom{\delta}{\beta}}\right)$, *is a positive definite matrix of measures.*

Proof of the lemma. Consider any function $f : A \times A \to \mathbf{C}$; the measure

$$\mu = \sum_{\alpha,\beta,\gamma,\delta} f(\alpha,\gamma)\overline{f(\beta,\delta)}\sigma_{\alpha\beta}^{\gamma\delta}$$

is a positive measure, since, with notations of the previous proof,

$$\hat{\mu}(k) = \lim_{N\to\infty} \frac{1}{q^N}\langle T^k\Big(\sum_{\alpha,\gamma} f(\alpha,\gamma)V_N\binom{\gamma}{\alpha}\Big), \sum_{\beta,\delta} f(\beta,\delta)V_N\binom{\delta}{\beta}\rangle;$$

putting $\hat{\mu}(-k) = \overline{\hat{\mu}(k)}$, $k \mapsto \hat{\mu}(k)$ becomes a positive definite sequence on $\mathbf{Z}$ and the lemma is proved (Bochner's theorem). □

Proof of the proposition. Once more, it is sufficient to consider the matrix whose entries are

$$\mu_{\alpha\beta} = \sum_{\gamma,\delta} \sigma_{\alpha\beta}^{\gamma\delta} v_\gamma \overline{v_\delta}$$

for any $v = (v_\gamma)$ in $\mathbf{C}^s$. But if $t = (t_\alpha)$ is another vector of $\mathbf{C}^s$, the measure

$$\sum_{\alpha,\beta} t_\alpha \overline{t_\beta}\mu_{\alpha\beta} = \sum_{\alpha,\beta,\gamma,\delta} t_\alpha \overline{t_\beta} v_\gamma \overline{v_\delta}\sigma_{\alpha\beta}^{\gamma\delta}$$

is seen to be a positive measure by applying the lemma with $f(\alpha,\gamma) = t_\alpha v_\gamma$. □

When this property holds, we write $Z \gg_{\text{op}} 0$ as in the scalar case (if it does not lead to some ambiguity). Observe that this property is preserved by product of a scalar matrix with a matrix of measures, both being positive definite in this sense.

We derive the first estimate on the maximal spectral type of (X_ζ, T) when ζ is primitive and aperiodic.

Corollary 10.2. *For every* $\alpha,\beta,\gamma,\delta \in A$, *the measure* $\sigma_{\alpha\beta}^{\gamma\delta}$ *is absolutely continuous with respect to* σ_α *and* σ_β; *it follows that the maximal spectral type of* (X_ζ, T) *is generated by the measure* $\sum_{\alpha,\beta,\gamma,\delta}\left|\sigma_{\alpha\beta}^{\gamma\delta}\right|$ *as well.*

Proof. Since $\left(\sigma_{\binom{\gamma}{\alpha}\binom{\delta}{\beta}}\right)$ is a positive definite matrix of measures, it follows readily that

$$\sigma_{\binom{\gamma}{\alpha}\binom{\delta}{\beta}} \ll \sigma_{\binom{\gamma}{\alpha}\binom{\gamma}{\alpha}} \text{ and } \sigma_{\binom{\delta}{\beta}\binom{\delta}{\beta}},$$

respectively equal to σ_α and σ_β. □

Remark 10.4. It would be interesting to manage to compare $\mathrm{Tr}(\Sigma)$ and $\mathrm{Tr}(Z) = \sum_{\alpha,\beta}\sigma_{\alpha\beta}^{\alpha\beta}$, so that, by the corollary, every entry $\sigma_{\alpha\beta}^{\gamma\delta}$ of Z could be compared to

$\mathrm{Tr}(Z)$. Clearly, $\mathrm{Tr}(Z) \ll \mathrm{Tr}(\Sigma)$ and more precisely, $\mathrm{Tr}(Z) \leq s^2\, \mathrm{Tr}(\Sigma)$, but we obtained no remarkable inequality in the other direction. Nevertheless, Z will be preferred to Σ for its strong mixing property, and some others that we prove first.

Proposition 10.6. *For every $a \in \mathbf{Z}$ we have*

$$\widehat{Z}(aq) = \frac{1}{q} C\, \widehat{Z}(a) \tag{10.10}$$

and

$$\lim_{n\to\infty} \widehat{Z}(aq^n) = P\, \widehat{Z}(a). \tag{10.11}$$

Proof. Property (10.11) is an immediate consequence of (10.10) that we establish now. Fix $a \in \mathbf{N}$. In order to prove that

$$\sigma_{\alpha\beta}^{\gamma\delta}\hat{}\,(aq) = \frac{1}{q} \sum_{\alpha',\beta'} C_{\alpha\beta}^{\alpha'\beta'}\, \sigma_{\alpha'\beta'}^{\gamma\delta}\hat{}\,(a)$$

we need to estimate

$$A_N := \mathrm{Card}\left\{n + aq < q^N,\ \zeta^N(\gamma)_{n+aq} = \alpha,\ \zeta^N(\delta)_n = \beta\right\}.$$

As usual, we decompose n into $mq + r$ with $0 \leq r \leq q-1$, so that $m+a$ must be $< q^{N-1}$, and, using identity (5.1), we get

$$\begin{aligned}
A_N &= \mathrm{Card}\left\{(m+a)q + r < q^N,\ \zeta(\zeta^{N-1}(\gamma)_{m+a})_r = \alpha,\ \zeta(\zeta^{N-1}(\delta)_m)_r = \beta\right\} \\
&= \sum_{\alpha',\beta'} \mathrm{Card}\left\{(m+a) < q^{N-1},\ \zeta^{N-1}(\gamma)_{m+a} = \alpha',\ \zeta^{N-1}(\delta)_m = \beta'\right\} \\
&\quad \times \mathrm{Card}\left\{r < q,\ \zeta(\alpha')_r = \alpha,\ \zeta(\beta')_r = \beta\right\} \\
&= \sum_{\alpha',\beta'} C_{\alpha\beta}^{\alpha'\beta'}\ \mathrm{Card}\left\{(m+a) < q^{N-1},\ \zeta^{N-1}(\gamma)_{m+a} = \alpha',\ \zeta^{N-1}(\delta)_m = \beta'\right\}.
\end{aligned}$$

Now, $\sigma_{\alpha\beta}^{\gamma\delta}\hat{}\,(aq) = \lim_{N\to\infty} A_N/q^N = \frac{1}{q} \sum_{\alpha',\beta'} C_{\alpha\beta}^{\alpha'\beta'}\, \sigma_{\alpha'\beta'}^{\gamma\delta}\hat{}\,(a)$ which was to be proved. □

Recall that property (10.10) is already shared by Σ as observed in (10.4), but the following has been proved for $s\Sigma$ in the commutative case only (see the remark at the end of chapter 8).

Proposition 10.7. *The matrix Z is q-strongly mixing, in other words,*

$$\lim_{p\to\infty} \widehat{Z}(aq^p + b) = \widehat{Z}(b)\, \widehat{Z}(a), \tag{10.12}$$

for every $a, b \in \mathbf{Z}$.

Proof. This proof is quite similar to the previous one. Fix a, b, p in $\mathbf{N}$ and consider

$$B_N := \text{Card}\left\{n + aq^p + b < q^N,\ \zeta^N(\gamma)_{n+aq^p+b} = \alpha,\ \zeta^N(\delta)_n = \beta\right\};$$

as above, we factorize B_N into

$$B_N = \sum_{\alpha',\beta'} \text{Card}\left\{(m+a) < q^{N-p},\ \zeta^{N-p}(\gamma)_{m+a} = \alpha',\ \zeta^{N-p}(\delta)_m = \beta'\right\}$$
$$\times \text{Card}\left\{r + b < q^p,\ \zeta^p(\alpha')_{r+b} = \alpha,\ \zeta^p(\beta')_r = \beta\right\} + \varepsilon_{N,p}$$

where the error, $\varepsilon_{N,p}$, results from omitting in B_N the terms of index $n = mq^p + r$, $q^p - b \le r < q^p$; thus $\varepsilon_{N,p} = O(bq^{N-p})$, and we get

$$\sigma_{\alpha\beta}^{\gamma\delta}\widehat{\ }(aq^p + b) = \lim_{N\to\infty} B_N/q^N$$

$$= \sum_{\alpha',\beta'} \sigma_{\alpha'\beta'}^{\gamma\delta}\widehat{\ }(a)\frac{1}{q^p}\text{Card}\left\{r + b < q^p,\ \zeta^p(\alpha')_{r+b} = \alpha,\ \zeta^p(\beta')_r = \beta\right\} + O\left(\frac{b}{q^p}\right).$$

The proposition is obtained by taking the limit over p. □

With help of the following proposition, the form of Z becomes clear.

Proposition 10.8. *The matrix Z satisfies*

$$Z = ZP \tag{10.13}$$

Proof. From (10.7) one deduces that $\widehat{Z}(0) = \lim_{N\to\infty} C^N/q^N = P$ and the identity follows from (10.12) with $a = 0$.

Note that

$$\sigma_{\alpha\beta}^{\gamma\delta} = \sum_j q_j(\gamma,\delta)\Big(\sum_{\alpha',\beta'} p_j(\alpha',\beta')\sigma_{\alpha\beta}^{\alpha'\beta'}\Big) = \sum_j q_j(\gamma,\delta)\sigma_{\alpha\beta}^j$$

where $\sigma_{\alpha\beta}^j$ denotes the common measure $\sigma_{\alpha\beta}^{\gamma\delta}$ when $(\gamma,\delta) \in E_j$. □

10.4 Main Theorem on Maximal Spectral Type

We are now in a position to describe the maximal spectral type of the system (X_ζ, T) associated with the primitive and aperiodic substitution ζ of length q. We start with a description of the matrix PZ, that we deduce from the above properties of Z.

Theorem 10.1. *The matrix of measures PZ can be diagonalized with respect to a basis consisting of $\gg 0$ vectors of $\mathbf{C}^{s^2}$, and the measures $\lambda_0, \ldots, \lambda_{k-1}$, which appear on the diagonal, are q-strongly mixing probability measures.*

Proof. From (10.13), it is clear that $PZw = 0$ for every $w \in \ker P$. It remains to prove that PZ can be diagonalized on Im P. This will be a consequence of the results established in chapter 7, summarized in corollary 7.1.

For every $a \in \mathbf{Z}$, we denote by χ_a the character in $\overline{\Gamma}(Z)$, obtained as the limit of the sequence $(\gamma_{aq^n})_n$ on Z. Then, from (10.12),

$$\chi_a\,(PZ) = P\,(\chi_a Z) = PZ\,\widehat{Z}(a)$$

and, since $Z = ZP$ (10.13), we get

$$\chi_a\,(PZ) = (PZ)\,\widehat{PZ}(a). \tag{10.14}$$

Also, $P\widehat{Z}(0) = P^2 = P$ and $\widehat{PZ}(0)$ restricted to Im P reduces to I. Corollary 7.1 may be applied and asserts that PZ can be diagonalized on Im P; there exist k vectors $w_0, w_1, \ldots, w_{k-1} \in$ Im P, and k measures $\lambda_0, \ldots, \lambda_{k-1}$ such that

$$PZ\,w_j = \lambda_j\,w_j, \quad j = 0, \ldots, k-1. \tag{10.15}$$

Let λ be one of the λ_j's, and w be a corresponding eigenvector. From (10.14) and (10.15) we have

$$\begin{aligned}\chi_a \lambda w = \chi_a PZw &= PZ\,\widehat{PZ}(a)w \\ &= PZ\,\widehat{\lambda}(a)w \\ &= \widehat{\lambda}(a)\,\lambda w\end{aligned}$$

and $\lim_{p\to\infty} \widehat{\lambda}(aq^p + b) = \widehat{\lambda}(a)\widehat{\lambda}(b)$ for $a, b \in \mathbf{Z}$. Such a measure needs to be a probability measure, and we are left with the choice of the eigenvectors to finish the proof.

Fix again λ, one of the λ_j's, and consider the L-space defined by

$$L = \{\mu \in M(\mathbf{T}),\ \chi_a \mu = \widehat{\lambda}(a)\,\mu, \quad \text{for all } a \in \mathbf{Z}\}$$

(see section 1.4 for all the sequel). Of course, $\lambda \in L$, but every $\lambda_j \neq \lambda$ is in $L^\perp$: actually, if $\lambda' \neq \lambda$, there exists some $a \in \mathbf{Z}$ such that $\widehat{\lambda'}(a) \neq \widehat{\lambda}(a)$; but, λ' being q-strongly mixing, $\chi_a = \widehat{\lambda'}(a)$ on λ' while χ_a is equal to $\widehat{\lambda}(a)$ on every $\mu \in L$. This remark, together with lemma 1.2, implies that $\lambda' \in L^\perp$. Let now $Z = Z_L + Z_{L^\perp}$ be the decomposition of Z relative to L. By a slight modification of proposition 7.11, it is easy to prove that Z_L inherits the property we established for Z in proposition 10.5 : $Z_L \gg_{\text{op}} 0$; and, P being $\gg_{\text{op}} 0$ too, the product PZ_L in turn must be $\gg_{\text{op}} 0$, as already observed.

But, from the choice of L, PZ_L is similar to the following

$$\begin{pmatrix} \begin{pmatrix} \lambda & & 0 \\ & \ddots & \\ 0 & & \lambda \end{pmatrix} & 0 \\ 0 & 0 \end{pmatrix}.$$

One can thus find a scalar projection operator P_λ such that

$$PZ_L = \lambda P_\lambda, \quad \text{with} \quad P_\lambda \underset{\text{op}}{\gg} 0,$$

and Im P_λ is the eigenspace corresponding to λ. Since $\mathscr{L}(\mathbf{C}^s)$ is generated by the positive definite elements it contains, it follows from the definition of $P_\lambda \gg_{\text{op}} 0$ that Im P_λ, itself, is generated by positive definite operators on $\mathbf{C}^s$ and the eigenvectors w may be chosen $\gg 0$. The proof is complete. □

From this study of the matrix PZ, we know really more about the form of Z :

Corollary 10.3. *There exits a basis of s^2-dimensional $\gg 0$ vectors, with respect to which the matrix Z is similar to the following :*

$$\left(\begin{array}{c|c} \begin{pmatrix} \lambda_0 & & 0 \\ & \ddots & \\ 0 & & \lambda_{k_1} \end{pmatrix} & 0 \\ \hline M & 0 \\ \underbrace{\qquad\qquad}_{\text{Im}P} & \underbrace{\qquad}_{\ker P} \end{array}\right)$$

In particular, we notice that $\text{Tr}(Z) = \text{Tr}(PZ) = \lambda_0 + \cdots + \lambda_{k-1}$. If $\text{Tr}(\Sigma)$ could be compared to $\text{Tr}(Z)$ as indicated in a foregoing remark, the description of the spectrum of ζ would be achieved. Nevertheless, we shall succeed in proving by another means that these λ_j are generating σ_{max} (in the sense defined in chapter 7). Recall the notations : if D_n is the subgroup of $\mathbf{T}$ generated by $2\pi/q^n$, ω_n is its Haar measure and $D = \cup_{n\geq 1} D_n$, we denote by $\tilde{\lambda}$, for any $\lambda \in M(\mathbf{T})$, the D-quasi-invariant measure

$$\tilde{\lambda} := \sum_{n\geq 1} 2^{-n}(\lambda * \omega_n).$$

Since the λ_j are q-strongly mixing already, the best expected result we could get would be an equivalence between σ_{max} and $\widetilde{\text{Tr}Z}$. This holds in fact and provides the main result of this chapter.
In the following, we write S for the q-shift S_q, $S(\mu)$, for the pullback under S of a measure μ and $S(M)$, for the pullback under S of a matrix of measures M.

Theorem 10.2. *The maximal spectral type of the primitive and aperiodic substitution ζ of length q is equivalent to $\sum_{j=0}^{k-1} \tilde{\lambda}_j$.*

Proof. We have to compare every measure in the block M with the $\tilde{\lambda}_j$. According to theorem 10.1, let $w \in \operatorname{Im} P$ be a strongly positive vector corresponding to the diagonal measure λ. Then $PZw = \lambda w$ and with the notations of corollary 10.3,

$$Zw = \lambda w + Mw \qquad \text{so that } PMw = 0. \tag{10.16}$$

Since, equivalently, $Cw = qw$, one has

$$\frac{1}{q}CZw = \lambda w + \frac{1}{q}CMw.$$

We deduce from (10.16) that

$$S(Zw) = \lambda w + S(Mw)$$

since λ is q-invariant. But, by (10.10), $S(Zw) = \frac{1}{q}CZw$, so that

$$\frac{1}{q}CMw = S(Mw). \tag{10.17}$$

The matrix of measures Zw is $\gg 0$ but this property never holds for Mw unless $Mw = 0$, since

$$\widehat{Mw}(0) = \widehat{Z}(0)w - w = Pw - w = 0.$$

However, this will be true in restriction to some L-space. Indeed, consider the L-space $L = \{\mu \in M(\mathbf{T}),\ |\mu| \perp \tilde{\lambda}\}$; then, by (10.16),

$$(Zw)_L = (Mw)_L$$

and a new application of proposition 7.11 yields $(Zw)_L \gg 0$, hence $R := (Mw)_L \gg 0$. We have proved that R is a $\gg 0$ matrix consisting in measures singular with respect to $\tilde{\lambda}$. If $\widehat{R}(0) = 0$, then R must be $\equiv 0$ (since $\gg 0$), in other words, $(Mw)_L$ must be zero. This would imply that every μ in M is absolutely continuous with respect to the $\tilde{\lambda}_j$'s and the theorem.

We need for that an additional property of the matrix R.

Lemma 10.4. $S(R) = \frac{1}{q}CR.$

Proof of the lemma. If the L-spaces L and $L^{\perp}$ happen to be invariant under the shift S, then, by using (10.17), we do get two decompositions of the matrix $S(Mw)$, namely

$$\begin{aligned} S(w) &= S((Mw)_L) + S((Mw)_{L^{\perp}}) \\ &= \tfrac{1}{q}C\,(Mw)_L + \tfrac{1}{q}C\,(Mw)_{L^{\perp}}. \end{aligned}$$

The identity of the lemma then derives from the unicity of such a decomposition.

We are thus reduced to prove the S-invariance of both spaces L and $L^{\perp}$.

$\star$ With our notations, observe that $S(\omega_n) = \omega_{n-1}$ if $n \geq 1$ so that $S(\tilde{\lambda}) = \sum_{n\geq 1} 2^{-n-1}(\lambda * \omega_n)$ is equivalent to $\tilde{\lambda}$. If the positive measure $\mu \in L^{\perp}$, which simply means $\mu \ll \tilde{\lambda}$, then

$$S(\mu) \ll S(\tilde{\lambda}) \ll \tilde{\lambda},$$

and $L^{\perp}$ is invariant under S.

$\star$ Let now $\mu \in L$ be a positive measure. By the definition of L, there exists a Borel set A such that $\mu(A) = ||\mu||$ and $\tilde{\lambda}(A) = 0$. It follows that $S(\mu)$ is supported by the Borel set $B = S(A)$; indeed, $S^{-1}(B) = S^{-1}(S(A)) = \cup_{j=0}^{q-1}(A + j/q)$ contains A so that

$$S(\mu)(B) = \mu(S^{-1}(B)) \geq \mu(A) = ||\mu|| = ||S(\mu)||.$$

On the other hand, we shall see that B is annihilated by $\tilde{\lambda}$; since $\tilde{\lambda} * \delta_{j/q} \ll \tilde{\lambda}$, and remembering that $\tilde{\lambda}(A) = 0$, we get

$$S(\tilde{\lambda})(B) \leq \sum_{j=0}^{q-1} \tilde{\lambda}(A + \frac{j}{q}) = 0;$$

but also $\tilde{\lambda} \ll S(\tilde{\lambda})$, so that $\tilde{\lambda}(B) = 0$ in turn and the measures $S(\mu)$ and $\tilde{\lambda}$ are still mutually singular. This proves that L is S-invariant and gives the lemma. □

Turning back to the proof of the theorem and to the matrix R, we obtain, by iteration of the lemma, that

$$S^n(R) = \frac{1}{q^n} C^n\, R = \frac{1}{q^n}(C^n Mw)_L,$$

which tends in norm to PMw. But, by (10.16), PMw=0 and thereby, $\lim_{n\to\infty} S^n(R)\widehat{\ }(0) = 0$; now, $S^n(R)\widehat{\ }(0) = \hat{R}(0)$ and the $\gg 0$ matrix R must be identical to zero.

Finally, the matrix Mw consists in measures absolutely continuous with respect to $\tilde{\lambda}$, if w is associated to λ. This result, combined with theorem 10.1, gives the description of $\sigma_{max} \sim \sum_{j=0}^{k-1} \tilde{\lambda}_j$, which was our purpose. □

Remark 10.5. 1. We can prove a little bit more precise result : keeping the notations of the proof, if $\mu \in Mw$, then

$$\mu \ll \tilde{\lambda} \text{ and } \mu \perp \tilde{\lambda}_j \text{ for any } \tilde{\lambda}_j \neq \lambda.$$

Indeed, one can easily check the relation

$$\chi_a \mu = \widehat{\lambda}(a)\, \mu, \quad a \in \mathbf{Z} \tag{10.18}$$

and from this one, $\chi_a \tilde{\mu} = \widehat{\lambda}(a)\, \tilde{\mu}$. Let now λ' be one of the measures λ_j distinct from λ. There must exist $a \in \mathbf{Z}$ such that $\widehat{\lambda}(a) \neq \widehat{\lambda'}(a)$, so that χ_a takes two different constant values on $\tilde{\mu}$ and λ'. As already seen, this implies $\tilde{\mu} \perp \lambda'$ or, equivalently, $\mu \perp \tilde{\lambda}'$.

2. What can one say about pairs of positive measures satisfying identity (10.18) more generally, i.e. about (μ, λ) such that $\lim_{n\to\infty} \widehat{\mu}(aq^n + b) = \widehat{\lambda}(a)\widehat{\mu}(b)$ for every $a, b \in \mathbf{Z}$?
   Of course, λ must be q-invariant, and, if $\lambda = m$, then every $\mu \in M_0(\mathbf{T})$ can be so associated with λ. Note that the stronger property

$$\lim_{n\to\infty} \mu(S^{-n}A \cap B) = \nu(A)\mu(B), \quad \text{for every Borel sets } A, B$$

   implies the S-ergodicity of both involved probability measures μ and ν, and also $\nu \ll \sum_{n\geq 1} 2^{-n} S^n(\mu)$. For discussions about these properties see [171].
3. This theorem is constructive since, as for Σ, Z can be obtained as weak-star limit point of a matrix Riesz product and the diagonal measures λ_j, theoretically, can be explicitly computed. One may expect that the description of those "generalized Riesz products" can be carried out as in the commutative case.

Chapter 11
Spectral Multiplicity of General Automata

In the previous chapter, we made use of the bi-correlation matrix Z to get the maximal spectral type of the system (X_ζ, T) associated with a primitive and aperiodic substitution of length q (and height equal to one). The diagonal measures λ_j being q-strongly mixing, they must be equal or mutually singular and a better knowledge of the distinct ones is needed to estimate the spectral multiplicity of the system. However, the computation of Z is rather intricate and we begin by showing how to get these measures easily from the correlation matrix $\Sigma = (\sigma_{\alpha\beta})$. In the next section, we deduce the spectral multiplicity from Σ only and we close the chapter with examples.

11.1 More About the Spectrum of Constant-Length Substitutions

11.1.1 The Convex Set $\mathcal{K}$

We recall the few notations we need from the previous chapter. We considered the operators C and P in $\mathcal{L}(\mathcal{L}(\mathbf{C}^s), \mathcal{L}(\mathbf{C}^s))$, that we endow with the Hilbert structure associated to the inner product

$$\langle v, w\rangle = \mathrm{Tr}(v^* w), \quad \text{if } v, w \in \mathcal{L}(\mathbf{C}^s).$$

The eigenspace F, corresponding to the eigenvalue q for C, has been identified with Im P^*, the space of $v \in \mathcal{L}(\mathbf{C}^s)$ satisfying $vP = v$; also, $v = (v(\alpha,\beta)) \in F$ is quite well-determined by its values on the ergodic classes $E_0, \ldots, E_{k-1}$ of C, that we denote by $(v(j))$ if $v = v(j)$ on E_j. We are now interested in F_+ consisting in those $v \in F$ which are $\gg 0$ as operators on $\mathbf{C}^s$ and we assume $k \geq 2$.

Definition 11.1. We denote by $\mathcal{K}$ the set of measures

$$\{v\Sigma := \langle v, \Sigma\rangle = \sum_{\alpha,\beta} v(\alpha,\beta)\sigma_{\alpha\beta}, \ v \in F_+, \ v(0) = 1\}.$$

M. Queffélec, *Substitution Dynamical Systems – Spectral Analysis: Second Edition*, Lecture Notes in Mathematics 1294, DOI 10.1007/978-3-642-11212-6_11,
© Springer-Verlag Berlin Heidelberg 2010

Proposition 11.1. *$\mathcal{K}$ is a non-trivial convex set, consisting of q-invariant probability measures.*

Proof. Obviously $\mathcal{K}$ is a convex set containing $\delta = \sum_{\alpha,\beta} \sigma_{\alpha\beta}$ since $v = (1\cdots 1) \in F_+$. For every $v \in \mathcal{L}(\mathbf{C}^s)$ with $v \gg 0$, the measure $\sum_{\alpha,\beta} v(\alpha,\beta)\sigma_{\alpha\beta}$ is a nonnegative one, for Σ is a $\gg 0$ matrix of measures (chapter 7); moreover,

$$||v|| = \widehat{v}(0) = \sum_{\alpha,\beta} v(\alpha,\beta)\widehat{\sigma}_{\alpha\beta}(0) = \sum_{\alpha} v(\alpha,\alpha)\mu([\alpha])$$

$$= v(0)\sum_{\alpha} \mu([\alpha])$$

$$= 1$$

so that v is a probability measure. Now, denoting the q-shift by S again, we have

$$S(v) = \langle v, S(\Sigma)\rangle = \langle v, \tfrac{1}{q}C\Sigma\rangle \quad \text{by (10.4)}$$

$$= \tfrac{1}{q}\langle C^* v, \Sigma\rangle$$

$$= v \quad \text{since } C^* v = qv,$$

and v is q-invariant.

It remains to prove that $\mathcal{K}$ is not reduced to $\{\delta\}$. But, since $k \geq 2$, F_+ is not reduced to the only vector $(1\cdots 1)$ and the proposition will follow from the following lemma.

Lemma 11.1. *Let $v \in \mathcal{L}(\mathbf{C}^s)$ with $v \gg 0$ be such that $\sum_{\alpha,\beta} v(\alpha,\beta)\sigma_{\alpha\beta} = \delta$. Then $v(\alpha,\beta) = 1$ for every (α,β).*

Proof of the lemma. The operator v being $\gg 0$, there exist finitely many nonnegative numbers C_z and a complex s-dimensional vector $(z_\alpha)_{\alpha\in A}$ such that $v(\alpha,\beta) = \sum_{\text{finite}} C_z\, z_\alpha \overline{z_\beta}$ for every (α,β). Now, if

$$\delta = \sum_{\text{finite}} C_z (\sum_{\alpha,\beta} z_\alpha \overline{z_\beta} \sigma_{\alpha\beta})$$

$$= \sum_{\text{finite}} C_z\, \sigma_{\sum_\alpha z_\alpha \mathbf{1}_{[\alpha]}}$$

then, necessarily, each measure $\sum_\alpha z_\alpha \mathbf{1}_{[\alpha]}$ must be of the form $a\delta$ for some function $a(z)$. But, remember that for an ergodic dynamical system (X,μ,T), the property $\sigma_f \ll \delta$ imposes the L^2-function f to be constant (chapter 2). Thus, each step function $\sum_\alpha z_\alpha \mathbf{1}_{[\alpha]}$ must be constant, which implies $z_\alpha = z_\beta$ for each z, α, β, and finally $v(\alpha,\beta) = 1$ for every (α,β). □

The proof of the proposition is complete. □

We turn back our attention to the diagonal measures λ_j, appearing in the diagonal form of the matrix PZ. By their strong-mixing property, they must be equal or mutually singular and from now on, we write $(\lambda_j)_{j\in J}$ for the *distinct*, thus mutually singular, such measures which generate the spectrum of the substitution system.

Theorem 11.1. *The probability measures $\lambda_j, j \in J$, are the extreme points of the convex set $\mathcal{K}$.*

Proof. We interpret the theorem 10.1 in the following way : the matrix PZ can be decomposed into $PZ = \sum_{j\in J} \lambda_j P_j$, where P_j is a scalar projection operator such that $P_j \gg_{\text{op}} 0$ for every $j \in J$. Actually, if L_j is the L-space $\{\mu,\ \mu \ll \lambda_j\}$,

$$(PZ)_{|L_j} = \lambda_j P_j \tag{11.1}$$

and the property follows (see the proof of theorem 10.1). Moreover, we deduce from the mutual singularity of the $\lambda_j, j \in J$, that

$$P_i P_j = 0 \quad \text{if } i \neq j.$$

The proof goes as follows : we first show that every $\nu \in \mathcal{K}$ is a convex combination of λ_j's, and then, that *all* the λ_j, in fact, belong to $\mathcal{K}$.

$\star$ Let $e = (e(\alpha,\beta))$ be the $\gg 0$ element of $\mathcal{L}(\mathbf{C}^s)$ defined by

$$e(\alpha,\beta) = \begin{cases} 0 & \text{if } \alpha \neq \beta \\ \mu([\alpha]) & \text{if } \alpha = \beta \end{cases}$$

and put $e_j = P_j e$. Each e_j is a $\gg 0$ operator in Im P_j and, thanks to the identity $\Sigma = Ze$, we get

$$P\Sigma = PZe = \sum_{j\in J} \lambda_j e_j. \tag{11.2}$$

Now, if $\nu \in \mathcal{K}$, there exists $v \in F_+$ with $v(0) = 1$, such that

$$\nu = \langle v, \Sigma \rangle = \langle v, P\Sigma \rangle$$

since $F = \text{Im}\, P^*$; we deduce from (11.2) that

$$\nu = \sum_{j\in J} \lambda_j \Big(\sum_{\alpha,\beta} e_j(\alpha,\beta) v(\alpha,\beta) \Big) =: \sum_{j\in J} c_j \lambda_j$$

where the c_j are nonnegative numbers, the sum of which being

$$\begin{aligned} \textstyle\sum_{j\in J} c_j &= \textstyle\sum_{\alpha,\beta} v(\alpha,\beta) \big(\sum_{j\in J} e_j(\alpha,\beta) \big) \\ &= \textstyle\sum_{\alpha} v(\alpha) \mu([\alpha]) \\ &= v(0) = 1, \end{aligned}$$

by noticing that $\sum_{j\in J} e_j = P\widehat{\Sigma}(0) = \widehat{\Sigma}(0)$. This establishes that $\mathscr{K}$ is contained into the convex hull of the set of measures $\lambda_j, j \in J$.

$\star$ We shall see that all the $\lambda_j, j \in J$, are involved; since they are mutually singular, they need to be extreme points of $\mathscr{K}$ (see subsection 3.5.1), and the only ones, according to the first part of the proof.

First of all, observe that every e_j is nonzero. Let us assume, on the contrary, that $e_j = 0$ for some j. Since $\mu([\alpha]) > 0$ for every α, any $\gg 0$ operator w is dominated by e in the sense : $w \ll Ce$ for some positive constant C, so that $P_j w$ would be zero; this would imply $P_j = 0$, whence a contradiction.

Fix now λ_i; we shall exhibit some $v \in F_+$ such that $v(0) = 1$ and $\lambda_i = v\Sigma$.

For every $j \in J$, we consider $f_j = P_j^* e_j$; by (11.1) and the property $ZP = Z$, we clearly have $P_j P = P_j$ and thus $P^* f_j = f_j$; moreover, $P_j^* \gg_{\text{op}} 0$, just like P_j, so that, finally, f_j is a $\gg 0$ operator in F, i.e. $f_j \in F_+$. We compute the constant value $f_j(0)$ of f_j on the ergodic class E_0. On the one-hand

$$f_j(0) = \sum_\alpha f_j(\alpha,\alpha)\mu([\alpha]) = \langle f_j, e\rangle,$$

on the other-hand

$$\langle f_j, e\rangle = \langle P_j^* e_j, e\rangle = \langle e_j, e_j\rangle$$

which proves that $f_j(0) \neq 0$. We are thus led to put

$$w_j = \frac{f_j}{\langle e_j, e_j\rangle}, \quad \text{for every } \ j \in J$$

and we deduce $\langle w_j, e_j\rangle = 1$, while, for $i \neq j$,

$$\langle w_i, e_j\rangle = \langle P_i^* w_i, e_j\rangle = \langle w_i, P_i e_j\rangle = 0.$$

It follows from this and from (11.2) that

$$w_i P\Sigma = \sum_{j\in J} \lambda_j \langle w_i, e_j\rangle = \lambda_i$$

so that $\lambda_i = v\Sigma$ for a suitable $v := w_i P$; this means that $\lambda_i \in \mathscr{K}$ and the proof is complete. □

Observe that the extreme points of $\mathscr{K}$ are exactly the measures $v\Sigma$ where v is an extremal vector in the convex set $K = \{v \in F_+,\ \ v(0) = 1\}$.

11.1.2 Examples

1. *Rudin-Shapiro automaton.* We proved in the previous chapter that F_+ consists in $v = (v(0), v(1))$ with $|v(1)| \leq v(0)$. It follows that the extreme points of $\mathscr{K}$ are δ_0 corresponding to $(1,1)$ and m corresponding to $(1,-1)$, as already observed.

2. *Q-mirror automaton.* Here also, F_+ consists in $\nu = (\nu(0), \nu(1))$ with $|\nu(1)| \leq \nu(0)$ so that $\mathscr{K}$ possesses two extreme points, δ_0 corresponding to $(1,1)$ and ν corresponding to $(1,-1)$. With help of matrix Riesz products, we shall identify ν with the generalized Riesz product $\prod_{n\geq 0}(1-\cos 2^n t)$, as expected since the system arising from this automaton and the Thue-Morse system are isomorphic.
We write

$$\begin{aligned}
\nu &= \textstyle\sum_{(\alpha,\beta)\in E_0\cup T_0} \sigma_{\alpha\beta} - \sum_{(\alpha,\beta)\in E_1\cup T_1} \sigma_{\alpha\beta} \\
&= \textstyle\sum_{\alpha,\beta} q_0(\alpha,\beta)\sigma_{\alpha\beta} - \sum_{\alpha,\beta} q_1(\alpha,\beta)\sigma_{\alpha\beta} \\
&=: M - N
\end{aligned}$$

where M and N are 2-invariant measures (proposition 11.1). We claim that, for every $a \in \mathbf{Z}$,

$$\begin{cases} \widehat{M}(2a+1) = \frac{1}{2}\big(\widehat{N}(a) + \widehat{N}(a+1)\big) \\ \widehat{N}(2a+1) = \frac{1}{2}\big(\widehat{M}(a) + \widehat{M}(a+1)\big) \end{cases} \tag{11.3}$$

From theorem 8.1, Σ is easily seen to be of the following form

$$\begin{pmatrix} \sigma & \eta & \rho & \tau \\ \check{\eta} & \sigma' & \check{\tau} & \rho' \\ \rho & \tau & \sigma & \eta \\ \check{\tau} & \rho' & \check{\eta} & \sigma' \end{pmatrix} \tag{11.4}$$

where σ and ρ are positive measures, $\check{\mu}(E) = \overline{\mu(-E)}$ as usual and μ' is defined by

$$\widehat{\mu}'(2n+1) = 0, \quad \widehat{\mu}'(2n) = \frac{1}{2}\widehat{\mu}(n), \ \ n \in \mathbf{Z}.$$

Since $M = 2(\sigma + \sigma') + 2(\tau + \check{\tau})$ and $N = 2(\rho + \rho') + 2(\eta + \check{\eta})$, we need to know $\widehat{\Sigma}(2a+1)$ to state the claim. By using identity (8.7) with $p = 1$ and $b = 1$, we get

$$\widehat{\Sigma}(2a+1) = \frac{1}{2}\big({}^t R_0 \widehat{\Sigma}(a) R_1\big) + \frac{1}{2}\big({}^t R_1 \widehat{\Sigma}(a+1) R_0\big) \tag{11.5}$$

where

$$R_0 = \begin{pmatrix} 1 & 0 & 0 & 0 \\ 0 & 0 & 1 & 0 \\ 0 & 0 & 1 & 0 \\ 1 & 0 & 0 & 0 \end{pmatrix} \quad \text{and} \quad R_1 = \begin{pmatrix} 0 & 1 & 0 & 0 \\ 1 & 0 & 0 & 0 \\ 0 & 0 & 0 & 1 \\ 0 & 0 & 1 & 0 \end{pmatrix}$$

We deduce from (11.4) and (11.5) the following identities.

$$\widehat{\sigma}(2a+1) = \tfrac{1}{2}\big(\widehat{\eta}(a)+\widehat{\rho}'(a)+\widehat{\check{\eta}}(a+1)+\widehat{\rho}'(a+1)\big)$$

$$\widehat{\tau}(2a+1) = \tfrac{1}{2}\big(\widehat{\rho}(a)+\widehat{\check{\eta}}(a)\big)$$

$$\widehat{\check{\tau}}(2a+1) = \tfrac{1}{2}\big(\widehat{\rho}(a+1)+\widehat{\eta}(a+1)\big)$$

which give rise to

$$\begin{aligned}\widehat{M}(2a+1) &= 2\widehat{\sigma}(2a+1)+2(\tau+\check{\tau})^{\widehat{}}(2a+1)\\ &= (\rho+\rho'+\eta+\check{\eta})^{\widehat{}}(a)+(\rho+\rho'+\eta+\check{\eta})^{\widehat{}}(a+1)\\ &= \tfrac{1}{2}\big(\widehat{N}(a)+\widehat{N}(a+1)\big)\end{aligned}$$

and the analogous for N, which proves the claim. The probability measure $\nu := M-N$ thus satisfies the relation

$$\widehat{\nu}(2a+1) = -\frac{1}{2}\big(\widehat{\nu}(a)+\widehat{\nu}(a+1)\big)$$

which is, together with $\widehat{\nu}(0)=1$, the characteristic equation of the generalized Riesz product $\prod_{n\geq 0}(1-\cos 2^n t)$.
3. *Bijective automaton.* We restrict our analysis to the example that we detailed in chapter 9 :

$$\zeta(0)=001,\ \zeta(1)=122,\ \zeta(2)=210,$$

and for which $F_+ = \{(v_0,v_1),\ -v_0/2 \leq v_1 \leq v_0\}$. We thus deduce from theorem 11.1 that the spectrum is generated by the two extreme measures of $\mathscr{K}$, $\delta_0 = \sum_{\alpha,\beta}\sigma_{\alpha\beta}$ and the 3-invariant measure

$$\nu = \sum_{\alpha}\sigma_\alpha - \frac{1}{2}\sum_{\alpha\neq\beta}\sigma_{\alpha\beta}$$

since the second ergodic class E_1 is $\{(\alpha,\beta),\ \alpha\neq\beta\}$. It is convenient to let $T := \sum_\alpha \sigma_\alpha = \mathrm{Tr}(\Sigma)$, which is not 3-invariant, but leads to

$$\nu = T - \frac{1}{2}(\delta - T) = (3T-\delta)/2.$$

Starting from

$$\Sigma = \begin{pmatrix}\sigma & \rho & \check{\tau}\\ \check{\rho} & \eta & \pi\\ \tau & \check{\pi} & \lambda\end{pmatrix}$$

we deduce from identity (8.7) that the positive measure $M := \eta + \tau + \check{\tau}$ satisfies $\widehat{M}(3a) = 1/3$ for every $a \in \mathbf{Z}$; in particular, $\widehat{M}(0) = 1/3$ which, combined with $M\{0\} = 1/3$, gives $M = \delta_0/3$. In the same way, we get $\sigma + \pi + \check{\pi} = \lambda + \rho + \check{\rho} = \delta_0/3$. Using further identity on Σ, namely, $\sum_\alpha \sigma_{\alpha\beta} = \mu([\beta])\delta_0 = \delta_0/3$ in our case, we deduce that

$$\tau + \check{\tau} = \rho + \check{\pi} = \check{\rho} + \pi$$

$$\rho + \check{\rho} = \pi + \check{\tau} = \check{\pi} + \tau$$

$$\pi + \check{\pi} = \rho + \check{\tau} = \check{\rho} + \tau$$

which reduces to $\rho = \pi = \tau$ and then $\sigma = \eta = \lambda$. At this stage,

$$\Sigma = \begin{pmatrix} \sigma & \rho & \check{\rho} \\ \check{\rho} & \sigma & \rho \\ \rho & \check{\rho} & \sigma \end{pmatrix}$$

and $\nu = (9\sigma - \delta)/2$. Finally, taking the trace in identity (8.7) once more, with $q = 3$ and $p = 1$, we derive

$$T := 3\sigma = \frac{2}{3}m + \frac{\delta}{3} \quad \text{hence} \quad \nu = m.$$

We can precise the expression of Σ

$$\Sigma = \frac{1}{9}\begin{pmatrix} \delta + 2m & \delta - m & \delta - m \\ \delta - m & \delta + 2m & \delta - m \\ \delta - m & \delta - m & \delta + 2m \end{pmatrix}$$

In particular, this substitution has a Lebesgue spectrum in the orthocomplement of eigenfunctions. Also, the correlation measure denoted by σ_{χ_2} at the end of chapter 9 is thus identified with the Lebesgue measure m.

Question 11.1. 1. Does the automaton, more generally associated to the symmetric group of order n, $G = S_n$, admit a *Lebesgue component* as continuous spectrum ?
2. Is it possible to describe the automata providing a dynamical system with a Lebesgue component ?

11.2 Spectral Multiplicity of Constant-Length Substitutions

The computation of the spectral multiplicity of dynamical systems has raised many questions. The most famous one, generally attributed to Banach, asks for the possible existence of an ergodic dynamical system with a simple Lebesgue (reduced) spectrum; more generally, which essential ranges of the multiplicity function (chapter 2) are available ? J. Mathew and M. G. Nadkarni constructed in [177] a transformation with a Lebesgue component of multiplicity 2, and, by generalizing their example [178], they exhibit a transformation with a Lebesgue component of multiplicity $N\phi(N)$ for every integer $N \geq 2$ (ϕ, the Euler function). On the other hand, E. A. Robinson [214] provides, for every $m \geq 1$, a transformation with a singular component of multiplicity m.

It would be interesting to obtain such results with help of automata, and we discuss now this question before quoting additional results about spectral multiplicity.

11.2.1 Main Theorem on Spectral Multiplicity

As usual, we consider ζ a substitution of length q defined on the alphabet A with s letters. In chapter 7, we gave a first estimate of the spectral multiplicity $m(\zeta)$ of ζ : by considering for every $n \geq 0$ the invariant spaces

$$H_n = \Big[U, (\mathbf{1}_{\zeta^n[\alpha]}, \alpha \in A)\Big], \tag{11.6}$$

we proved that $\overline{\bigcup_{n\geq 0} H_n} = L^2(X,\mu)$, and we deduced $m(\zeta) \leq s$.

Later, we provided examples of commutative substitutions with $m(\zeta) = 1$ or 2, and, in the non-commutative case, denoting by θ the associated substitution on the group G generated by the instructions (ϕ_j) (chapters 8–9), we observed that $m(\theta)$ is always greater than 1.

Notation : We denote by $r(\Sigma)$ the *rank* of the matrix of measures Σ, which, in fact, is a function in $L^\infty(\sigma_{max})$: actually, writing $\Sigma = F \cdot \sigma_{max}$ with $F = (f_{ij})$, where the entries are functions in $L^1(\sigma_{max})$, $r(\Sigma)$ is a.s. equal to the rank of F, which is, indeed, a function in $L^\infty(\sigma_{max})$.

We have the following description of the spectral multiplicity.

Theorem 11.2. *The spectral multiplicity of ζ is the essential supremum of the function $r(\Sigma)$.*

We decompose the proof in two lemmas.

Lemma 11.2. *The spectral multiplicity of U on H_n does not depend on $n \geq 1$.*

Proof of the Lemma. According to the definition (11.6), $f \circ \zeta \in H_n$ if $f \in H_{n+1}$. Conversely, consider the map $f \in H_n \mapsto \widetilde{f} \in H_{n+1}$ where $\widetilde{f} = f \circ \zeta^{-1}$ on $\zeta(X)$ and $\widetilde{f} = 0$ elsewhere. From identity (7.3) we deduce that

$$\widehat{\sigma}_{\widetilde{f},\widetilde{g}}(k) = \begin{cases} 0 & \text{if } k \notin q\mathbf{Z} \\ \dfrac{1}{q}\widehat{\sigma}_{f,g}(k/q) & \text{if } q \text{ divides } k. \end{cases}$$

It follows that $\sigma_{f,g} = 0$ implies $\sigma_{\widetilde{f},\widetilde{g}} = 0$ so that the map $f \mapsto \widetilde{f}$ preserves the U-orthogonality. Assuming now that $H_n = [U, (f_j)]$ for some finite family of functions (f_j), we claim that $H_{n+1} = [U, (\widetilde{f_j})]$, which will give the lemma. Let us fix $g \in H_{n+1}$; g can be decomposed into $g = \sum_{k<q} g_k$ where g_k vanishes out of $T^k\zeta(X)$, so that, without loss of generality, we may assume g to be zero out of $\zeta(X)$. Now, $G = g \circ \zeta$ is a member of H_n and G can be approximated in L^2 by some combination $\sum_j R_j(U) f_j$, where the R_j are trigonometric polynomials. Thanks to the obvious identity

$$(R(U)f) \circ \zeta^{-1} = R(U^q) f \circ \zeta^{-1} \text{ on } \zeta(X),$$

we see that g itself can be approximated in L^2 by $\sum_j R_j(U^q)\widetilde{f}_j$, whence the claim. U has the same multiplicity on both H_n and H_{n+1}. □

We are thus led to compute the spectral multiplicity of U on H_1. Before that, we note a consequence of lemma 11.2, and we introduce a definition.

Definition 11.2. For any probability measure λ, we denote by $n(\lambda)$ and call *multiplicity of* λ, the maximal number of U-orthogonal functions in $L^2(X, \mu)$ whose spectral measure is λ.

With this definition, we deduce the following from the previous lemma and theorem 11.1.

Corollary 11.1. *The spectral multiplicity* $m(\zeta)$ *is equal to* $\sup_j n(\lambda_j)$ *where the* (λ_j) *are the distinct diagonal probability measures of the diagonalized matrix PZ (which generate the maximal spectral type).*

Lemma 11.3. *The spectral multiplicity of U on* H_1 *is equal to ess sup* $r(\Sigma)$.

Proof of the Lemma. Since the operator U restricted to H_d has simple spectrum, we focus on the matrix Σ_c, whose entries are the continuous part of measures of Σ. If, now, λ_j runs through the distinct and continuous diagonal probability measures of the diagonalized PZ (remember the purity law), Σ_c may be decomposed into

$$\Sigma_c = \sum_j E_j \cdot \lambda_j \quad \text{with} \quad E_j = \frac{d\Sigma_c}{d\lambda_j}.$$

It follows from proposition 7.11 that each matrix of functions E_j is positive definite λ_j-a.e, and satisfies via (10.4)

$$E_j \circ S = \frac{1}{q} C\, E_j.$$

Also, we have

$$r(\Sigma_c) = \sup_j r(E_j)$$

and, in view of corollary 11.1, it remains to establish that $r(E_j) = n(\lambda_j)$ (a.e.) for those λ_j.

For sake of simplicity, we assume that each E_j is a scalar operator, which condition is equivalent to the following : $P\Sigma_c = \Sigma_c$. Let λ be one of those measures λ_j and let w be an associated scalar positive definite operator. If L is the L-space $L(\lambda) := \{\mu,\ |\mu| \ll \lambda\}$, then $(\Sigma_c)_L = \Sigma_L = \lambda w$. Suppose the rank of w to be equal to r; we first prove that $n(\lambda) \geq r$.

Since w is a positive definite operator, there exists a unitary operator V diagonalizing w, so that, if $\theta_0, \theta_1, \ldots, \theta_{r-1}$ are the positive eigenvalues of w, we have

$$V^*\Sigma_L V = (V^*\Sigma V)_L = V^*wV \cdot \lambda$$

$$= \begin{pmatrix} \begin{pmatrix} \theta_0 & & 0 \\ & \cdot & \\ & & \cdot \\ 0 & & \theta_{r-1} \end{pmatrix} & 0 \\ 0 & 0 \end{pmatrix}. \tag{11.7}$$

We now exhibit U-orthogonal functions $f_0, \ldots, f_{r-1}$ in $L^2(X,\mu)$, satisfying $\sigma_{f_j} = \lambda$ for $0 \leq j \leq r-1$. For every $\gamma \in A$, we consider

$$F_\gamma = \sum_{\alpha \in A} \overline{v(\alpha,\gamma)} \mathbf{1}_{[\alpha]}$$

where $V := (v(\alpha,\beta))$. Then, for every $\gamma, \delta \in A$,

$$\begin{aligned} \sigma_{F_\gamma,F_\delta} &= \Sigma_{\alpha,\beta \in A} \overline{v(\alpha,\gamma)} v(\beta,\delta) \sigma_{\alpha\beta} \\ &= (V^*\Sigma V)_{\gamma\delta}. \end{aligned} \tag{11.8}$$

Finally, let E be a borelian support of the fixed measure λ, so that, for every $\sigma \in \Sigma$, we have both : $\mathbf{1}_E \sigma \in L$ and $\mathbf{1}_{E^c} \sigma \in L^\perp$. We claim that the functions

$$f_\gamma = \frac{1}{\sqrt{\theta_\gamma}} \mathbf{1}_E(U)\, F_\gamma,$$

with $0 \leq \gamma \leq r-1$, do the job. Actually, combining (11.7) and (11.8), it is easily checked that

$$\sigma_{f_\gamma} = \frac{1}{\theta_\gamma} \mathbf{1}_E \cdot \sigma_{F_\gamma} = \lambda,$$

and

$$\sigma_{f_\gamma, f_\delta} = \frac{1}{\sqrt{\theta_\gamma \theta_\delta}} \mathbf{1}_E \cdot \sigma_{F_\gamma, F_\delta} = 0 \quad \text{if } \gamma \neq \delta.$$

The functions (f_γ) are r U-orthogonal functions with identical spectral measure equal to λ, and thus $n(\lambda) \geq r = \text{rank}(w)$.

On the opposite direction, put $r = n(\lambda)$. One can thus find r U-orthogonal functions $f_0, \ldots, f_{r-1}$ in $L^2(X,\mu)$ such that $\sigma_{f_j} = \lambda$ for $0 \leq j \leq r-1$. In addition, keeping in mind lemma 11.2 and proposition 2.12, we see that those functions may be picked out of H_1. Consequently, for every $\varepsilon > 0$, there exists a finite family $(\phi_{i\alpha})_{\substack{0\leq i\leq r-1 \\ \alpha\in A}}$ of elements of $L^2(\sigma_{max})$, such that

$$\|f_i - \sum_{\alpha\in A} \phi_{i\alpha}(U)\mathbf{1}_{[\alpha]}\|_{L^2(X)} \leq \varepsilon, \quad \text{for } 0 \leq i \leq r-1.$$

It follows, for every $0 \leq i, j \leq r-1$, that

$$\|\sigma_{f_i,f_j} - \sum_{\alpha,\beta\in A} \phi_{i\alpha}\overline{\phi_{j\beta}}\sigma_{\alpha\beta}\|_{M(\mathbf{T})} \leq C\,\varepsilon \tag{11.9}$$

where the implied constant C only depends on $||f_i||$, $0 \leq i \leq r-1$.
Using now the assumption on the f_i, namely, that $(\sigma_{f_i,f_j})_L = \sigma_{f_i,f_j} = \delta_{ij}\,\lambda$ and $\Sigma_L = w\lambda$, we deduce from (11.9) that the matrix $I_r = (\delta_{ij})$ can be termwise approximated in $L^1(\lambda)$ by a matrix whose entries are the following $L^1(\lambda)$-functions

$$\Big(\sum_{\alpha,\beta\in A} \phi_{i\alpha}\overline{\phi_{j\beta}}w(\alpha,\beta)\Big)_{i,j}. \tag{11.10}$$

If the rank of w were less than or equal to $r-1$, each associated operator with entries given by (11.10), considered as a λ-a.e. defined scalar operator, would have a rank bounded by $r-1$. But, the map $A \to \text{rank}(A)$ being a l.s.c. function of A, the same property would hold for I_r, which leads to a contradiction. Therefore, the rank of w must be at least r and the lemma is proved. □
Proof of the Theorem. The theorem is a straightforward consequence of lemma 11.3, corollary 11.1 and proposition 2.12. □

11.2.2 Examples

We now give some applications of the previous theorem. In the following, "spectrum" always means "spectrum in the orthocomplement of the eigenfunctions".
1. *Generalized Rudin-Shapiro automata.* To be compared with [177], we first obtain

Theorem 11.3. *The Rudin-Shapiro automaton has a 2-fold Lebesgue spectrum.*

Proof. In chapter 8, we have computed the corresponding matrix Σ and obtained

$$\Sigma = \frac{1}{8}\begin{pmatrix} \omega+m & \gamma\omega & \gamma\omega & \omega-m \\ \overline{\gamma}\omega & \omega+m & \omega-m & \overline{\gamma}\omega \\ \overline{\gamma}\omega & \omega-m & \omega+m & \overline{\gamma}\omega \\ \omega-m & \gamma\omega & \gamma\omega & \omega+m \end{pmatrix},$$

where ω is the discrete measure $\frac{1}{2}(\delta_0 + \delta_\pi)$. It follows that $f = \sqrt{2}(\mathbf{1}_{[0]} - \mathbf{1}_{[3]})$ and $g = \sqrt{2}(\mathbf{1}_{[2]} - \mathbf{1}_{[1]})$ are two U-orthogonal functions with Lebesgue spectral measure and $m(\zeta)$ must be at least 2 : Actually,

$$\sigma_f = 2(\sigma_{00} + \sigma_{33} - \sigma_{03} - \sigma_{30}) = m,$$

and so is σ_g, while

$$\sigma_{f,g} = 2(\sigma_{02} + \sigma_{31} - \sigma_{32} - \sigma_{01}) = 0.$$

But directly, it is easy to apply theorem 11.2 since $\Sigma_c = P\Sigma_C = m \cdot w$, where the scalar operator

$$w = \begin{pmatrix} 1 & 0 & 0 & -1 \\ 0 & 1 & -1 & 0 \\ 0 & -1 & 1 & 0 \\ -1 & 0 & 0 & 1 \end{pmatrix}$$

has rank 2. □

In [203], following D. Rider [211], we consider generalized Rudin-Shapiro sequences with components in $\{e^{2\pi ik/q}, 0 \le k \le q-1\}$ and $q \ge 2$. For any fixed $q \ge 2$, this sequence can be derived from a q-automaton on the alphabet $\mathscr{A} = \mathbf{Z}/q\mathbf{Z} \times \mathbf{Z}/q\mathbf{Z}$ whose h^{th} instruction is given by

$$\phi_h(j) = ((j_2 + hj_1) \bmod q, h) \quad \text{if } j = (j_1, j_2) \in \mathscr{A} \text{ and } 0 \le h \le q-1.$$

The spectrum of the associated system is described in [203] and we quote the result (to be compared to [178]).

Theorem 11.4. *Let $q \ge 2$ and let φ denote the Euler indicator. The spectrum of the generalized Rudin-Shapiro q-automaton consists of a $q\varphi(q)$-fold Lebesgue component, and, for every non-trivial divisor d of q, of a $d\varphi(d)$-fold singular component (given by a generalized Riesz product).*

2. *Bijective automaton.* Turning back to our emblematic bijective substitution ζ on the alphabet $\{0,1,2\}$, defined by

$$\zeta(0) = 001,\ \zeta(1) = 122,\ \zeta(2) = 210,$$

we can now precise the spectral multiplicity of the associated system.

Proposition 11.2. *The previous bijective substitution has a 2-fold Lebesgue spectrum.*

Proof. This is a direct consequence of the description of the corresponding correlation matrix Σ (see subsection 11.1.2) which leads to

$$\Sigma_c = \frac{1}{9}\begin{pmatrix} 2m & -m & -m \\ -m & 2m & -m \\ -m & -m & 2m \end{pmatrix}$$

whose rank is equal to 2. □

Remark 11.1. The spectral multiplicity is not known in the general case where $G = S_n$ (chapter 9); as already observed, it would be interesting to explicit the relationship between the spectral multiplicity $m(\zeta)$ and the number $\sup\dim\rho$ where $\rho \in \mathscr{R}$, the set of irreducible unitary representations.

3. *Commutative automaton.* We try now to apply theorem 11.2 to general commutative automata (or substitutions), in which case, generating measures were easy to exhibit. In chapter 8 indeed, we showed that the correlation matrix Σ can be diagonalized, with generalized Riesz products on the diagonal : for each $\gamma \in \Gamma = \widehat{G}$,

$$\sigma_\gamma = \prod_{n\geq 0} P_\gamma(q^n t),$$

where $P_\gamma(t) = \frac{1}{q}\left|\gamma(\phi_0) + \gamma(\phi_1)e^{it} + \cdots + \gamma(\phi_{q-1})e^{i(q-1)t}\right|^2$ and $\phi_0 = id$, are those measures.
We first notice the following easy consequence of the results obtained in chapter 1.

Proposition 11.3. *For a commutative substitution ζ, $m(\zeta)$ is the number of distinct $\gamma \in \Gamma$ giving rise to identical polynomials P_γ.*

Proof. From the proof of lemma 11.3, $n(\sigma_\gamma)$ is exactly the number of γ leading to identical σ_γ. Now, according to corollary 1.1, $\sigma_\gamma = \sigma'_\gamma$ if and only if $P_\gamma = P'_\gamma$; otherwise, the measures are mutually singular. The proposition is proved. □

We already pointed out that the identity : $P_\gamma = P'_\gamma$ may occur with $\gamma \neq \gamma'$. Indeed, a condition such as $\gamma'(\phi_j) = \overline{\gamma(\phi_{q-j-1})}$ infers the coincidence between P_γ and P'_γ; in particular, if the automaton is *symmetrical (or palindromic)*, in other words, if $\phi_j = \phi_{q-j-1}$ for every j, then, obviously,

$$P_\gamma = P_{\bar{\gamma}}, \quad \sigma_\gamma = \sigma_{\bar{\gamma}},$$

so that

$$m(\zeta) \geq 2 \quad \text{if} \quad \gamma^2 \neq 1.$$

Now, the question naturally arises, whether multiplicity larger than 2 can be obtained with some commutative automaton. We do not think so and in this direction, J. Kwiatkowski and A. Sikorski in [156] proved the following.

Theorem 11.5. *Let ζ be a commutative substitution such that the so-called group G, generated by the instructions (ϕ_j), has a prime order p. Then $m(\zeta) \leq 2$ and $m(\zeta) = 2$ if and only if the automaton is symmetrical.*

Proof. We identify G with the additive group $\mathbf{Z}/p\mathbf{Z}$ and we put $\omega = e^{2\pi i/p}$; each character $\gamma \in \Gamma = \widehat{G}$ can thus be identified with some integer $k \mod p$ and the duality equation becomes

$$\gamma(g) = \omega^{kg} = e^{2\pi ikg/p}, \text{ for } \gamma \in \Gamma, \ g \in G.$$

For every $g \in G$ and $0 \leq r \leq q-1$, we denote by $S_r(g)$ the number of decompositions of g into $g_{j+r} - g_j$ with $j \geq 0$ and $j + r \leq q - 1$, the g_j being now the instructions of G.

Lemma 11.4. *Let $\gamma \neq \gamma'$ in Γ, identified with k and $k' =: hk$ for some $h \in \mathbf{Z}/p\mathbf{Z}$, $h \neq 1$. Then, $P_\gamma = P'_\gamma$ if and only if $S_r(g) = S_r(gh)$ for every $g \in G$ and $0 \leq r \leq q-1$.*

Proof of the Lemma. By expanding $P_\gamma = \frac{1}{q}\left|\sum_{j=0}^{q-1} \omega^{kg_j} e^{ijt}\right|^2$, we obtain

$$\widehat{P_\gamma}(r) = \frac{1}{q} \sum_{j=0}^{q-1-r} \omega^{k(g_{j+r}-g_j)} \quad \text{for } 0 \leq r < q.$$

Therefore, $P_\gamma = P'_\gamma$ means that for such r,

$$\sum_{j=0}^{q-1-r} \omega^{k(g_{j+r}-g_j)} = \sum_{j=0}^{q-1-r} \omega^{hk(g_{j+r}-g_j)},$$

or, equivalently,

$$\sum_{g \in G} \omega^{kg} S_r(g) = \sum_{g \in G} \omega^{hkg} S_r(g),$$

which may be written

$$\sum_{g \in G} \omega^{hkg}(S_r(hg) - S_r(g)) = 0, \tag{11.11}$$

since, $p = |G|$ being a prime number, G is also $\{hg, g \in G\}$.
Let us fix r. The degree of the minimal polynomial of ω^{hk} over $\mathbf{Q}$ is equal to $n-1$, while in (11.11), ω^{hk} is annihilating a polynomial $\in \mathbf{Q}[X]$ of degree $\leq n-2$. It follows that

$$S_r(hg) = S_r(g) \quad \text{for every } g \in G.$$

This is valid for any $0 \leq r \leq q-1$ and the lemma is proved. □
Turning back to the theorem, we shall prove successively that :

–if ζ is not symmetrical, then $m(\zeta) = 1$;

–if ζ is symmetrical, then $m(\zeta) = 2$.

We suppose now that $P_\gamma = P'_\gamma$ with $\gamma, \gamma' \in \Gamma$ identified with j and hj. We first remark that the identity $P_\gamma = P'_\gamma$ with $\gamma \neq \gamma'$ must imply $g_{q-1} = 0 = g_0$ since $\widehat{P_\gamma}(q-1) = \frac{1}{q}\omega^{k(g_{q-1})}$.

$\star$ Suppose the automaton to be non-symmetrical. There must exist $k \geq 1$ such that $g_0 = g_{q-1}, g_1 = g_{q-2}, \ldots, g_{k-1} = g_{q-k}$ but $g_k \neq g_{q-1-k}$. Put $\ell = q-k-1$. Then we have

$$S_\ell(g) - S_\ell(-g) = \begin{cases} 1 & \text{if } g = g_\ell \quad \text{or } g = -g_k \\ -1 & \text{if } g = g_k \quad \text{or } g = -g_\ell \\ 0 & \text{otherwise,} \end{cases}$$

and, using the previous lemma with the above defined h, we deduce that $S_\ell(hg) - S_\ell(-hg)$ is equal to 1, -1 or 0 for the same g's. If $h \neq 1$, this imposes

$$hg_\ell = -g_k \quad \text{and} \quad hg_k = -g_\ell;$$

combining these two identities, we derive that $h^2 = 1$, thus $h = -1$; this would entail $g_k = g_\ell$ and a contradiction. We have shown that $P_\gamma = P'_\gamma$ implies $\gamma = \gamma'$ and $m(\zeta) = 1$ in this case.

$\star$ Suppose now the automaton to be symmetrical, and set

$$k = \inf\{j,\ g_j \neq 0\},$$

so that $g_0 = g_1 = \cdots = g_{k-1} = g_{q-k} = \cdots = g_{q-2} = g_{q-1} = 0$, and $g_k = g_{q-1-k} \neq 0$. For $\ell = q-k-1$ we have $S_\ell(g_k) = 1$, $S_\ell(g) = 0$ if $g \neq \pm g_k$ and, applying lemma 11.4 with the same h again, we get simultaneously both $g = \pm g_k$ and $hg = \pm g_k$. Finally $h = \pm 1$, that is $\gamma' = \gamma$ or $\overline{\gamma}$, and $m(\zeta) = 2$.

The proof is complete. □

Remark 11.2. In the same paper, the authors study the commutative substitutions with associated $G = \mathbf{Z}/4\mathbf{Z}$ and prove in this case that $m(\zeta) = 1$ or 2. Actually, commutative substitutions with $m(\zeta) = 2$ can easily be exhibited : for instance, so is ζ defined on $\{0,1,2,3\}$ by $\zeta(0) = 0132$, $\zeta(1) = 1203$, $\zeta(2) = 2310$, $\zeta(3) = 3021$. (See also [102]).

11.2.3 More About Lebesgue Multiplicity

In this last paragraph, we give an account of the problems and recent results on spectral multiplicity. As already seen in [178, 203], a Lebesgue component may appear with an even multiplicity, but not every even number can be reached through these classes of examples. Whence the following questions about the Lebesgue multiplicity.

1. Can we exhibit, for every k, a dynamical system arising from some automaton (or substitution) with a $2k$-fold Lebesgue component ?

2. Can we exhibit a dynamical system arising from some automaton (or substitution) with an odd Lebesgue multiplicity ?
3. Can we exhibit a dynamical system with a simple Lebesgue component ?

In [160], M. Lemańczyk gives a positive answer to the first question. He deals with generalizations of Morse sequences, including previous generalizations introduced in [143] and in [55]. More precisely, let $\omega = \omega_0 \cdots \omega_{\ell-1}$ be some fixed word in A^* where $A = \{0,1,\infty\}$ and $\omega_{\ell-1} \neq \infty$. (Actually, symbols "$\infty$" represent "holes" in the sequence). Now, we define an infinite sequence $x_\omega \in \{0,1\}^{\mathbf{N}}$ by putting

$$x_\omega(n) = \mathrm{Card}\{i \geq 0,\ \omega_j = n_{i+j} \text{ for all } j \leq \ell - 1 \text{ such that } \omega_j \neq \infty\} \mod 2$$

if $n = \sum_{j\geq 0} n_j 2^j$; this means that $x_\omega(n)$ is the number modulo 2 of occurrences of ω in the binary expansion of n (up to the holes). For example, x_1 is the classical Thue-Morse sequence and x_{11} the classical Rudin-Shapiro sequence (see [12]). As a particular result, M. Lemańczyk proved the following :

Theorem 11.6 (Lemańczyk). *The dynamical system generated by the sequence x_ω with $\omega = 1\infty\cdots\infty 1$, or $\omega = 1\infty\cdots\infty\, 0$ has a Lebesgue spectrum of multiplicity $2^{|\omega|-1}$ in the orthocomplement of the eigenfunctions, and admits every dyadic number as an eigenvalue.*

Now, for any odd integer s, the transformation T^s is still ergodic and has a $2^k s$-fold Lebesgue component. The problem of "even Lebesgue multiplicity" is thus solved.

On the opposite, simple and partly continuous spectrum may appear with generalized Morse sequences [154] but the problem whether a Lebesgue component may appear in those cases is open. In [107], it is proved that the existence of a Lebesgue component depends on whether a construction of flat trigonometric polynomials with coefficients ± 1 is possible. More precisely,

Theorem 11.7 (Guénais). *There exists a Morse sequence for which the spectrum of the associated dynamical system is simple and contains a Lebesgue component if and only if there exists a sequence of trigonometric polynomials $P_n(x) = \sum_{j=0}^{p_n-1} \varepsilon_j(n) e^{2\pi i j x}$ on the circle ($p_n \geq 2$, $\varepsilon_j(n) = \pm 1$ for $j = 0,\ldots,p_n - 1$) such that $\lim_{n\to\infty} ||P_n||_1/\sqrt{p_n} = 1$.*

But the problem of whether such a sequence of polynomials exists remains an open problem in harmonic analysis [32, 205].

Chapter 12
Compact Automata

If now A is a compact metric set and if $\phi_0, \phi_1, \ldots, \phi_{q-1}$ are q continuous maps from A into A ($q \geq 2$), it is easy to define the automatic action of the family (ϕ_j) on A, and to exhibit, in this way, automatic sequences taking their values in the compact set A. Such an example is given by the sequence $(e^{2i\pi\alpha S_q(n)})$, where $S_q(n)$, as usual, denotes the sum of digits in the q-adic expansion of n, and where α is any irrational number.

In this chapter, we intend to extend the previous analysis to the associated dynamical systems, as far as possible.

12.1 Strictly Ergodic Automatic Flows

Let A be a compact metric set and let $\phi_0, \phi_1, \ldots, \phi_{q-1}$ be q continuous maps from A into A, that we still call *instructions*; the family (ϕ_j) gives rise to a semi-group of continuous maps from A into A, in the following way :
If $k \geq 1$ and $n < q^k$, we put as before

$$\phi_n^{(k)} = \phi_{n_0} \circ \phi_{n_1} \circ \cdots \circ \phi_{n_{k-1}} \tag{12.1}$$

if $n = n_0 + n_1 q + \cdots + n_{k-1} q^{k-1}$ is the q-adic expansion of n. According to the still valid relation

$$\phi_n^{(k)} \circ \phi_m^{(p)} = \phi_{n+mq^k}^{(k+p)} \quad \text{for} \quad n < q^k \quad \text{and} \quad m < q^p, \tag{12.2}$$

we see that the $(\phi_n^{(k)})$ form a semi-group Φ. We shall refer to Φ as the *automatic semi-group* generated by the instructions $(\phi_j)_{0 \leq j \leq q-1}$.

Definition 12.1. The pair $(A, (\phi_j))$ is called a *compact automaton of length q* or, in short, a *compact q-automaton*.

M. Queffélec, *Substitution Dynamical Systems – Spectral Analysis: Second Edition*, Lecture Notes in Mathematics 1294, DOI 10.1007/978-3-642-11212-6_12,
© Springer-Verlag Berlin Heidelberg 2010

If $q^{j-1} \leq n < q^j$ for some j, or equivalently, if $n = n_0 + n_1 q + \cdots + n_{j-1} q^{j-1}$ with $n_{j-1} \neq 0$, we then put, without ambiguity,

$$\phi_n = \phi_{n_0} \circ \phi_{n_1} \circ \cdots \circ \phi_{n_{j-1}}.$$

Definition 12.2. The sequence $t = (t_n)_{n \geq 0}$ in $A^{\mathbf{N}}$ is called *Φ-automatic* if there exists $x_0 \in A$ such that

$$t_n = \phi_n(x_0) \quad \text{for every} \quad n \geq 0.$$

Note that, for every $n < q^k$, $\phi_n(x_0)$ is nothing but $\phi_n^{(k)}(x_0)$ as soon as $\phi_0(x_0) = x_0$.

Actually, we shall study the *dynamical flow* (A, Φ), that is the action of the semi-group Φ on the compact set A, and deduce from that the properties of any dynamical system associated with a Φ-automatic sequence t. We need an additional assumption on Φ to go further :

Definition 12.3. We say that the dynamical flow (A, Φ) is *primitive* if, for every open set Ω in A, one can find an integer $k := k(\Omega)$, and, for any $x \in A$, an other integer $n < q^k$ such that

$$\phi_n^{(k)}(x) \in \Omega. \tag{12.3}$$

If $\phi_0 = id$ is the identity on A, this property of primitivity exactly means the density of the trajectories $(\phi_n(x))_{n \geq 0}$, for every $x \in A$, in other words, the *minimality* of the flow.

We denote by $C(A)$ the space of all continuous complex-valued functions on A endowed with the sup norm $||\cdot||$ and we consider the bounded linear operator M defined on $f \in C(A)$ by

$$Mf(x) = \frac{1}{q} \sum_{r=0}^{q-1} f(\phi_r(x)) \quad \text{for } x \in A.$$

We shall now prove our main result.

Theorem 12.1. *Let Φ be the automatic semi-group generated by the instructions $(\phi_j)_{0 \leq j \leq q-1}$. We assume that Φ is equicontinuous and that (A, Φ) is a primitive flow. Then, for every $f \in C(A)$, the sequence*

$$M^k f = \frac{1}{q^k} \sum_{r=0}^{q^k-1} f \circ \phi_n^{(k)}$$

converges in $C(A)$ to the limit $\int_A f \, d\mu$, where μ is the unique M-invariant Borel measure on A.

Remark 12.1. When A is a finite set of cardinality s, this theorem establishes the existence of frequency for each letter in any automatic sequence, under the primitivity assumption. In this case, indeed, the operator M is easily seen to be the

s-dimensional endomorphism whose matrix (in the canonical basis) is $M(\zeta)/q$ ($\zeta = \phi_0\phi_1\cdots\phi_{q-1}$). The unique invariant left vector of $M(\zeta)$, $(\mu([\alpha]))_{\alpha\in A}$, defines then a unique invariant linear form on $\mathbf{C}^s$.

Proof. We first show, for every $f \in C(A)$, that the sequence $(M^n f)$ converges uniformly on A to some constant function.

Since the semi-group Φ is assumed to be equicontinuous, so is the sequence $(M^n f)$ and therefore, $(M^n f)$ is a totally bounded thus relatively compact set in the closed ball centered at 0, with radius $||f||$. Let g be some cluster point of $(M^n f)$ in $C(A)$. If $k > \ell$ we have

$$\begin{aligned}||M^{k-\ell}g - g|| &\le ||M^{k-\ell}g - M^k f|| + ||M^k f - g|| \\ &\le ||g - M^\ell f|| + ||M^k f - g||\end{aligned}$$

since, obviously, M is a contraction on $C(A)$. If now $g = \lim_i M^{k_i} f$ where (k_i) is non-decreasing, we deduce from the above inequality that $g = \lim_i M^{k_{i+1}-k_i} g$, and g is as well a uniform cluster point of the sequence $(M^n g)$.

Under the primitivity assumption on Φ, we shall prove that g must be constant. We need, to that effect, an estimate of the visiting number : Card $\{n < q^{k+\ell},\ \phi_n^{(k+\ell)}(x) \in \Omega\}$, for every open set Ω and $k = k(\Omega)$ given by (12.3). More precisely, if $x \in A$ and $\ell \ge 0$, we shall see that

$$\text{Card}\ \{n < q^{k+\ell},\ \phi_n^{(k+\ell)}(x) \in \Omega\} \ge q^\ell. \tag{12.4}$$

Fix x in A and let $n = a + bq^k$ with $a < q^k$ and $b < q^\ell$, so that, from (12.2), $\phi_n^{(k+\ell)}(x) = \phi_a^{(k)} \circ \phi_b^{(\ell)}(x)$. Thanks to the primitivity and the choice of $k = k(\Omega)$, for any fixed $b < q^\ell$ one can find some $a := a_b < q^k$ such that $\phi_a^{(k)}(\phi_b^{(\ell)}(x)) \in \Omega$ whatever $x \in A$. Therefore, each of the q^ℓ integers : $n_0 = a_0,\ n_1 = a_1 + q^k, \ldots, n_{q^\ell-1} = a_{q^\ell-1} + (q^\ell - 1)q^k$ satisfies $\phi_n^{(k+\ell)}(x) \in \Omega$ and (12.4) is proved.

Let now $x_0 \in A$ be such that $||g|| = |g(x_0)|$, and, for every $\varepsilon > 0$, let us choose for Ω, an open set on which $|g| < \inf_A |g(x)| + \varepsilon$. Fix k and ℓ, and denote by (n_j) the finite sequence of integers satisfying

$$n_j < q^{k+\ell} \quad \text{and} \quad \phi_{n_j}^{(k+\ell)}(x_0) \in \Omega,$$

the cardinality of which being at least q^ℓ by (12.4). We may then write

$$|M^{k+\ell}g(x_0)| \le M^{k+\ell}|g|(x_0) = \frac{1}{q^{k+\ell}}\Big(\sum_{\substack{n<q^{k+\ell}\\ n\in(n_j)}} |g|(\phi_n^{(k+\ell)}(x_0)) + \sum_{\substack{n<q^{k+\ell}\\ n\notin(n_j)}} |g|(\phi_n^{(k+\ell)}(x_0))\Big),$$

and, according to the choice of Ω and referring to (12.4) again, we have

$$|M^{k+\ell}g(x_0)| \leq \frac{q^\ell}{q^{k+\ell}}(\inf_A |g| + \varepsilon) + \frac{q^{k+\ell} - q^\ell}{q^{k+\ell}}||g||$$

$$\leq \frac{1}{q^k}(\inf_A |g| + \varepsilon) + (1 - \frac{1}{q^k})|g(x_0)|.$$

Letting ℓ run over the sequence $k_{i+1} - k_i - k$ where (k_i) is such that $\lim_i M^{k_i} f = g$ (and $\lim_i M^{k_{i+1}-k_i} g = g$ in turn), we get, taking the limit on i,

$$|g(x_0)| \leq \frac{1}{q^k}(\inf_A |g| + \varepsilon) + (1 - \frac{1}{q^k})|g(x_0)|$$

and

$$\sup_A |g(x)| \leq \inf_A |g(x)| + \varepsilon.$$

This holds for every $\varepsilon > 0$ and $|g|$ must be constant. If we start with a nonnegative function f, g itself is constant, and we classically decompose $f = f_1 + f_2$ into $f_1^+ - f_1^- + i(f_2^+ - f_2^-)$ to get the claim for a complex function f.

We have proved that the only cluster points of the sequence $(M^n f)$ in $C(A)$ are the constants. It remains to prove that there exists a unique such one.

If M^* is the adjoint operator of M, acting on the space of complex regular measures supported by A, we consider the sequence of probability measures $\mu_N = \frac{1}{N} \sum_{n<N} M^{*n}(\delta_a)$, where a is some point in A. Let now μ be a weak-star cluster point of $(\mu_N)_N$; then, for every $f \in C(A)$ and $p \geq 0$,

$$\int_A M^p f \, d\mu = \int_A f \, d\mu$$

so that,

$$\int_A g \, d\mu = \int_A f \, d\mu$$

for every cluster point g of $(M^p f)_p$, and the constant g must be $\int_A f \, d\mu$. This proves that $M^n f$ converges uniformly on A to the constant $\int f \, d\mu$, and it is well-known that this property for every $f \in C(A)$ induces the existence of a unique M-invariant Borel measure on A. The proof is now complete. □

In the finite case (chapter 5), we proved that each word occurring in the substitutive sequence $u = \zeta(u)$ occurs with a positive frequency when the automaton is primitive. In the compact case (under same assumption), we deduce the following result.

Corollary 12.1. *Let (A, Φ) and M be chosen as in the previous theorem. Then, for every $\ell > 0$ and all functions $f_0, f_1, \ldots, f_{\ell-1}$ in $C(A)$, the sequence*

$$\frac{1}{q^k}\sum_{n=0}^{q^k-\ell-1} f_0(\phi_n^{(k)}(x))f_1(\phi_{n+1}^{(k)}(x))\cdots f_{\ell-1}(\phi_{n+\ell-1}^{(k)}(x))$$

converges to a constant, uniformly in $x\in A$.

Proof. Fix $k\geq 0$ and put

$$F_k(x)=\frac{1}{q^k}\sum_{n=0}^{q^k-\ell-1} f_0(\phi_n^{(k)}(x))f_1(\phi_{n+1}^{(k)}(x))\cdots f_{\ell-1}(\phi_{n+\ell-1}^{(k)}(x))$$

for $x\in A$. From the semi-group property (12.2), we derive

$$\begin{aligned} MF_k(x) &:= \tfrac{1}{q}\textstyle\sum_{r=0}^{q-1} F_k(\phi_r(x)) \\ &= \tfrac{1}{q^{k+1}}\textstyle\sum_{r=0}^{q-1}\sum_{n=0}^{q^k-\ell-1} f_0(\phi_{n+rq^k}^{(k+1)}(x))\cdots f_{\ell-1}(\phi_{n+\ell-1+rq^k}^{(k+1)}(x)), \end{aligned}$$

so that, for every $p\geq 1$ and $k\geq 1$,

$$||M^pF_k-F_{k+p}||\leq ||f_0||\cdots||f_{\ell-1}||\,\frac{\ell}{q^k}.$$

If μ is the unique M-invariant probability measure on A, we see that the sequence $(\int F_k\,d\mu)_k$ is a Cauchy sequence in $\mathbf{C}$: indeed

$$\begin{aligned} \left|\int_A F_{k+p}\,d\mu-\int_A F_k\,d\mu\right| &= \left|\int_A F_{k+p}\,d\mu-\int_A M^pF_k\,d\mu\right| \\ &\leq ||F_{k+p}-M^pF_k|| \\ &= O(\frac{1}{q^k}). \end{aligned}$$

For any fixed k, the sequence $(M^pF_k)_p$ converges uniformly on A to $C_k:=\int_A F_k\,d\mu$ by the previous theorem, and $C=\lim_{k\to\infty}C_k$ exists by the above remark. From those convergences and from the following inequality

$$||F_{k+p}-C||\leq ||F_{k+p}-M^pF_k||+||M^pF_k-C_k||+|C_k-C|,$$

we deduce the uniform convergence of F_k to the constant C. This proves the corollary. □

Corollary 12.2. *Let (A,Φ) be chosen as in the previous theorem, with moreover $\phi_0=id$. Then, for any Φ-automatic sequence t, the associated topological system (X,T) is uniquely ergodic, where T, as usual, denotes the one-sided shift acting on the closed orbit X of t in the compact set $A^{\mathbf{N}}$.*

Proof. It is sufficient to establish the pointwise convergence to a constant of the sequence $(\frac{1}{q^k}\sum_{n<q^k} F\circ T^n)_k$, for a complete class of continuous functions on X. Corollary 12.1, combined with the assumption $\phi_0 = id$, settles the convergence for every function of the form $F = f_0 \otimes f_1 \otimes \cdots \otimes f_{\ell-1}$ and corollary 12.2 follows. □

Example 12.1. We consider $A = \mathbf{U}$ and the compact automaton of length q, with instructions

$$\phi_0 = id,\ \ \phi_1 = \cdots = \phi_{q-1} = R_\alpha,$$

where R_α is the irrational rotation $z \mapsto ze^{2\pi i\alpha}$. For any $x_0 \in \mathbf{R}/\mathbf{Z}$, we define the unimodular sequence $t = (t_n)$ by

$$t_n = \phi_n(e^{2\pi i x_0}) = e^{2\pi i x_0} e^{2\pi i S_q(n)\alpha}$$

as easily checked. By applying theorem 12.1, we rediscover Kamae's result [134] stating the unique ergodicity of the system generated by t, and, by the way, the uniform distribution modulo 1 of the sequence $(S_q(n)\alpha)$; recall that this latter result has been proved by M. Mendes France by using the correlation measure (chapter 4) (see also [58, 179]).

12.2 Application to Bounded Remainder Sets

Let $u = (u_n)_{n\geq 0}$ be a real sequence.

Definition 12.4. If u is uniformly distributed mod 1, we call *discrepancy* of the interval $I \subset [0,1)$ of length $|I|$, the sequence

$$D_N(I) = Card\{n < N,\ u_n \in I \bmod 1\} - N|I|.$$

For any such sequence u, it is well-known [222] that infinitely many N can be found for which $\sup_I |D_N(I)| > C\log N$. But $D_N(I)$ may happen to be very small for some I; consider indeed $u_n = n\alpha$ where $\alpha \in [0,1)$ is an irrational number, and $I = [1-\alpha, 1)$. Then $\mathbf{1}_I(\{n\alpha\}) = [(n+1)\alpha] - [n\alpha]$ and

$$D_N(I) = \sum_{n<N} \mathbf{1}_I(\{n\alpha\}) - N\alpha = [N\alpha] - N\alpha$$

so that $|D_N(I)| < 1$. Actually, Kesten proved in [146] a conjecture from Erdòs and Szüsz concerning this sequence : $D_N(I) = O(1)$ if and only if $|I| \in \mathbf{Z}\alpha\,(\mathrm{mod}\,1)$. His proof involves the continuous fraction expansion of α. A few years later, an ergodic proof of this characterization was given by Furstenberg, Keynes and Shapiro [96], and by Petersen simultaneously [199]; they observed that, putting $f = \mathbf{1}_I - |I|$, then $D_N(I) = \sum_{n<N} f\circ R_\alpha(0)$ is obtained from a cocycle over the irrational rotation.

Under the assumption : $D_N(I) = O(1)$, we derive that $||\sum_{n<N} f \circ R_\alpha||_{L^2}$ must be bounded : Actually, for any $x = j\alpha$, $j \in \mathbf{N}$, we have that

$$\Big|\sum_{n=0}^{N-1} f \circ R_\alpha^n(j\alpha)\Big| = \Big|\sum_{k=j}^{j+N-1} f(k\alpha)\Big| \leq \Big|\sum_{k=0}^{N+j-1} f(k\alpha)\Big| + \Big|\sum_{k=0}^{j-1} f(k\alpha)\Big| = O(1).$$

Since the sequence $(j\alpha)_{j\in\mathbf{N}}$ is dense and since each function $\sum_{n=0}^{N-1} f \circ R_\alpha^n$ has at most a finite number of jumps, we get a uniform bound

$$\Big|\Big|\sum_{n=0}^{N-1} f \circ R_\alpha^n\Big|\Big|_\infty = O(1).$$

A fortiori the L^2-norm must be bounded. Finally, we conclude with theorem 2.6 : f is a coboundary, that is $f(x) = g(x+\alpha) - g(x)$ for some $g \in L^2(\mathbf{T})$. Also

$$e^{2i\pi f(x)} = G(x+\alpha)\overline{G(x)} = e^{2i\pi|I|}.$$

which means that $e^{2i\pi|I|}$ is an eigenvalue for the irrational rotation R_α, whence the conclusion. (See also [199]).

We turn back to the subsequence $(S_q(n))$ where $q \geq 2$ and we ask for estimates of the discrepancy for the real sequence $u = (u_n)$ with $u_n = S_q(n)\alpha$ and $\alpha \notin \mathbf{Q}$. But now, $\mathrm{Card}\{n < N, S_q(n)\alpha \in I \bmod 1\}$ no longer appears as a cocycle on $(\mathbf{T}, R, m)$. Nevertheless, following P. Liardet [167], we consider the strictly ergodic dynamical system (X, T, μ) where $X = \overline{O(u)}$ and T the one-sided shift on X (see example 12.1). The discrete factor of this system can easily be identified to the q-odometer. For every bounded and μ-a.e. continuous function F on X,

$$\lim_{N\to\infty} \frac{1}{N} \sum_{n<N} F(T^n u) = \int_X F\, d\mu.$$

Putting then $\pi(x) = x_0$, the projection of $x = (x_n)_{n\geq 0} \in X$ onto $\mathbf{T}$, and letting $F := f \circ \pi$ with $f = \mathbf{I}_I - |I|$, we see that $D_N = \sum_0^{N-1} F(T^n u)$, therefore D_N is a cocycle on (X, T, μ). The previous ergodic method may be applied.

Suppose that $D_N = O(1)$. First of all, observe that theorem 2.6 implies that

$$F = f \circ \pi = G - G \circ T \quad \text{where} \quad G \in L^2(X, \mu).$$

Thus, $e^{2\pi i G(Tx)} = e^{2\pi i G(x)} e^{2\pi i|I|}$ and $e^{2\pi i|I|}$ is an eigenvalue of $U = U_T$ on $L^2(X, \mu)$ so that $|I|$ needs to be a q-adic rational number : $|I| = p/q^r \pmod 1$ [134].

To go further, we shall make use of the correlation measures.

Lemma 12.1. *Let U be an isometry of the Hilbert space H. Then $F \in H$ is a coboundary*

$$F = G - UG, \quad \text{with } G \in H$$

if and only if

$$\frac{1}{\sin^2 \pi t} \in L^1(\sigma_F)$$

where σ_F is the spectral measure of F defined, for $n \geq 0$, by

$$\widehat{\sigma}_F(n) = \langle U^n F, F \rangle_H \quad \text{and } \widehat{\sigma}_F(-n) = \overline{\widehat{\sigma}_F(n)}.$$

Proof. It is easy, for $F = G - UG$, to check that

$$\begin{aligned}
\widehat{\sigma}_F(n) &= 2\widehat{\sigma}_G(n) - \widehat{\sigma}_G(n-1) - \widehat{\sigma}_G(n+1) \\
&= \int_{\mathbf{T}} (2e^{2\pi int} - e^{2\pi i(n-1)t} - e^{2\pi i(n+1)t})\, d\sigma_G(t) \\
&= \int_{\mathbf{T}} e^{2\pi int} \cdot 4\sin^2 \pi t\, d\sigma_G(t)
\end{aligned}$$

i.e. $\sigma_F = 4\sin^2 \pi t \cdot \sigma_G$; in particular, $\sigma_F\{0\} = 0$ and

$$\| \sigma_G \| = \| G \|_{L^2}^2 = \int_{\mathbf{T}} \frac{d\sigma_F(t)}{4\sin^2 \pi t}.$$

Conversely, suppose that $\dfrac{1}{\sin^2 \pi t} \in L^1(\sigma_F)$. Using (2.1), we have

$$\| F + UF + \cdots + U^{n-1} \|_2^2 = \| \sigma_{F+UF+\cdots+U^{n-1}F} \| .$$

But, recall that $\sigma_{\varphi(F)} = | \varphi |^2 \sigma_F$, thanks to the isometry : $[U,F] \simeq L^2(\sigma_F)$ (theorem 2.1); thus

$$\begin{aligned}
\sigma_{F+UF+\cdots+U^{n-1}F} &= | 1 + e^{2\pi it} + \cdots + e^{2\pi i(n-1)t} |^2 \sigma_F \\
&= \left| \frac{e^{2\pi int} - 1}{e^{2\pi it} - 1} \right|^2 \sigma_F .
\end{aligned}$$

The density functions $\left| \dfrac{e^{2\pi int} - 1}{e^{2\pi it} - 1} \right|^2$ are dominated by $\dfrac{1}{\sin^2 \pi t}$; therefore, this sequence is uniformly bounded in L^1 and the sums $F + UF + \cdots + U^{n-1}F$, in turn, are bounded in L^2, which gives the equivalence via theorem 2.6. □

Applying this lemma with $F \in L^2(\mu)$, we deduce from the construction of μ that

$$\begin{aligned}\widehat{\sigma}_F(k) &= \int_X F \circ T^k \cdot \overline{F}\, d\mu \\ &= \lim_{N\to\infty} \frac{1}{N} \sum_{n<N} F(T^{n+k}u)\, \overline{F(u)} \\ &= \lim_{N\to\infty} \frac{1}{N} \sum_{n<N} f(u_{n+k})\, \overline{f(u_n)} \\ &= \widehat{\sigma}_{f(u)}(k) \quad \text{for } k \geq 0.\end{aligned}$$

We thus have to compute the correlation measure of the sequence $f(u)$ in order to decide whether $\dfrac{1}{\sin^2 \pi t} \in L^1(\sigma_{f(u)})$. The Fourier expansion of $f \in L^2(\mathbf{T})$,

$$f(t) = \sum_{m\in\mathbf{Z}^*} \widehat{\mathbf{1}}_I(m) e^{2\pi imt} =: \sum_{m\in\mathbf{Z}^*} c_m e_m(t)$$

leads to $\sigma_{f(u)} = \sum_{m\in\mathbf{Z}^*} |c_m|^2\ \sigma_{e_m}$ where σ_{e_m} is nothing but the correlation measure of the sequence $(e^{2\pi im\alpha S_q(n)})$. Fortunately, in chapter 3, we already identified this measure with a generalized Riesz product : putting $z = e^{2\pi im\alpha}$, in fact we have

$$\sigma_{e_m} = \rho_z := \prod_{n\geq 0} P_z(q^n t),$$

with $P_z = \frac{1}{q}\ |1 + z\, e^{2\pi it} + z^2 e^{2\pi i2t} + \cdots + z^{q-1}\, e^{2\pi i(q-1)t}|^2$.

Lemma 12.2. *Let ρ_z be the above generalized Riesz product and $\beta := m\alpha$; then*

$$\frac{1}{\sin^2 \pi t} \in L^1(\rho_z) \ \textit{if and only if} \ \left|\frac{\sin \pi q\beta}{\sin \pi\beta}\right| < 1.$$

Proof. We simply write ρ instead of ρ_z. For $k \geq 1$, consider

$$J_k := \int_{1/q^k}^{1/q^{k-1}} \frac{1}{|t|^2}\, d\rho(t) =: \int_{I_k} \frac{1}{|t|^2}\, d\rho(t).$$

Clearly,

$$\frac{1}{\sin^2 \pi t} \in L^1(\rho) \quad \text{if and only if} \quad \sum_{k\geq 1} J_k < +\infty.$$

We are left to compare J_k and J_{k+1}. Remember that ρ is q-invariant and D_q-quasi-invariant (section 3.5), so that we may write

$$J_k = \sum_{j=0}^{q-1} \int_{1/q^{k+1}+j/q}^{1/q^k+j/q} \frac{1}{|\, qt \,|^2}\, d\rho(t)$$

$$= \frac{1}{q^2} \sum_{j=0}^{q-1} \int_{I_{k+1}} \frac{1}{|\, t \,|^2}\, d(\rho * \delta_{j/q})(t)$$

$$= \frac{1}{q} \int_{I_{k+1}} \frac{1}{|\, t \,|^2} \frac{1}{P(t)}\, d\rho(t).$$

For k going to infinity, $P(t) \sim P(0)$ on I_{k+1} and $J_k \sim \frac{1}{q\,P(0)} J_{k+1}$. But

$$P(0) = \frac{1}{q} \,|\, 1 + z + \cdots + z^{q-1} \,|^2 = \frac{1}{q} \left| \frac{e^{2\pi i q\beta} - 1}{e^{2\pi i\beta} - 1} \right|^2$$

$$= \frac{1}{q} \frac{\sin^2 \pi q\beta}{\sin^2 \pi\beta}.$$

We deduce that $\frac{J_{k+1}}{J_k} \sim_{k\to\infty} (\frac{\sin \pi q\beta}{\sin \pi\beta})^2$ and $\sum_{k\geq 1} J_k$ converges if and only if $|\sin \pi q\beta \,| < |\, \sin \pi\beta \,|$ (note that equality never holds since $\beta \notin \mathbf{Q}$). □

Definition 12.5. Given a sequence $u = (u_n)_{n\geq 0}$ of elements of a compact metric space X, a subset B of X is called a *bounded remainder set* for u if there exists a $b \in [0,1]$ such that $\mathrm{Card}\{n < N,\, u_n \in B\} - Nb$ is bounded as a function of N.

As a consequence of those lemmas, one can deduce [167]

Theorem 12.2. *: Let $\alpha \notin \mathbf{Q}$, $q \geq 2$ and $u = (S_q(n)\alpha)_{n\geq 0}$. The unique bounded remainder intervals for u are the trivial ones.*

Proof. If $m\alpha$ is close to zero for some m, so is $qm\alpha$ and thereby, the condition : $\left|\frac{\sin \pi q m\alpha}{\sin \pi m\alpha}\right| < 1$ does not hold. By applying lemma 12.2 with $\beta = m\alpha$, we get that $\int \frac{1}{\sin^2 \pi t}\, d\sigma_{e_m}$ diverges for at least one m. Now recall that

$$\sigma_{f(u)} = \sum_{m\neq 0} |\, \widehat{\mathbf{I}}_I(m) \,|^2 \, \sigma_{e_m} \text{ with } |\, \widehat{\mathbf{I}}_I(m) \,| = \left| \frac{\sin \pi m|I|}{\pi m} \right|,$$

and $|I|$ being a q-adic rational number. Suppose $|I|$ to be different from zero or one. For any irrational number α, one can find a sequence (m_k) satisfying $m_k\alpha \to 0$ mod 1 and m_k prime to q. Thus, one can find $m \in (m_k)$ such that $\widehat{\mathbf{I}}(m) \neq 0$ and $\int \frac{1}{\sin^2 \pi t} d\sigma_{e_m} = \infty$; lemma 12.1 is thus failing and I is not a bounded remainder set. □

Remark 12.2. 1. Precise estimates of $D_N = \sup_I D_N(I)$ have been worked out by Drmota and Gräbner for the sequence $(S_q(n)\alpha)$, in terms of rational approximations of α.

2. For other results on bounded remainder sets see [83, 207].
3. In the wake of its proof of Kesten's result, Petersen studied particular cocycles over the irrational rotation with source $\varphi = \mathbf{1}_{[0,\beta[}$, and he deduced in [199]:

Proposition 12.1. *The following properties are equivalent :*

(i) $\sum_{n\geq 1} \frac{||n\beta||^2}{n^2||n\alpha||^2} < +\infty$;

(ii) $e^{2\pi i\varphi}$ *is a multiplicative coboundary over the irrational rotation;*

(iii) $\beta \in \mathbf{Z}\alpha$.

Appendix A
Schrödinger Operators with Substitutive Potential

The Schrödinger equation on $\mathbf{R}^n$: $(-\Delta + V(x))\psi(x) = E\psi(x)$, where Δ is the Laplace operator, appears, in many problems related to Quantum Mechanics, as the eigenvalue equation for a one-electron energy operator H, $H\psi = -\Delta\psi + V \cdot \psi$, when V describes the potential energy of the electron.

The discovery of Quasicrystals, in 1984, has motivated a great interest in models provided by discrete Schrödinger operators acting on $\ell^2(\mathbf{Z}^\nu)$. The one-dimensional case with potentials taking finitely many values on the infinite chain $\mathbf{Z}$, or on the semi-infinite chain $\mathbf{N}$, has attracted particular attention and has been considerably studied. It is, of course, impossible to cite all the contributions and progresses. We would just like to explain the role plaid by substitutions in this area.

A.1 Classical Facts on 1D Discrete Schrödinger Operators

A.1.1 Preliminaries

We consider $\mathscr{L}$, the space of bilateral complex sequences, and H, the *discrete Schrödinger operator* defined on $\varphi = (\varphi(n))_{n\in\mathbf{Z}} \in \mathscr{L}$ by

$$(H\varphi)(n) = \varphi(n+1) + \varphi(n-1) + v(n)\varphi(n), \quad n \in \mathbf{Z}, \tag{A.1}$$

where the *potential* $v = (v(n))_{n\in\mathbf{Z}}$ is a bounded and real sequence. It is easily checked that H restricted to $\ell^2(\mathbf{Z})$ is a self-adjoint bounded operator.

We are interested in the description of the spectral invariants of $H \in \mathscr{B}(\ell^2(\mathbf{Z}))$ according to the properties of the sequence v, more precisely, we wish to have a good knowledge of

1. $sp(H) = \{\lambda \in \mathbf{R},\ H - \lambda I \text{ non-invertible}\}$;

2. σ, maximal spectral type of H.

Recall that the *spectral measure* $\sigma_{f,g}$ of $f,g \in \ell^2(\mathbf{Z})$ is defined by

$$\int_{\mathbf{R}} x^k \, d\sigma_{f,g}(x) = \langle H^k f, g \rangle_{\ell^2(\mathbf{Z})}, \ k \geq 0$$

($\sigma_{f,f}$ is denoted by σ_f) and the maximal spectral type σ is a bounded positive measure on $\mathbf{R}$, defined up to equivalence, such that

$$\sigma_{f,g} \ll \sigma \quad \text{for every } f,g \in \ell^2(\mathbf{Z}).$$

As in the unitary case, $sp(H)$ is the topological support of σ and the possible eigenvalues of H are the discrete point masses of σ.

First Observations

1. *If* $v \equiv 0$*, then* $H = \Delta$ *and* σ *is equivalent to the Lebesgue measure on* $[-2,2]$.
   To prove this claim, we observe that H is conjugate to M_ω, the multiplication operator by $\omega : t \mapsto 2\cos t$ on $L^2(\mathbf{T})$, through the canonical isometric isomorphism $\varphi \in \ell^2(\mathbf{Z}) \mapsto \Phi \in L^2(\mathbf{T})$ with $\Phi(t) = \sum_{n\in\mathbf{Z}} \varphi(n) e^{int}$. It is clear that $sp(M_\omega) = [-2,2]$ and so is $sp(H)$ by conjugation. The maximal spectral type σ of H is generated by the spectral measures σ_{e_k} with $e_k(n) = \delta_{nk}$ for $n,k \in \mathbf{Z}$; but, for every $k \in \mathbf{Z}$, σ_{e_k} is the pull-back under ω of the Lebesgue measure on $\mathbf{T}$. In particular,
   $$\begin{aligned} \langle H^n e_0, e_0 \rangle &= \tfrac{1}{2\pi} \textstyle\int_0^{2\pi} (2\cos t)^n dt \\ &= \textstyle\int_{-2}^{2} \frac{x^n}{\pi\sqrt{4-x^2}} dx, \end{aligned}$$
   and σ_{e_0} is absolutely continuous with respect to the Lebesgue measure on $[-2,2]$ with $1/\pi\sqrt{4-x^2}$ as its density.
2. On the opposite, the operator $\varphi \mapsto (v(n)\varphi(n))_{n\in\mathbf{Z}}$ has a pure point spectrum since the $e_k, k \in \mathbf{Z}$, form a complete family of eigenvectors.
   Thus, in the general case, H is a perturbation of Δ by this operator and all is possible!
3. *General case.* For $n,m \in \mathbf{Z}$, we denote by $\sigma_{n,m}$ the spectral measure σ_{e_n,e_m} and we put $\sigma_n = \sigma_{n,n}$.

   Proposition A.1. *The maximal spectral type (up to equivalence) is*
   $$\sigma = \sigma_0 + \sigma_{-1}.$$

   This is an easy consequence of the following lemma.

   Lemma A.1. *For every* $n \in \mathbf{Z}$*,* $e_n = p_n(H)e_0 + q_n(H)e_{-1}$ *where* p_n *and* q_n *are polynomials in* $\mathbf{R}[X]$.

   Proof of the Lemma. This can be established by induction on n, starting with
   $$\begin{cases} p_{-1}(t) = 0, & q_{-1}(t) = 1 \\ p_0(t) = 1, & q_0(t) = 0 \end{cases}$$

Then,

$$\begin{aligned} He_n(k) &= e_n(k+1) + e_n(k-1) + v(k)e_n(k) \\ &= e_{n-1}(k) + e_{n+1}(k) + v(k)e_n(k) \end{aligned}$$

so that, $e_{n+1} = He_n - v(n)e_n - e_{n-1} = p_{n+1}(H)e_0 + q_{n+1}(H)e_{-1}$ by induction on $n \geq 0$, with

$$\begin{cases} p_{n+1}(t) = (t - v(n))p_n(t) - p_{n-1}(t) \\ q_{n+1}(t) = (t - v(n))q_n(t) - q_{n-1}(t) \end{cases} \tag{A.2}$$

Writing $e_{n-1} = He_n - v(n)e_n - e_{n+1}$ when $n \leq 0$ and proceeding in the same way, we see that p_n and q_n have degree n if $n \geq 0$ and $-n-1$ if $n \leq -1$. □

Proof of the Proposition. If we put $\mathbf{S} = \begin{pmatrix} \sigma_0 & \sigma_{0,-1} \\ \sigma_{-1,0} & \sigma_{-1} \end{pmatrix}$, we deduce from the above lemma that

$$\sigma_{n,m} = (p_n, q_n)\, \mathbf{S} \begin{pmatrix} p_m \\ q_m \end{pmatrix};$$

in particular, we have $\sigma_{n,m} \ll \sigma_0 + \sigma_{-1}$ for every $n, m \in \mathbf{Z}$, which gives the proposition. □

Remark A.1. A system with absolutely continuous spectrum behaves like a conductor while a system with pure point spectrum behaves like an insulator. We shall see that periodic potentials lead to absolutely continuous spectrum; on the opposite, random potentials give rise (almost surely) to pure point spectrum and, for potentials intermediate between those two extreme cases, one may expect the presence of a singular continuous component.

A.1.2 Schrödinger Equation

We begin by investigating the possible eigenvalues of H. A good reference for this section is [22].

Definition A.1. $E \in \mathbf{R}$ is an *eigenvalue* of H if there exists $\varphi \in \ell^2(\mathbf{Z})$, called an *eigenvector* of H, such that $H\varphi = E\varphi$; in other words φ is a solution of the (tight-binding) *Schrödinger equation*

$$\psi(n+1) + \psi(n-1) + v(n)\psi(n) = E\psi(n), \quad n \in \mathbf{Z}. \tag{A.3}$$

Equation (A.3) is nothing but an order-two linear recurrence equation that involves matrices in $SL(2, \mathbf{R})$ and the solutions form a two-dimensional vector space of $\mathscr{L}$. Actually, equation (A.3) for ψ means,

$$\begin{pmatrix} \psi(n+1) \\ \psi(n) \end{pmatrix} = g_n \begin{pmatrix} \psi(n) \\ \psi(n-1) \end{pmatrix} \quad \text{for every } n \in \mathbf{Z}, \tag{A.4}$$

with $g_n = \begin{pmatrix} E - v(n) & -1 \\ 1 & 0 \end{pmatrix} \in SL(2, \mathbf{R})$. By iterating (A.4), we get

$$\begin{pmatrix} \psi(n+1) \\ \psi(n) \end{pmatrix} = S_n \begin{pmatrix} \psi(0) \\ \psi(-1) \end{pmatrix} \text{ if } n \geq 0 \text{ with } S_n = g_n g_{n-1} \cdots g_0$$

and

$$\begin{pmatrix} \psi(n) \\ \psi(n-1) \end{pmatrix} = S_n \begin{pmatrix} \psi(0) \\ \psi(-1) \end{pmatrix} \text{ if } n \leq -1 \text{ with } S_n = g_n^{-1} g_{n-1}^{-1} \cdots g_{-1}^{-1}.$$

Note that, by (A.2),

$$S_n := S_n(E) = \begin{pmatrix} p_{n+1}(E) & q_{n+1}(E) \\ p_n(E) & q_n(E) \end{pmatrix} \quad \text{if } n \geq 0$$

$$= \begin{pmatrix} p_n(E) & q_n(E) \\ p_{n-1}(E) & q_{n-1}(E) \end{pmatrix} \quad \text{if } n \leq -1.$$

Definition A.2. The matrices (g_n) are called the *transfer matrices.*

Proposition A.2. *The eigenvalues of H are simple : for every $E \in \mathbf{R}$, there exists at most one solution $\varphi \in \ell^2(\mathbf{Z})$ satisfying (A.3) (up to a multiplicative constant).*

Proof. Indeed, for every $E \in \mathbf{C}$, equation (A.3) admits two solutions in $\mathscr{L}$, and if φ and ψ are two such solutions, the wronskian

$$W_n(\varphi, \psi) = \varphi(n)\psi(n-1) - \varphi(n-1)\psi(n) \quad \text{if } n \geq 0$$

$$= \det \begin{pmatrix} \varphi(n) & \psi(n) \\ \varphi(n-1) & \psi(n-1) \end{pmatrix} = \det S_{n-1} \cdot \det \begin{pmatrix} \varphi(0) & \psi(0) \\ \varphi(-1) & \psi(-1) \end{pmatrix}$$

$$= W_0(\varphi, \psi)$$

and the analogue for $n \leq -1$. This proves that the wronskian must be constant and, if φ and ψ are in $\ell^2(\mathbf{Z})$, this constant must be zero. The eigenvectors are proportional. □

We recall some classical facts on the spectrum of a self-adjoint operator H on a Hilbert space $\mathscr{H}$. If $E \in sp(H)$ is an isolated point of $\sigma(H)$ then E is an eigenvalue. On the other hand, a theorem due to H. Weyl states that a point of $\sigma(H)$ which is not an eigenvalue is an *approximate eigenvalue* in the following sense :

Definition A.3. E is an *approximate eigenvalue* of H if there exists a sequence (x_k) of elements in $\mathscr{H}$ such that $||x_k|| = 1$ and

$$\lim_{k\to\infty} ||Hx_k - Ex_k|| = 0.$$

Let us get back to the Schrödinger operator.

Definition A.4. A solution $\varphi \in \mathscr{L}$ of (A.3) is said to be *polynomially bounded* if there exists $k \geq 1$ such that $\varphi(n) = O(1+|n|^2)^k$ as $n \to \pm\infty$.

One can prove the following by constructing an approximate sequence with help of a truncation.

Proposition A.3. *Let $E \in \mathbf{R}$. If the equation (A.3) has a polynomially bounded solution φ, then E belongs to $sp(H)$.*

If $\varphi \in \mathscr{L}$ is such a solution, E is called a *generalized eigenvalue* and φ a *generalized eigenvector* of H.

This provides a first description of the spectrum of H that we admit [22].

Corollary A.1. *$sp(H)$ is the closure of the set*

$$\{E \in \mathbf{R},\ \textit{for which (A.3) has a polynomially bounded solution}\}$$

To decide whether E is an eigenvalue of H or not, we are thus led to study the behaviour of the sequence $(S_n(E))$ as $n \to \pm\infty$.

Before continuing, we need some notations and properties of $SL(2,\mathbf{R})$ or $SL(2,\mathbf{C})$, that we summarize below.

1. If $g \in SL(2,\mathbf{R})$, we note $||g||$ the operator norm of g, i.e. $||g|| = \sup_{||x||_2=1} ||gx||_2$ where $||\cdot||_2$ is the euclidian norm on $\mathbf{R}^2$. If $r(g)$ denotes the *spectral radius* of g, $r(g) = \lim_{n\to\infty} ||g^n||^{1/n}$.
2. If $g = \begin{pmatrix} a & b \\ c & d \end{pmatrix} \in SL(2,\mathbf{C})$, $||g||$ can be computed in terms of the coefficients :
$$||g||^2 = \frac{1}{2}(K^2 + \sqrt{K^2-4}) \quad \text{where } K^2 = |a|^2+|b|^2+|c|^2+|d|^2,$$
and $||g|| = ||g^{-1}||$.
3. We shall be needing the Cayley-Hamilton identity
$$g^2 = Tr(g)\cdot g - I \tag{A.5}$$
which infers
$$Tr(g^2) = Tr^2(g) - 2 \tag{A.6}$$
and $Tr(g) = Tr(g^{-1})$. By (A.5) indeed, $g^{-1} = Tr(g)\cdot I - g$ and $Tr(g^{-1}) = 2Tr(g) - Tr(g) = Tr(g)$.
4. Another consequence of the Cayley-Hamilton identity is the following inequality : If $x \in \mathbf{R}^2$,
$$||x|| \leq 2\sup\{|Tr(g)|\cdot||gx||, ||g^2x||\} \tag{A.7}$$

5. **Fricke Formula :** We shall also need the following generalization of (A.6).

$$\text{Let} \quad g,\, h \in SL(2,\mathbf{C}); \quad \text{then we have} \tag{A.8}$$

$$Tr^2(g)+Tr^2(h)+Tr^2(gh) = Tr(ghg^{-1}h^{-1})+2+Tr(g)+Tr(h)+Tr(gh).$$

6. **Lyapunov exponents.** If $S_n = T_1 \cdots T_n$ is a product of elements of $SL(2,\mathbf{C})$, $(||S_n||)_{n\geq 1}$ is a sequence of numbers ≥ 1 by item 2. and we put

$$\gamma_n = \frac{1}{n}\log||S_n||.$$

If the limit γ of (γ_n) exists as $n \longrightarrow \infty$ (for instance, in the ergodic case as we shall see), then γ is called the *Lyapunov exponent* of this product of matrices.

Theorem A.1 (Oseledeč). *Let $(g_n)_{n\in\mathbf{N}^*}$ be a sequence of matrices in $SL(2,\mathbf{C})$ satisfying.*

a)
$$\lim_{n\longrightarrow+\infty} \frac{1}{n} \log ||g_n g_{n-1} \cdots g_1|| = \gamma$$

b)
$$\lim_{n\longrightarrow+\infty} \frac{1}{n} \log ||g_n|| = 0.$$

Then, there exists a nonzero vector x such that

$$\lim_{n\longrightarrow\infty} \frac{1}{n} \log ||g_n \cdots g_1 x|| = -\gamma$$

and, for any w independent of x,

$$\lim_{n\longrightarrow\infty} \frac{1}{n} \log ||g_n \cdots g_1 w|| = \gamma.$$

Consider now the sequence $(S_n)_{n\in\mathbf{Z}}$ where $S_n = S_n(E)$. If φ and ψ are two fundamental solutions of (A.3), then, for every $n \in \mathbf{Z}$,

$$||S_n||^2 = |\varphi(n)|^2 + |\varphi(n+1)|^2 + |\psi(n)|^2 + |\psi(n+1)|^2,$$

since $\begin{pmatrix} \varphi(0) & \psi(0) \\ \varphi(-1) & \psi(-1) \end{pmatrix} = I$, and the asymptotic behaviour of the solutions follows from the asymptotic behaviour of $||S_n||$. Suppose, for instance, that the positive $\gamma_n := \gamma_n(E)$ admit a limit $\gamma := \gamma(E) > 0$ as $n \to +\infty$. By the previous Oseledeč theorem, one of the two vectors $(\varphi(n+1), \varphi(n))$ and $(\psi(n+1), \psi(n))$ goes to infinity while the other goes to zero, with exponential rate.

A.1.3 Periodic Potential

The Floquet theory for periodic differential equations can be transposed in this context of difference equation and thus leads to a complete spectral description of H in this case. Those systems with periodic potential will be useful later, when we aim to study substitutive Schrödinger operators by means of periodic approximations.

Suppose v to be a periodic sequence with period-length $p : v_{n+p} = v_n$ for every $n \in \mathbf{Z}$. With our notations,

$$g_{n+p} = g_n \quad \text{for every } n \in \mathbf{Z} \quad \text{where} \quad g_n := g_n(E, v).$$

The block $S_p = g_p g_{p-1} \cdots g_1$ will play a specific role and we put $\mathbf{t}(E) = Tr(S_p(E))$. Note that $\mathbf{t}$ is a polynomial function of E, of degree p, with leading coefficient equal to one. By the Cayley-Hamilton identity, S_p is a solution of the equation

$$X^2 - \mathbf{t}(E)X + 1 = 0, \quad X \in SL(2, \mathbf{R}).$$

Proposition A.4. *$E \in sp(H)$ if and only if $|\mathbf{t}(E)| \leq 2$.*

Proof. $\triangleleft$ let E be such that $|\mathbf{t}(E)| \leq 2$. Then S_p has two complex conjugate eigenvalues and it is easily checked that $|Tr(S_p^k)| \leq 2$ for all $k \in \mathbf{Z}$.

Assume now that E is not in $sp(H)$. For every $k \in \mathbf{Z}$, one can find $\varphi \in \ell^2(\mathbf{Z})$, $\varphi \neq 0$, such that $(H - EI)\,\varphi = e_k$. In particular, choosing $k = 0$, there must exist $\varphi \in \ell^2(\mathbf{Z})$ such that

$$\varphi(n+1) + \varphi(n-1) + (v(n) - E)\,\varphi(n) = \begin{cases} 0 & \text{if } n \neq 0 \\ 1 & \text{if } n = 0. \end{cases}$$

This implies that one of the two vectors $(\varphi(1), \varphi(0))$ or $(\varphi(0), \varphi(-1))$ is nonzero. Suppose, for instance, that this is the first one, denoted by x. Since $S_p^k x = \begin{pmatrix} \varphi(kp+1) \\ \varphi(kp) \end{pmatrix}$ for every $k \geq 0$, we get, by applying inequality (A.7) with $g = S_p^k$, that

$$\sup\left\{2(|\varphi(kp+1)|^2 + |\varphi(kp)|^2),\ |\varphi(2kp+1)|^2 + |\varphi(2kp)|^2\right\} \geq \|x\|$$

which is incompatible with $\varphi \in \ell^2(\mathbf{Z})$. (If we need to consider the second vector, we would get a contradiction by looking at the negative k).

$\triangleleft$ We now assume that $|\mathbf{t}(E)| > 2$; the eigenvalues of S_p are the real numbers λ and $1/\lambda$, where, say, $|\lambda| > 1$. There exists $w \neq 0$ with $\|S_{np}w\| = \|S_p^n w\| = O\left(\frac{1}{|\lambda|^n}\right)$ and $\|S_p^n x\| = O(|\lambda|^n)$ for any x non-proportional to w. It follows that every solution of the Schrödinger equation must increase with exponential rate in one direction at least; E cannot be in $sp(H)$ in view of corollary A.1. □

Using the spectral radius formula, one easily sees that the Lyapunov exponent exists and is equal to :

$$\gamma(E) = \frac{1}{p} \log r(S_p(E)).$$

The following alternative to the proposition is clear.

Proposition A.5. *$E \in sp(H)$ if and only if $\gamma(E) = 0$.*

Remark A.2. We can precise the previous proposition in the following way : E is in $sp(H)$ if and only if there exists $\varphi \in \ell^\infty(\mathbf{Z})$ and θ in $\mathbf{R}$ such that

$$H\varphi = E\varphi, \quad Tr(S_p(E)) = 2\cos\theta \quad \text{and} \quad \tau^p(\varphi) = e^{i\theta}\varphi$$

where τ is the backward shift on $\ell^\infty(\mathbf{Z})$. This is the discrete version of the *Floquet theorem.*

Corollary A.2. *The spectrum of the periodic Schrödinger operator H is the union of p closed intervals with disjoint interiors :*

$$sp(H) = \mathbf{t}^{-1}([-2,2]) =: \bigcup_{m=1}^{p} J_m.$$

Proof. $H\varphi = E\varphi$ formally means :

$$\begin{pmatrix} \cdot & \cdot & \cdot & \cdot & \cdot & \cdot \\ \cdots & 1 & E - v(-1) & 1\cdots & & \\ \cdots & \cdot & 1 & E - v(0) & 1\cdots & \\ \cdots & \cdot & \cdot & 1 & E - v(1) & 1\cdots \\ \cdot & \cdot & \cdot & \cdot & \cdot & \cdot \end{pmatrix} \varphi = 0$$

where the matrix is of the Jacobi type. Now, previous Floquet's theorem says that $E \in sp(H)$ if and only if $\det(E - V(\theta)) = 0$ for some real θ, where

$$V(\theta) = \begin{pmatrix} v(1) & 1 & \cdot & \cdot & \cdot & e^{-i\theta} \\ 1 & v(2) & 1 & \cdot & \cdot & \cdot \\ 0 & 1 & \cdot & \cdot & \cdot & \cdot \\ \cdot & \cdot & \cdot & \cdot & \cdot & 1 \\ e^{i\theta} & \cdot & \cdot & \cdot & 1 & v(p) \end{pmatrix}$$

But $V(\theta)$ is symmetric, and $\det(E - V(\theta)) = \mathbf{t}(E) - 2\cos\theta$. It follows that, for every $t \in [-2,2]$, the equation $\mathbf{t}(E) = t$ admits p real roots, whence the description of $\mathbf{t}^{-1}([-2,2])$. □

The following theorem gives a final answer to the spectral problem in the periodic case.

Theorem A.2. *Let H be the Schrödinger operator with a periodic potential v. Then H has an absolutely continuous spectrum.*

Proof. $\star$ Following the same scheme, we see that H admits no eigenvalues : a non-trivial solution φ of $(H-EI)\,\varphi=0$ cannot belong to $\ell^2(\mathbf{Z})$.

$\star$ In order to describe the maximal spectral type σ of H, we interpret H as an operator acting on $L^2(\mathbf{T},\mathbf{C}^p)$. Consider indeed the canonical isomorphism between $\ell^2(\mathbf{Z})$ and $L^2(\mathbf{T})$, $\varphi\mapsto\Phi$, with $\Phi(t)=\sum_{n\in\mathbf{Z}}\varphi(n)\,e^{int}$. By the Fourier inversion formula, the periodic potential v may be written under the form

$$v(n)=\sum_{j=1}^{p}a_j\,e^{2\pi i n j/p}=\hat{\nu}(n)$$

where ν is the discrete measure $\sum_{j=1}^{p}a_j\,\delta_{\frac{2\pi j}{p}}$. Therefore, H is conjugate to the operator $\widehat{H}$ acting on $L^2(\mathbf{T})$ in this way :

$$\begin{aligned}(\widehat{H}\Phi)(t)&=2\cos t\,\Phi(t)+\textstyle\sum_{n\in\mathbf{Z}}\sum_{j=1}^{p}a_j e^{2\pi i n j/p}\,\varphi(n)\,e^{int}\\ &=2\cos t\,\Phi(t)+\textstyle\sum_{j=1}^{p}a_j\Phi\left(t+\frac{2\pi j}{p}\right).\end{aligned}$$

Putting more generally $F_j(t)=F\left(t+\dfrac{2\pi j}{p}\right)$ and denoting by $\mathbf{F}$ the vector-valued function with components $(F_j), 1\le j\le p$, $\widehat{H}$ extends to a multiplication operator on $L^2(\mathbf{T},\mathbf{C}^p)$ by the following formula :

$$\mathbf{F}(t)\mapsto M(t)\cdot\mathbf{F}(t),\quad \text{where } M(t)=D(t)+G,$$

D being the diagonal matrix $(d_{jj})_{1\le j\le p}$ with $d_{jj}(t)=2\cos\left(t+\dfrac{2\pi j}{p}\right)$ and G the circulant matrix

$$\begin{pmatrix}0 & a_1 & 0 & \cdot & \cdot & 0\\ 0 & 0 & a_2 & \cdot & \cdot & 0\\ \cdot & & & & & \cdot\\ \cdot & & & & & a_{p-1}\\ a_p & 0 & 0 & \cdot & \cdot & 0\end{pmatrix}.$$

In particular, $\widehat{H}$ has an absolutely continuous spectrum and the same holds for H. $\square$

Example A.1. Consider $v(n)=(-1)^n$. Then $\mathbf{t}=E^2-3$ and

$$sp(H)=\{E,\;-2\le E^2-3\le 2\}=[-\sqrt{5},-1]\cup[1,\sqrt{5}].$$

In addition, following the proof of the theorem, if we consider

$$\mathbf{F}(t) = (F(t), F(t+\pi)),$$

then $\widehat{H}\mathbf{F}(t) = M(t) \cdot \mathbf{F}(t)$ with

$$M(t) = \begin{pmatrix} 2\cos t & 1 \\ 1 & -2\cos t \end{pmatrix}$$

The spectrum $sp(H)$ can be rediscovered as the union of the images of the continuous version of the eigenvalues :

$$sp(H) = \bigcup_{i=1}^{2} \lambda_i(\mathbf{T}),$$

$\lambda_1(t)$, $\lambda_2(t)$ being, for each t, the eigenvalues of $M(t)$; in particular, in our example,

$$\begin{aligned} \lambda_1(t) &= \sqrt{1+4\cos^2 t} \\ \lambda_2(t) &= -\sqrt{1+4\cos^2 t} \end{aligned}$$

and $sp(H) = [-\sqrt{5}, -1] \cup [1, \sqrt{5}]$ again.

Remark A.3. 1. Note that the action of $\widehat{H}$ may be expressed by :

$$\widehat{H}(F) = 2\cos t \cdot F + F * \nu,$$

and clearly, $\widehat{H}$ extends to an operator on $M(\mathbf{T})$ by the same formula : $\widehat{H}(\rho) = 2\cos t \cdot \rho + \rho * \nu$, if $\rho \in M(\mathbf{T})$, as soon as $v(n) = \hat{\nu}(n)$, for some $\nu \in M(\mathbf{T})$.

2. In [108], a different point of view is developed, centered on the dynamical system induced by $\mathbf{t}$.

A.2 Ergodic Family of Schrödinger Operators

Henceforth, one will be interested in a potential $v = (v(n))_{n\in\mathbf{Z}} \in \mathscr{L}$ generating an ergodic subshift (Ω, T, μ), and in some cases, a strictly ergodic one, that is a compact, uniquely ergodic and minimal dynamical system under the shift's action. To each $\omega \in \Omega$, we associate the Schrödinger operator H_ω with potential $\omega = (\omega(n))$ and we put $H := H_v$. Before obtaining general results for the specific operators H_v, μ-almost sure results on the family (H_ω), $\omega \in \Omega$, will be deduced from the ergodic theory. Complements can be found in [22].

A.2.1 General Properties

We can already make these preliminary simple observations :
1. In all cases, we readily have :

Proposition A.6. *If (Ω, T, μ) is a subshift, then*

$$H_{T\omega} = U^* H_\omega U$$

where $U := U_\tau$ is the unitary operator associated with τ, the shift on $\ell^2(\mathbf{Z})$.

Thus $sp(H_\omega) = sp(H_{T\omega})$. If σ_ω is the maximal spectral type of H_ω and, for $n \in \mathbf{Z}$, if $\sigma_{\omega,n}$ denotes the spectral measure of e_n, relative to H_ω, then $\sigma_{T\omega,n} = \sigma_{\omega,n+1}$ for every $n \in \mathbf{Z}, \omega \in \Omega$.

We denote by Σ the spectrum of H and $G = \{\omega \in \Omega,\ sp(H_\omega) = \Sigma\}$. G is a non-empty and T-invariant subset of Ω by the previous proposition A.6. If (Ω, T, μ) is ergodic, $\mu(G) = 0$ or 1 and, finally,

$$sp(H_\omega) = \Sigma \quad \text{for } \mu - a.e.\ \omega$$

since $\mu(G) > 0$.
2. Suppose now the system (Ω, T) to be minimal. We have :

Proposition A.7. *If (Ω, T, μ) is a minimal subshift, then $sp(H_\omega) = sp(H_{\omega'})$ for all $\omega, \omega' \in \Omega$ i.e. the spectrum of H_ω does not depend on ω.*

This will be a consequence of the following, interesting in itself, lemma.

Lemma A.2. *Let H_n, H be self-adjoint operators on the Hilbert space $\mathscr{H}$, such that H_n tends to H in the strong operator topology. Then*

$$sp(H) \subset \limsup \mathrm{top}(sp(H_n)) := \bigcap_n \overline{\bigcup_{m \geq n} sp(H_m)}.$$

Proof of the lemma. Let $E \notin \overline{\bigcup_{m \geq n} sp(H_m)}$ for some fixed n. Then, the distance $d_n := \mathrm{dist}(E, \bigcup_{m \geq n} sp(H_m))$ must be positive. As $(H_m - EI)^{-1}$, in turn, is self-adjoint,

$$\|(H_m - EI)^{-1}\| = \sup_{\mu \in sp(H_m)} \left| \frac{1}{\mu - E} \right| \leq \frac{1}{d_n} \qquad (m \geq n).$$

by applying the spectral image theorem. Thus, $\|H_m f - Ef\| \geq d_n \|f\|$ for every $f \in \mathscr{H}$. Taking the strong limit on m, we obtain that

$$\|Hf - Ef\| \geq d_n \|f\|;$$

$(H - EI)$ being itself self-adjoint, we conclude that $E \notin sp(H)$. □

Proof of the proposition. As usual, Ω is endowed with the metric $d(\omega,\omega') = \sum_{\mathbf{Z}} \frac{1}{2^{|n|}} |\omega(n) - \omega'(n)|$. If $d(\omega,\omega') \leq \varepsilon$, there exists N such that $\omega_{[-N,N]} = \omega'_{[-N,N]}$, and for every $f \in \mathscr{H}$,

$$\|H_\omega f - H_{\omega'} f\|^2 \leq C \sum_{|n|>N} |f(n)|^2.$$

This proves that, for any fixed $f \in \mathscr{H}$, the map $\omega \mapsto \|H_\omega f\|$ is continuous on Ω. But, by definition of Ω, $\omega \in \Omega$ if $\omega = \lim_{i\to\infty} T^{n_i} v$ for some sequence (n_i), so that, $H_{T^{n_i} v}$ converges strongly to H_ω. The lemma, combined with proposition A.6, implies that

$$sp(H_\omega) \subset \limsup_{i\to\infty} \operatorname{top}(sp(H_{T^{n_i} v})) \subset \Sigma.$$

Since the system is assumed to be minimal, Ω is the closed orbit of any of its points, and for every $\omega \in \Omega$, $v = \lim_{j\to\infty} T^{m_j} \omega$ for a suitable sequence (m_j). Finally,

$$\Sigma \subset sp H_\omega \quad \text{for every } \omega \in \Omega$$

and the proposition is proved. □

3. We do not know whether the spectral measure σ_ω depends on ω, and if it does, how. But one can prove the following.

Proposition A.8. *Assume that the system (Ω, T, μ) is ergodic and fix $E \in \mathbf{R}$. Then $\mu\{\omega \in \Omega,\ \sigma_\omega\{E\} > 0\} = 0$. As a consequence, $\kappa = \int_\Omega \sigma_\omega \, d\mu(\omega)$ is a continuous measure.*

Proof. Put $\Omega_E = \{\omega \in \Omega,\ \sigma_\omega\{E\} > 0\}$. The complement set Ω_E^c is clearly invariant since $\sigma_{T\omega}\{E\} = \sigma_{T\omega,0}\{E\} + \sigma_{T\omega,-1}\{E\} = \sigma_{\omega,1}\{E\} + \sigma_{\omega,0}\{E\}$ by proposition A.1; hence $\sigma_\omega\{E\} = 0$ implies $\sigma_{T\omega}\{E\} = 0$ in turn. It follows that $\mu(\Omega_E) = 0$ or 1 by ergodicity of the system. We claim that $\mu(\Omega_E) = 0$.
Consider, indeed, $P_{\omega,E}$ the projection of rank ≤ 1 onto the eigenspace corresponding to the eigenvalue E. We have

$$\begin{aligned}\sum_n \sigma_{n,\omega}\{E\} = \sum_n \langle P_{\omega,E} e_n, e_n\rangle = Tr(P_{\omega,E}) &= 1 \text{ if } \omega \in \Omega_E \\ &= 0 \quad \text{otherwise.}\end{aligned}$$

Thus

$$\mu(\Omega_E) = \int_\Omega Tr(P_{\omega,E}) \, d\mu(\omega) = \sum_n \int_\Omega \sigma_{n,\omega}\{E\} \, d\mu(\omega).$$

But

$$\int_\Omega \sigma_{n,\omega}\{E\} \, d\mu(\omega) = \int_\Omega \sigma_{n,T\omega}\{E\} \, d\mu(\omega) = \int_\Omega \sigma_{n+1,\omega}\{E\} \, d\mu(\omega);$$

hence the sequence $(\int_{\Omega_E} \sigma_{n,\omega}\{E\}\, d\mu(\omega))_n$ must be constant, thus identically zero. This implies that Ω_E is negligible, and the claim follows. □

Remark A.4. The periodic case may appear as a particular case of the ergodic one. Since the involved measure μ has a finite support, the almost-sure results are thus replaced by deterministic ones.

A.2.2 Lyapunov Exponents

In this section, the system (Ω, T, μ) is supposed to be ergodic. The following subadditive ergodic theorem will be useful [142, 148].

Theorem A.3 (Kingman). *Let (X, T, μ) be an ergodic dynamical system and let (f_n) be a subadditive cocycle in $L^1(X, \mu)$, i.e. be such that*

$$f_{n+m}(x) \leq f_n(x) + f_m(T^n x), \quad \mu - a.e., \quad \text{for } n, m \geq 0.$$

Then, $f_n(x)/n$ converges, for μ-almost all x and in $L^1(X, \mu)$, to the constant $\inf_n \frac{1}{n} \int_X f_n \, d\mu$.

We introduced the Lyapunov exponent as a measuring intrument, when existing, of the asymptotic behaviour of the solutions to the Schrödinger equation. Since we are now dealing with an ergodic family of Schrödinger equations, the involved product of matrices

$$g_n g_{n-1} \cdots g_0 = S_n := S_n(E, \omega) \quad (n \geq 0)$$

and the anologous for $n \leq 0$, also depend on ω. We are thus led to distinguish two kinds of exponents.

Notation : We consider, for $n \in \mathbf{Z}$ and $E \in \mathbf{R}$ (or $\mathbf{C}$) , the two quantities :

$$\gamma_n(E, \omega) = \log \|S_n(E, \omega)\|,$$

and

$$\gamma_n(E) = \int_\Omega \log \|S_n(E, \omega)\| \, d\mu(\omega).$$

Observe that $\gamma_n(E) \geq 0$ since $\det S_n = 1$ and $\|S_n\| \geq 1$. From the multiplicative cocycle equation in $SL(2, \mathbf{R})$ satisfied by $(S_n(E, \omega))$:

$$S_n(E, T^m\omega) \cdot S_m(E, \omega) = S_{n+m}(E, \omega),$$

it follows that $\|S_{n+m}(E, \omega)\| \leq \|S_n(E, T^m\omega)\| \cdot \|S_m(E, \omega)\|$ for every $n, m \geq 0$. Therefore, $(\gamma_n(E))$ is a subadditive ergodic sequence for $n \geq 0$. In addition, note that $\|S_{-n}(E, T^n\omega)\| = \|S_{n-1}(E, \omega)\|$ if $n \geq 1$ and

$$\gamma_{-n}(E) = \gamma_{n-1}(E) \quad \text{for } n \geq 1.$$

We easily deduce that

$$\gamma(E) = \lim_{|n|\to\infty} \frac{1}{|n|} \gamma_n(E) = \inf_{n\geq 1} \frac{1}{n} \gamma_n(E)$$

exists and is nonnegative for every $E \in \mathbf{R}$ or $\mathbf{C}$.

Definition A.5. We call *mean Lyapunov exponent* the nonnegative number

$$\gamma(E) = \lim_{|n|\to\infty} \frac{1}{|n|} \int_\Omega \log \|S_n(E,\omega)\| \, d\mu(\omega)$$

which is defined for every $E \in \mathbf{R}$ or $\mathbf{C}$.

It has been remarked that the function $E \in \mathbf{C} \mapsto \gamma(E)$ is subharmonic, that is uppersemicontinuous and submean; indeed, $S_n(E)$ is an holomorphic matrix in E, therefore $\log\|S_n(E,\omega)\|$ is subharmonic as well as $\int_\Omega \log\|S_n(E,\omega)\| \, d\mu(\omega)$; finally, $\gamma(E)$ is subharmonic as the limit of such ones.

Now, since we have, for every $n, m \geq 0$ and fixed E,

$$\log\|S_{n+m}(E,\omega)\| \leq \log\|S_m(E,\omega)\| + \log\|S_n(E,T^m\omega)\|$$

we may apply the subadditive ergodic theorem to $\omega \mapsto \log\|S_n(E,\omega)\|$ and get the following.

Proposition A.9. *Let E be a fixed real or complex number. Then, for μ-almost ω, the sequence $\frac{1}{|n|}\log\|S_n(E,\omega)\|$ converges to $\gamma(E)$ as $|n| \to \infty$.*

Definition A.6. We could speak of the *individual Lyapunov exponent* for

$$\lim_{|n|\to\infty} \frac{1}{|n|} \operatorname{Log} \|S_n(E,\omega)\| = \gamma(E,\omega)$$

when the limit exists.

Remark A.5. 1. This proposition is already contained in the following well-known multiplicative ergodic theorem, due to Furstenberg and Kesten [95], which is a noncommutative generalization of Birkhoff's theorem and can also be obtained as a corollary of the posterior Kingman theorem.

Theorem A.4 (Furstenberg & Kesten). *Let (X,T,μ) be an ergodic dynamical system and let $A : X \to GL_d(\mathbf{R})$ be a measurable function, with both $\log\|A\|$ and $\log\|A^{-1}\|$ in $L^1(X,\mu)$. Then, the sequence $\frac{1}{n}\log\|A(T^{n-1}x)\cdots A(x)\|$ converges to a constant $\Lambda(A)$, for μ-almost all x and in $L^1(X,\mu)$.*

2. When the system is uniquely ergodic, one can ask about the uniform convergence in both Kingman and Furstenberg-Kesten theorems. In this direction we have :

Theorem A.5 (Furman). *Let (X,T,μ) be a uniquely ergodic system and let $A : X \to GL_d(\mathbf{R})$ be a continuous function; then, for every $x \in X$ and uniformly on X*

$$\limsup_{n\to\infty} \frac{1}{n} \log \|A(T^{n-1}x)\cdots A(x)\| \leq \Lambda(A)$$

where $\Lambda(A)$ is the almost-sure constant limit in the previous theorem.

The following will be used [19].

Theorem A.6 (Avron-Simon). *When the involved system is uniquely ergodic, $\int_{\mathbf{R}} |\gamma_n(E,\omega) - \gamma(E)|^2 dE \underset{|n|\to\infty}{\longrightarrow} 0$ uniformly in $\omega \in \Omega$.*

A.2.3 Results from Pastur, Kotani, Last and Simon

The set $\mathcal{N} = \{E \in \mathbf{R},\ \gamma(E)$ exists and is zero$\}$ is particularly meaningful. The relation between the positivity of γ and the nature of the spectrum of H has been first pointed out by Pastur. The following theorem is proved in [22].

Theorem A.7 (Pastur). *If $m(\mathcal{N}) = 0$, then, for μ-almost all ω, H_ω has no absolutely continuous part in its spectrum. Also, if there exists a positive measure ν on $\mathbf{R}$ such that $\nu(\mathcal{N}) = 0$ then $\sigma_\omega \perp \nu$ for μ-almost all ω.*

Note the following improvement due to Kotani [149].

Theorem A.8 (Kotani). *For μ-almost all ω, the topological support of $(\sigma_\omega)_{ac}$ is the essential closure of $\mathcal{N}$.*

This theorem admits an important consequence in view to the next section, where substitutive potentials and, more generally, uniformly recurrent ones are involved.

Theorem A.9 (Kotani). *Let v be a non-periodic sequence taking finitely many values. If (Ω,T,μ) is an ergodic dynamical system, then, for μ-almost all ω, H_ω has no absolutely continuous part in its spectrum.*

Proof. We just sketch the proof. Suppose on the contrary that $m(\mathcal{N}) > 0$; one can show, using first Kotani's theorem, that every point ω in the topological support of μ is entirely determined by the knowledge of $(\omega(n))_{n\leq 0}$ or $(\omega(n))_{n\geq 0}$ and that the map $(\omega(n))_{n\geq 0} \mapsto (\omega(n))_{n\leq -1}$ (resp. $n \leq -1 \mapsto n \geq 0$) is continuous on the support of μ. In particular, one can find an index K such that the equalities : $\omega(n) = \omega'(n)$ for $n = -1,-2,\ldots,-K$ imply $\omega(0) = \omega'(0)$, so that, by T-invariance of the support of μ, $\omega(n) = \omega'(n)$ for every $n \geq 0$. It follows that the support of μ is finite, with Card Supp$(\mu) \leq K^s$ if v is supposed to take s distinct values. This exactly means that v is a periodic sequence, whence the theorem. □

We shall invoke later the following consequence of those two fundamental results A.7 and A.9.

Corollary A.3. *Let v be a sequence taking finitely many values such that (Ω, T, μ) is a strictly ergodic dynamical system. If the individual Lyapunov exponent $\gamma(E, v)$ exists and is zero for every $E \in sp(H)$, then $m(sp(H)) = 0$.*

Proof. Applying theorem A.6, we know that $\int_{\mathbf{R}} |\gamma_n(E,v) - \gamma(E)|^2 dE \to 0$ as $|n| \to +\infty$, thus, for a subsequence (n_j), $\gamma_{n_j}(E,v)$ tends to $\gamma(E)$ for almost all E. Since $\gamma_{n_j}(E,v)$ tends to zero for every $E \in sp(H)$, $\gamma(E) = 0$ for $E \in sp(H)\backslash N$ where $m(N) = 0$. Suppose now that $m(sp(H)) > 0$; $m(sp(H)\backslash N)$ is still positive and $\gamma(E) = 0$ on a set of positive measure. It follows from Pastur's theorem that, for μ-almost all ω, σ_ω must have an absolutely continuous part. But, the aperiodic sequence v taking finitely many values, Kotani's theorem applies : for μ-almost all ω, H_ω has no absolutely continuous part in its spectrum. This contradiction proves that $m(sp(H)) = 0$. □

Thus, if the aperiodic potential takes finitely many values, an absolutely continuous spectrum is excluded for μ-almost all ω and if, moreover, the system (Ω, T, μ) is minimal, the spectrum $sp(H_\omega)$ is independent of ω.

The problem to know whether the absolutely continuous spectrum itself does not depend on ω for strictly ergodic systems has been definitively solved by Last and Simon in [157], after many partial results (see also [115]).

Theorem A.10 (Last & Simon). *Let v be a sequence taking finitely many values such that (Ω, T, μ) is a strictly ergodic dynamical system. Then, the essential support of the absolutely continuous part of the spectrum is independent of ω.*

In case of an aperiodic potential v, it follows from theorems A.9 and A.10 that the absolutely continuous spectrum of H_ω is empty for every $\omega \in \Omega$.

A.3 Substitutive Schrödinger Operators

We restrict our attention to potentials v with $v(n) = V \cdot u(n)$ where u is a fixed point of a primitive substitution ζ defined on the alphabet A, and where V is some real constant. Recall that bilateral substitution subshifts can be constructed so as to obtain a strictly ergodic dynamical system.

Proposition A.10. *Let ζ be a primitive substitution on A. If α and β are letters in A such that*

$$\beta \text{ is the last letter of } \zeta(\beta) \text{ and } \alpha \text{ is the first letter of } \zeta(\alpha),$$

then there exists a unique fixed point x of ζ such that $x_{-1} = \beta$ and $x_0 = \alpha$. Moreover, the system (X, T), with T the bilateral shift and X the closed orbit of x under T in $A^{\mathbf{Z}}$, is strictly ergodic.

Definition A.7. Such a fixed point will be called *admissible*.

Note that x is admissible if and only if $x_{-1}x_0 \in \mathcal{L}_\sigma$.

Basic Examples.

1. Let $u = (u_n)_{n\geq 1}$ be the Fibonacci fixed point $u = \zeta^\infty(a)$ where ζ is defined on $\{a,b\}$ by :
$$\begin{aligned}\zeta(a) &= ab\\ \zeta(b) &= a\end{aligned}.$$
The two-word aa is an element of $\mathcal{L}_\zeta$, and a is the last letter of $\eta(a)$ where $\eta = \zeta^2$. We thus extend u to the left by $\lim_{n\to\infty} \eta^n(a)$, and u becomes a fixed point of η which generates the same dynamical system. Note that $u_{-n} = u_{n-1}$ for $n \geq 2$. Finally we put $a = 0$, $b = 1$ to get the potential v. By construction, the central terms of the sequence u are
$$\begin{array}{c}\cdots 1\,0\,0\,1\,0\,1\,0\,0\,1\,0\,1 \quad 0 \quad 0 \quad 1\,0\,0\,1\,0\,1\,0\,0\,1\,0\,0\,1\,0\,1\,0\cdots\\ \cdots\; u_{-1}\; u_0\; u_1\;\cdots\end{array}$$
2. Let $u = (u_n)_{n\geq 1}$ be the Toeplitz fixed point $u = \zeta^\infty(a)$ where ζ is defined on $\{a,b\}$ by
$$\zeta(a) = ab, \quad \zeta(b) = aa;$$
We extend u to the left by $\lim_{n\to\infty} \eta^n(a)$ with $\eta = \zeta^2$ again; thus $u_{-n} = u_n$ for $n \geq 1$ and we put $a = 1$, $b = -1$ to get v.
3. Let $u = (u_n)_{n\in\mathbf{Z}}$ be the bilateral Thue-Morse fixed point, that means : the right part of u is $\zeta^\infty(a)$ where ζ is defined on $\{a,b\}$ by
$$\zeta(a) = ab, \quad \zeta(b) = ba,$$
and u is extended to the left by $\eta^\infty(a)$ with the same η; finally we put $a = 1$, $b = -1$ to get a ± 1-valued potential.

Remark A.6. 1. The choice of the extension of u to the negative indices has no impact on the set $sp(H)$ (proposition A.9), but does have one on the spectral invariants.

2. The first two substitutions admit a pure point spectrum while the third one has a purely continuous singular reduced spectrum. But the Schrödinger operator associated with each of those three sequences has the same characteristics. The link between the spectral properties of the potential system and the spectral properties of the Schrödinger operator is not so sharp.

The substitution sequences being intermediate between periodic ones and random ones, the maximal spectral type of H is expected to be intermediate between absolutely continuous one and pure point one, that is to be continuous singular. In order to prove such a result, according to the observation at the end of the previous section, it remains to exclude eigenvalues.

Among the considerable amount of results, two types of them will be retained, giving sufficient conditions on the potential to ensure the absence of eigenvalues together with a purely singular spectrum. Although they largely exceed the framework of substitutions, they are not exhausting the question for those sequences.

A.3.1 The Trace Map Method

Historically, it seems that the Fibonacci potential has been first introduced and the interesting trace map first used, to work out the spectral problem. Actually, the three quoted examples (and a class of other ones) can be analyzed exactly in the same way : We begin with a description of $sp(H)$ with help of the trace map, then we check that eigenvalues cannot exist and finally, we conclude that $m(sp(H)) = 0$ by estimating the individual Lyapunov exponent. As we said, this scheme of proof can be shortened by using the Last-Simon theorem. But the description of the spectrum by means of the trace map remains interesting as we shall see.

The key point of the proof is an approximation argument by periodic Schrödinger operators whose spectrum is well-known.

Lemma A.3. *Let v be a sequence taking finitely many values and H, the Schrödinger operator corresponding to v. For every m we denote by v_m the periodic sequence coinciding with v on the segment $[a_m, b_m]$ and by H_m the corresponding operator. If $b_m \to +\infty$, $a_m \to -\infty$, then $sp(H) \subset \limsup \mathrm{top}\,(sp(H_m))$.*

Proof. Since $\|(H - H_m)\,\varphi\|^2 \leq C\sum_{|n|\geq \inf(b_m,|a_m|)} |\varphi(n)|^2$, the lemma is a consequence of lemma A.2. □

We choose to give the details for the founder Fibonacci case [231].

Theorem A.11 (A. Süto). *Let $v = V \cdot u$ be a Fibonacci potential, where u is the bilateral Fibonacci $0-1$-sequence and $V \in \mathbf{R}$. Then the maximal spectral type of $H := H_v$ is continuous singular.*

Recall the inductive property shared by the elementary words of u :

$$\zeta^n(0) = \zeta^{n-1}(0)\,\zeta^{n-1}(1) = \zeta^{n-1}(0)\,\zeta^{n-2}(0) \tag{A.9}$$

and thus $|\zeta^n(0)| = F_n$ where $F_{n+1} = F_n + F_{n-1}$, $F_0 = 1$, $F_1 = 2$.

This relation directly transposes to the transfer matrices g_n. Let us put

$$T_n = g_{F_n} g_{F_n-1} \cdots g_1 \quad \text{if} \quad n \geq 1$$

and

$$L_n = (g_1 g_2 \cdots g_{F_n})^{-1} = g^{-1}_{-F_n-1} \cdots g^{-1}_{-2}$$

by the symmetry of u. Then, if $\Phi_n = \begin{pmatrix} \varphi(n+1) \\ \varphi(n) \end{pmatrix}$,

$$\Phi_{F_n} = T_n \Phi_0 \quad \text{and} \quad \Phi_{-F_n-2} = L_n \Phi_{-2} \quad \text{for} \quad n \geq 1.$$

Finally, put, for $n \geq 1$,

$$x_n = Tr(T_n), \quad y_n = Tr(L_n).$$

From a relation between the matrices T_n, L_n, we shall deduce a relation on their traces.

Lemma A.4 (Trace Map). *The sequence (x_n) satisfies the equation :*

$$x_{n+2} = x_n x_{n+1} - x_{n-1} \quad \text{for} \quad n \geq 0, \tag{A.10}$$

if we set $x_0 = Tr(g_2)$ and $x_{-1} = 2$. Moreover, $y_n = x_n$ for $n \geq 1$.

Proof. On the transfer matrices, relation (A.9) can be read as follows.

$$\begin{aligned} T_{n+2} &= T_n T_{n+1} && \text{if} \quad n \geq 1 \\ &= T_n \cdot T_{n-1} T_n && \text{if} \quad n \geq 2 \end{aligned}$$

and

$$\begin{aligned} x_{n+2} &= Tr(T_{n+2}) = Tr(T_n^2 T_{n-1}) = Tr((x_n T_n - I)\, T_{n-1}) \\ &= x_n Tr(T_{n-1} T_n) - x_{n-1} \\ &= x_n x_{n+1} - x_{n-1} \quad \text{if} \quad n \geq 2. \end{aligned}$$

For $n = 0, 1$ this remains true provided that $x_0 = E - V = Tr(g_2)$ and $x_{-1} = 2$. Analogously, $L_{n+2} = L_n L_{n+1}$ and $y_{n+3} = y_{n+2} y_{n+1} - y_n$ if $n \geq 1$. By induction on n it is easily checked that $y_n = x_n$ for every $n \geq 1$. □

We shall establish the following description of the spectrum.

Theorem A.12. $sp(H) = B_\infty := \{E \in \mathbf{R},\ (x_n)\ \textit{is bounded}\}$.

Proof of theorem A.12. First, x_n being a continuous function of E for any $n \in \mathbf{Z}$, observe that B_∞ must be closed in $\mathbf{R}$.

We successively prove both inclusions.

$\star$ $B_\infty \subset sp(H)$. We shall need the identities to be proved

$$v(F_n + j) = v(j) \quad \text{if} \quad 1 \leq j \leq F_n \quad \text{and} \quad n \geq 3 \tag{A.11}$$

$$v(-F_{2n} + j) = v(j) \quad \text{if} \quad 1 \leq j \leq F_{2n+1} \quad \text{and} \quad n \geq 1 \tag{A.12}$$

Starting with $\zeta^{n+2}(0) = \zeta^{n+1}(0)\, \zeta^n(0) = \zeta^n(0)\, \zeta^{n-1}(0) \cdot \zeta^n(0)$, we find for $n \geq 3$,

$$\begin{aligned} \zeta^{n+2}(0) &= \zeta^n(0)\, \zeta^{n-1}(0) \cdot \zeta^{n-2}(0)\, \zeta^{n-3}(0)\, \zeta^{n-2}(0) \\ &= \zeta^n(0)\, \zeta^n(0) \zeta^{n-3}(0)\, \zeta^{n-2}(0) \end{aligned}$$

whence (A.11); now (A.12) can be obtained in the same way by using $\eta = \zeta^2$. It follows that

$$S_{2F_n} = g_{2F_n} g_{2F_n-1} \cdots g_1 = g_{F_n} \cdots g_1 \cdot g_{F_n} \cdots g_1 = T_n^2$$

and $S_{2F_n} = x_n T_n - I$.

Let now $E \in B_\infty$ and suppose that $E \notin sp(H)$. As in the periodic case, we prove that one of the two vectors Φ_0 or Φ_{-2} is non zero, if the sequence (Φ_n) derives from a non-trivial solution φ of $(H - EI)\varphi = e_0$.

If $\Phi_0 = w \neq 0$, then $\Phi_{2F_n} = T_n^2 w$, and, according to inequality (A.7), $0 < \|w\|^2 \leq \sup\{2c\|\Phi_{F_n}\|, \|\Phi_{2F_n}\|\}$ if $|x_n| \leq c$, so that $\varphi \notin \ell^2$.

If now $\Phi_{-2} = w \neq 0$, $0 < \|w\|^2 \leq \sup\{2c\|\Phi_{-F_n-2}\|, \|\Phi_{-2F_n-2}\|\}$ and $\varphi \notin \ell^2$ again.

In both cases, we get a contradiction. Note that the same argument may be used to prove that no point of B_∞ can be an eigenvalue of H.

$\star$ $sp(H) \subset B_\infty$. We approximate H by the periodic Schrödinger operators $H_m := H_{v_m}$, where v_m is F_m-periodic with period

$$\overleftarrow{\zeta^m(0)}\ \overrightarrow{\zeta^m(0)} \quad \text{if } m \text{ is even}$$

and

$$\overleftarrow{\zeta^{m-1}(0)}\ \overrightarrow{\zeta^{m-2}(0)} \quad \text{if } m \text{ is odd}$$

By lemma A.2, $sp(H) \subset \limsup\text{top}\, \Sigma_m$ where $\Sigma_m = sp(H_m) = \{E,\ |x_m| \leq 2\}$ according to the characterization established in the periodic case (proposition A.4). The proof of theorem A.12 will be complete if we prove the following :

Lemma A.5. *The sequence of subsets $(\Sigma_m \cup \Sigma_{m+1})_m$ is non-increasing and is contained in B_∞.*

Indeed, admitting lemma A.5, we deduce that

$$\Sigma_m \cup \Sigma_{m+1} \supset \bigcup_{k=m}^{\infty} \Sigma_k \quad \text{for every } m$$

and $B_\infty \supset \overline{\left(\bigcup_{k=m}^{\infty} \Sigma_k\right)}$ since B_∞ is closed. Hence, $B_\infty \supset \bigcap_m \overline{\left(\bigcup_{k=m}^{\infty} \Sigma_k\right)} \supset sp(H)$.

Thus both inclusions are established and theorem A.12 is proved. □

We turn back to lemma A.5 which happens to be a consequence of the trace map.

Lemma A.6. *Consider $(x_n)_{n\geq 0}$ with $x_{-1} = 2$, a solution of (A.10). Then, (x_n) is unbounded if and only if there exists a unique N such that*

$$|x_{N-1}| \leq 2, \quad |x_N| > 2 \quad \text{and} \quad |x_{N+1}| > 2 \tag{A.13}$$

In this case, $|x_n| > 2$ *for every* $n \geq N$ *and one can find* $C > 1$ *such that*

$$|x_n| > 2\, C^{F_{n-N}} \quad \text{for } n \geq N.$$

Otherwise,

$$|x_n| \leq 2 + |x_{-1} - x_0| = 2 + |V|.$$

Proof. ⋆ Suppose that (A.13) holds; then, from (A.10), we have

$$\begin{aligned}|x_{N+2}| &\geq |x_{N+1}|\,|x_N| - |x_{N-1}| \\ &\geq \tfrac{1}{2}\,|x_{N+1}|\,|x_N| + \left(\tfrac{1}{2}\,|x_{N+1}|\,|x_N| - |x_{N-1}|\right) \\ &\geq \tfrac{1}{2}\,|x_{N+1}|\,|x_N| > 2.\end{aligned}$$

It follows that $\log\frac{1}{2}\,|x_{n+2}| \geq \log\frac{1}{2}\,|x_{n+1}| + \log\frac{1}{2}\,|x_n|$ for $n \geq N$, and $\log\frac{1}{2}\,|x_n|$ grows faster than the Fibonacci sequence initiated at N.

⋆ Suppose now that (A.13) is not satisfied : if $|x_n| > 2$, then $|x_{n-1}|$ and $|x_{n+1}|$ must be ≤ 2. But the nontrivial invariant given by the Fricke formula (A.8) and expressed with $g = S_n$, $h = S_{n-1}$, leads to

$$x_{n+1}^2 + x_n^2 + x_{n-1}^2 - x_{n+1}\,x_n\,x_{n-1} = V^2 + 4 \qquad n \geq 0$$

that we write

$$V^2 + 4 = x_{n+1}^2 + x_{n-1}^2 + \left(x_n - \frac{1}{2}\,x_{n+1}\,x_{n-1}\right)^2 - \frac{1}{4}\,x_{n+1}^2\,x_{n-1}^2.$$

Hence, $\left(x_n - \frac{1}{2}\,x_{n+1}\,x_{n-1}\right)^2 = V^2 + \left(2 - \frac{x_{n+1}^2}{2}\right)\left(2 - \frac{x_{n-1}^2}{2}\right)$ and

$$|x_n| \leq \frac{1}{2}\,|x_{n+1}\,x_{n-1}| + \left(V^2 + \left(2 - \frac{x_{n+1}^2}{2}\right)\left(2 - \frac{x_{n-1}^2}{2}\right)\right)^{1/2}.$$

If $|x_{n-1}|$ and $|x_{n+1}|$ are ≤ 2, the maximum of $|x_n|$ is reached as $|x_{n-1}| = |x_{n+1}| = 2$ so that $|x_n| \leq 2 + |V| = 2 + |x_0 - x_{-1}|$.

We end this part by the proof of lemma A.5. From the above estimates, it is clear that

$$B_\infty^c \subset \bigcup_{m \geq 0} (\Sigma_m^c \cap \Sigma_{m+1}^c)$$

and that

$$\Sigma_m^c \cap \Sigma_{m+1}^c \subset \Sigma_{m+1}^c \cap \Sigma_{m+2}^c.$$

Finally, the remaining lemma A.5 is proved. □

Proof of theorem A.11. We already observed that H has no eigenvalue. Thus, Süto's theorem will be deduced for H, from the computation of the individual Lyapunov exponent, by a now standard process, as pointed out in section A.2.3.

Theorem A.13. *The individual Lyapunov exponent $\gamma(E,v)$ exists and is equal to zero for every $E \in sp(H)$.*

Proof. Since the potential satisfies the symmetry relation $v_{-n} = v_{n-1}$, we only consider the behaviour of $\frac{1}{n}\gamma_n(E,v) := \frac{1}{n}\log ||S_n(E,v)||$ as $n \to +\infty$. We first study (ommitting the dependence on E and v)

$$t_n = \frac{1}{F_n} \log ||T_n|| \quad \text{where} \quad T_n = g_{F_n} g_{F_n - 1} \cdots g_1.$$

From the relation

$$\begin{aligned} T_{n+1} &= T_{n-1} T_n = T_{n-2}^{-1} T_n^2 \\ &= x_n T_{n-2}^{-1} T_n - T_{n-2}^{-1} \\ &= x_n T_{n-1} - T_{n-2}^{-1} \end{aligned}$$

we obtain that

$$1 \leq ||T_{n+1}|| \leq |x_n| \, ||T_{n-1}|| + ||T_{n-2}||.$$

If $E \in B_\infty$, say, $|x_n| \leq C_0$ for every n, then $1 \leq ||T_n|| < C^n$ for some constant C. We already have the fact that

$$t_n \leq \frac{1}{F_n} n \cdot \log C \underset{n \to +\infty}{\longrightarrow} 0 \qquad \text{if} \quad E \in sp(H). \tag{A.14}$$

Now, consider more generally $S_n = g_n g_{n-1} \cdots g_1 g_0$. In order to exploit the previous step, we decompose n in the Fibonacci base :

$$n = F_{n_k} + F_{n_{k-1}} + \cdots + F_{n_0} \qquad (\text{actually with } n_k - n_{k-1} \geq 2)$$

and, reminding identity (A.11)

$$v(F_n + j) = v(j) \qquad \text{if} \quad 1 \leq j \leq F_n \qquad (n \geq 3)$$

we get,

$$\begin{aligned} S_n &= S_\ell T_{n_k} \quad \text{where} \quad n - F_{n_k} =: \ell < F_{n_k - 1} \\ &= T_{n_0} T_{n_1} \cdots T_{n_k} \qquad (\text{where } \mathrm{T}_{\mathrm{n}_0}, \mathrm{T}_{\mathrm{n}_1} \text{ may be distinct}) \end{aligned}$$

Finally, by (A.14),

$$0 \leq \log ||S_n|| \leq \sum_{j=0}^{k} \log ||T_{n_j}|| =: \sum_{j=0}^{k} F_{n_j} t_{n_j}$$

$$\leq \left(\sum_{0}^{k} n_j\right) \log C \quad \text{if} \quad E \in B_\infty$$
$$\leq C'(\log n)^2$$

and $\gamma(E,v) = 0$. □

Remark A.7. 1. This method has been worked out by Bovier and Ghez in [39] where they obtain sufficient combinatorial conditions on the involved sequence v to ensure singular continuous spectrum; this condition, in terms of *powers* of words, is rather technical.

2. Finally, the result holds for a dense G_δ set of ω : by minimality, the set of ω for which H_ω has no eigenvalues is dense since it contains the orbit $O(v)$ and it is a countable intersection of open sets [229]. So, by using the Last-Simon theorem, we get a purely continuous spectrum for a dense G_δ set of ω's.

A.3.2 *The Palindromic Density Method*

In [116], the authors exhibit a new combinatorial condition on the strictly ergodic orbit $X := \overline{O(v)}$ to exclude eigenvalues in the spectrum of $H := H_v$ (and, consequently, in the spectrum of H_ω for a dense G_δ set of ω's). Recall that a *palindrome* in the infinite word v on the alphabet A is a symmetric word. Following the authors, we say that :

Definition A.8. The infinite word $v \in A^{\mathbf{Z}}$ is *palindromic* if v contains arbitrarily long palindromes.

Theorem A.14 (Hof, Knill & Simon). *Let $v \in A^{\mathbf{Z}}$ be an aperiodic and palindromic sequence generating a strictly ergodic system (X,T). Then, the Schrödinger operator H_ω has a purely singular continuous spectrum for a dense G_δ set of values of ω.*

Proof. According to a previous remark, it is sufficient to exclude eigenvalues in the spectrum of *one* operator, for instance, $H := H_v$. The proof rests on the following combinatorial improvement that we admit.

Lemma A.7. *Under the assumptions of the theorem, one can find, for every constant $C > 1$, infinitely many palindromes w_i of length L_i, centered on m_i, such that : $|m_i| \to \infty$ and $C^{m_i}/L_i \to 0$.*

Let E be an eigenvalue of H and $H\varphi = E\varphi$ where $\varphi \in \ell^2(\mathbf{Z})$, $||\varphi|| = 1$.

$\star$ Suppose first that the palindromes $w_i := v_{[m_i-\ell_i,\, m_i+\ell_i]}$ have odd length $L_i := 2\ell_i + 1$ and are centered on m_i. To each w_i, we associate the element $\varphi_i \in \ell^2(\mathbf{Z})$ defined by $\varphi_i(n) = \varphi(2m_i - n)$, the reflecting word at m_i. Finally, we put $\Phi(n) = \begin{pmatrix} \varphi(n+1) \\ \varphi(n) \end{pmatrix}$ and $\Phi_i(n) = \begin{pmatrix} \varphi_i(n+1) \\ \varphi_i(n) \end{pmatrix}$ for $n \in \mathbf{Z}$. Without loss of generality, we may assume that the m_i are nonnegative indices.

The symmetry of the word w_i results into the following on the transfer matrices (g_n) :

$$g_{m_i-k} = g_{m_i+k} \quad \text{for } 0 \le k \le \ell_i;$$

thus,

$$A_i := g_1^{-1} \cdots g_{m_i}^{-1} = g_{2m_i-1}^{-1} \cdots g_{m_i}^{-1}$$

and since $\Phi_i(m_i) = \begin{pmatrix} \varphi(m_i-1) \\ \varphi(m_i) \end{pmatrix}$, we get

$$A_i \Phi(m_i) = \Phi(0) \quad \text{and} \quad A_i \Phi_i(m_i) = \Phi_i(0). \tag{A.15}$$

Moreover, because of the symmetry of w_i again, we check that

$$\varphi_i(n+1) + \varphi_i(n-1) = (E - v(n))\varphi_i(n)$$

for every $n \in [m_i - \ell_i,\ m_i + \ell_i]$, so that the wronskian $W := W(\varphi, \varphi_i)$ is constant on the interval $[m_i - \ell_i,\ m_i + \ell_i]$. From that property, we deduce an estimate of $W(m_i)$: since $||\varphi|| = ||\varphi_i|| = 1$, we derive

$$|W(m_i)| L_i \le \sum_{n \in \mathbf{Z}} |W(n)| \le 2$$

from the Cauchy-Schwarz inequality. But $W(m_i) = \varphi(m_i)(\varphi(m_i+1) - \varphi(m_i-1))$, thus $|\varphi(m_i)|$ and $|\varphi(m_i+1) - \varphi(m_i-1)|$ cannot be simultaneously greater than $(2/L_i)^{1/2}$. Whence the discussion :

If $|\varphi(m_i+1) - \varphi(m_i-1)| \le (2/L_i)^{1/2}$, then

$$||(\Phi - \Phi_i)(m_i)|| = |\varphi(m_i+1) - \varphi(m_i-1)| \le (2/L_i)^{1/2};$$

If $|\varphi(m_i)| \le (2/L_i)^{1/2}$, then, in turn,

$$||(\Phi + \Phi_i)(m_i)|| \le K(L_i)^{-1/2}$$

since $(\Phi + \Phi_i)(m_i) = \varphi(m_i) \begin{pmatrix} E - v(m_i) \\ 2 \end{pmatrix}$. In both cases, we deduce from (A.15) that

$$\begin{aligned} ||\Phi(0)|| - ||\Phi(2m_i)|| &\le K ||A_i|| (L_i)^{-1/2} \\ &\le C^{m_i} / L_i^{1/2} \end{aligned}$$

where $||g_n|| < C$ for all n and E. If C, m_i and L_i are such that $C^{m_i}/L_i^{1/2}$ tends to zero, then

$$\lim_i (||\Phi(0)|| - ||\Phi(2m_i)||) = ||\Phi(0)|| = 0$$

since $\varphi \in \ell^2(\mathbf{Z})$ and m_i tends to infinity. But φ must then be zero, which leads to a contradiction.

$\star$ Suppose now that the palindromes w_i have even length $L_i := 2\ell_i$ and are centered on $m_i + 1/2$. Then

$$g_{2m_i-n+1} = g_n \quad \text{for } m_i - \ell_i \leq n \leq m_i + \ell_i + 1,$$

and we consider $\varphi_i(n) = \varphi(2m_i - n + 1)$. The previous proof can be arranged in such a way to get the same contradiction. □

Examples and Problem : This last result covers many examples of potentials arising from primitive substitutions such as : Thue-Morse, Fibonacci, Period doubling, binary and ternary non-Pisot as well as sequences defined by circle maps, of the form $v_n = \mathbf{1}_{[0,\beta[}(n\alpha + \theta)$, with α irrational and suitable θ.

The Rudin-Shapiro sequence seems to escape those processes : Actually J.P. Allouche proved that neither the Rudin-Shapiro sequence, nor generalized ones in sense of [184], are palindromic [9]; also numerical experiments indicate that there might be eigenvalues in the spectrum of the associated Schrödinger operator [116]. Thus, the nature of the spectrum of H for the Rudin-Shapiro potential is an open question.

Appendix B
Substitutive Continued Fractions

Some algebraic properties of real numbers can be read on a suitable expansion, adic-expansion, continued fraction expansion or other one. In this chapter, we focus on the continued fractions viewed as sequences taking integer values, and, more specifically, on such sequences taking finitely many values, then we ask for algebraic properties of the underlain real numbers. In this framework too, the initial study of substitutive sequences has led to a more general interest in combinatorics of words and the parallel with the previous appendix is powerful. We just aim to gain a glimpse into this subject and we address the reader to [4, 11, 14] for more information, also on the adic-expansions [3, 86].

B.1 Overview on Continued Fraction Expansions

Let $a_1, a_2, \ldots$ be a sequence taking values in a finite alphabet $A = \{1, 2, \ldots, s\}$ which may be viewed as the *continued fraction expansion* of some positive real number; which algebraic properties does it enjoy? We already know that the rational numbers are real numbers with finite continued fraction expansion and that the numbers with eventually periodic continued fraction expansion are exactly the quadratic irrational ones by Lagrange's theorem. What else ? In particular, what can we say about irrational numbers whose continued fraction expansion is an automatic sequence on a finite alphabet ? We begin with usual definitions and general facts on the continued fractions.

It is well-known that every real number x is the limit of the sequence of rational numbers, called *convergents*,

$$\frac{p_n(x)}{q_n(x)} = a_0(x) + \cfrac{1}{a_1(x) + \cfrac{1}{a_2(x) + \cdots + \cfrac{1}{a_n(x)}}} =: [a_0(x); a_1(x), \ldots, a_n(x)],$$

where $a_0(x) = [x]$; the integers $a_n(x)$ are called the *partial quotients* of x. Rational numbers have a terminating continued fraction and we usually write for an irrational $x \in [0,1]$:

$$x = \cfrac{1}{a_1(x) + \cfrac{1}{a_2(x) + \cdots}} =: [0; a_1(x), a_2(x), \ldots].$$

This expansion involves linear recurrences of order two and by the way, matrices in $SL(2,\mathbf{Z})$: if we put

$$A_k(x) = \begin{pmatrix} a_k(x) & 1 \\ 1 & 0 \end{pmatrix} \quad \text{and} \quad M_n(x) = A_n(x) \ldots A_1(x),$$

then we get, by induction on $n \geq 1$,

$$M_n(x) = \begin{pmatrix} q_n(x) & p_n(x) \\ q_{n-1}(x) & p_{n-1}(x) \end{pmatrix}, \quad \text{with } q_0(x) = 1,\ p_0(x) = 0 \text{ if } x \in (0,1]. \tag{B.1}$$

The following identities can easily be obtained from (B.1).

1. By taking the determinant of $M_n(x)$,
$$q_n(x)p_{n-1}(x) - p_n(x)q_{n-1}(x) = (-1)^n.$$
2. By using the symmetry of the $A_k(x)$,
$$\frac{q_{n-1}(x)}{q_n(x)} = [0; a_n(x), \ldots, a_1(x)]. \tag{B.2}$$
3. Now, palindromes have a nice expression : if $[0; a_1, \ldots, a_n] = p_n/q_n$, then
$$[0; a_1, \ldots, a_n, a_n, \ldots, a_1] = \frac{p_n^2 + p_{n-1}^2}{p_n q_n + p_{n-1} q_{n-1}} \tag{B.3}$$
and
$$[0; a_1, \ldots, a_n, c, a_n, \ldots, a_1] = \frac{p_n(c p_n + 2p_{n-1})}{c p_n q_n + p_n q_{n-1} + p_{n-1} q_n}. \tag{B.4}$$
4. From above, one can deduce the striking identity [235]
$$[0; a_1, \ldots, a_{n-1}, a_n + 1, a_n - 1, a_{n-1}, \ldots, a_1] = \frac{p_n}{q_n} + \frac{(-1)^n}{q_n^2} \tag{B.5}$$

B.1.1 The Gauss Dynamical System

We consider $\mathscr{X} = [0,1]\backslash\mathbf{Q}$ and the *Gauss map* T defined by

$$Tx = \frac{1}{x} \bmod 1 \quad \text{for } x \in \mathscr{X}.$$

Then, $a_1(x) = [\frac{1}{x}]$ and $a_{n+1}(x) = a_1(T^n x)$ for every $n \geq 1$, so that x can be expressed in the form

$$x = \cfrac{1}{a_1(x) + \cfrac{1}{a_2(x) + \cdots + \cfrac{1}{a_n(x) + T^n x}}} =: [0; a_1(x), \ldots, a_n(x) + T^n x], \tag{B.6}$$

with

$$T^n x = [0; a_{n+1}(x), a_{n+2}(x), \ldots] \quad \text{for every } n \geq 0.$$

We recall that T on $\mathscr{X}$ preserves the so-called *Gauss measure* μ, which is absolutely continuous with respect to the Lebesgue measure on $[0,1]$, with density $\frac{1}{\log 2}\frac{1}{1+x}$. This probability measure is T-ergodic, even mixing [29] and it is the unique absolutely continuous T-invariant one; in fact, $\frac{1}{\log 2}\frac{1}{1+x}$ is the fixed point of the Perron-Frobenius operator on $L^1([0,1])$ associated with T [46].

Definition B.1. The ergodic dynamical system $(\mathscr{X}, \mu, T)$ is the *Gauss dynamical system.* This system is conjugate to the backward shift on $\mathbf{N}^{\mathbf{N}}$.

From identity (B.6), we derive that

$$x = \frac{p_n(x) + T^n x\, p_{n-1}(x)}{q_n(x) + T^n x\, q_{n-1}(x)}, \tag{B.7}$$

and

$$\begin{aligned} x q_{n-1}(x) - p_{n-1}(x) &= q_{n-1}(x)\frac{p_n(x) + T^n x\, p_{n-1}(x)}{q_n(x) + T^n x\, q_{n-1}(x)} - p_{n-1}(x), \\ &= \frac{(-1)^n}{q_n(x) + T^n x\, q_{n-1}(x)}. \end{aligned}$$

We deduce from above the important following bounds :

$$\frac{1}{2} < q_n(x)|x q_{n-1}(x) - p_{n-1}(x)| < 1. \tag{B.8}$$

On the other hand, from (B.7) again, we see that

$$T^n x = -\frac{x q_n(x) - p_n(x)}{x q_{n-1}(x) - p_{n-1}(x)}.$$

so that, for $n \geq 1$,

$$xTx\cdots T^n x = (-1)^n(xq_n(x) - p_n(x)) = |xq_n(x) - p_n(x)|. \tag{B.9}$$

Combining (B.8) and (B.9), we get that

$$\lim_{n\to\infty} \frac{1}{n}\log q_n(x) + \frac{1}{n}(\log x + \cdots + \log T^{n-1}x) = 0.$$

But now, according to the pointwise ergodic theorem, for almost all $x \in [0,1]$,

$$\lim_{n\to\infty} \frac{1}{n}(\log x + \cdots + \log T^{n-1}x) = \int_0^1 \log x \, d\mu(x) = -\frac{\pi^2}{12\log 2},$$

and we have proved the Paul Lévy's statistical estimate on the denominators $q_n(x)$:

Theorem B.1 (P. Lévy). *: For almost all $x \in [0,1]$,*

$$\lim_{n\to\infty} \frac{1}{n}\log q_n(x) = \frac{\pi^2}{12\log 2}, \tag{B.10}$$

where $(q_n(x))_n$ is the sequence of denominators of the convergents of x.

Definition B.2. We call a *Lévy number*, any $x \in [0,1]$ for which the limit in (B.10) exists and $\beta(x) := \lim_{n\to\infty} \log q_n(x)/n$ is then called the *Lévy constant* of x.

Note that the irrational quadratic numbers are Lévy numbers [82].

The behaviour of the sequence $(a_n(x))$ itself is less rigid but, as a consequence of Fatou's lemma, the following can be proved.

Proposition B.1. *For almost all $x \in [0,1]$, $\lim_{n\to\infty}(a_1(x) + \cdots + a_n(x))/n = +\infty$.*

B.1.2 Diophantine Approximation and BAD

From now on, we omit to mention the dependence on x in the continuous fraction expansion. The behaviour of the denominators $(q_n)_n$ is relevant for rational approximation and, consequently, for transcendence, as we briefly summarize. Recall the classical result of Dirichlet :

Theorem B.2 (Dirichlet). *For any $x \in [0,1]$, there exist infinitely many $q \in \mathbf{N}$ such that $||qx|| \leq 1/q$, where $||\cdot||$ denotes the distance to the nearest integer.*

From (B.8), we see that

$$\frac{1}{a_{n+1}+2} \leq q_n||q_n x|| \leq \frac{1}{a_{n+1}} \tag{B.11}$$

and the convergents provide explicit rational approximates satisfying Dirichlet's theorem. Moreover, they are the best ones in sense that $||qx|| \geq ||q_n x||$ for any $q < q_{n+1}$.

The inequality in Dirichlet's result is best possible (up to a multiplicative constant), more precisely, there exist real numbers x for which the inequality $||qx|| \leq c/q$ has at most a finite number of solutions for any given $c < 1/\sqrt{5}$. If we denote by BAD the set of *badly approximable numbers* that is

$$\mathrm{BAD} := \{x \in [0,1],\ ||qx|| > cq^{-1} \text{ for some constant } c := c(x)\},$$

the following can be deduced from (B.11) :

Proposition B.2. *The real number $x \in BAD$ if and only if the sequence of its partial quotients is bounded.*

All quadratic irrationals are in BAD since they have an ultimately periodic expansion. The problem whether an algebraic non-quadratic number could belong to BAD remains open, but in fact, the converse assertion is conjectured.

Note that BAD is uncountable with zero Lebesgue measure by proposition B.1. Whence, transcendental numbers do exist in BAD. Many explicit examples of such numbers have been constructed (see [6]) and we just mention a historical example [235].

Theorem B.3. *Let u be an integer ≥ 2. The number $g = \sum_{k=0}^{\infty} \frac{1}{u^{2^k}}$ is transcendental, with partial quotients bounded by $u+2$ if $u \geq 3$, and by 6 if $u = 2$.*

Proof. The transcendence of those numbers has been established by Mahler and we refer to [14] for a proof. It remains to show that they belong to BAD.

We fix $u \geq 2$ and we put $g_n = \sum_{k=0}^{n} u^{-2^k}$. The computation of the first terms gives

$$\begin{aligned} g_0 &= u^{-1}, \\ g_1 &= u^{-1} + u^{-2} = [0; u-1, u+1], \\ g_2 &= [0; u-1, u+2, u, u-1]. \end{aligned}$$

Writing $g_n = P_n/Q_n = [0; a_1, \ldots, a_N]$ with P_n/Q_n irreducible, then $Q_n = u^{2^n}$ so that $g_{n+1} = g_n + u^{-2^{n+1}}$. Now we remark by induction on $n \geq 1$ that $N = 2^n$ and

$$\frac{P_{n+1}}{Q_{n+1}} = \frac{P_n}{Q_n} + \frac{1}{Q_n^2} = [0; \underbrace{a_1, \ldots, a_{N-1}}, a_N + 1, a_N - 1, \underbrace{a_{N-1}, \ldots, a_1}],$$

using (B.5) for the last equality. Then the bounds on the partial quotients arise by induction on n again. □

B.2 Morphic Numbers

Aperiodic substitutive words constitute a class of sequences very close to periodic ones, and one may think reasonably that the arithmetical study of real numbers with such expansions could be carried out. Of course, according to the conjecture, the expected conclusion is the transcendence of those numbers, but, beyond this conclusion, growth estimates of the denominators q_n could be useful in view to Koksma's and Mahler's classifications (see [51]).

B.2.1 Schmidt's Theorem on Non-Quadratic Numbers

We shall consider numbers whose continued fraction expansion is the fixed point of some aperiodic primitive substitution, and call them *morphic numbers*. For such non-quadratic numbers, approximation by quadratic irrational numbers instead of rational ones proves to be more effective. The idea for getting transcendence goes as follows : suppose that the real number of $[0,1]$

$$x = [0; a_1, a_2, \ldots]$$

has bounded partial quotients. Besides suppose that the sequence $(a_n)_{n\geq 1}$ presents properties of "repetition". More precisely this sequence begins by arbitrarily long prefixes which are "almost-squares":

$$a_1 a_2 \ldots = V_n V_n' \ldots$$

where the length $|V_n|$ goes to infinity, and where V_n' is a "big" prefix of V_n. Approach the sequence $(a_n)_{n\geq 1}$ by the periodic one : $V_n V_n V_n \ldots$, and let ξ_n be the real (quadratic) number whose partial quotients are given by the letters of the periodic sequence $V_n V_n V_n \ldots$. Then, using the same notation for the sequence and the so-associated real number, the coincidence between x and ξ_n is apparently reduced to

$$\begin{aligned} x &= \underbrace{V_n} \mid \ldots \\ \xi_n &= \underbrace{V_n} \mid \underbrace{V_n}\, \underbrace{V_n} \ldots \end{aligned}$$

though in fact it is better (don't forget that V_n' is a prefix of V_n)

$$\begin{aligned} x &= \underbrace{V_n}\, \underbrace{V_n'} \mid \ldots \\ \xi_n &= \underbrace{V_n}\, \underbrace{V_n'} \mid \ldots \end{aligned}$$

The approximation of x by quadratic numbers is good enough to say something on x by using a powerful Schmidt's theorem just below.

Recall the following notation.

Definition B.3. Let ξ be is a root of the minimal equation $a\xi^2+b\xi+c=0$, with $a,b,c \in \mathbf{Z}$, and $\gcd(|a|,|b|,|c|)=1$. The *height* of ξ, denoted by $H(\xi)$, is defined by $H(\xi)=\max(|a|,|b|,|c|)$.

Theorem B.4 (W.M. Schmidt). *Let x be a real number in $[0,1]$. We suppose that x is neither rational, nor quadratic irrational. If there exist a real number $B>3$, and infinitely many quadratic irrational numbers ξ_k such that*

$$|x-\xi_k| < H(\xi_k)^{-B}$$

then x is transcendental.

We shall need the following classical lemmas on continued fraction expansions.

Lemma B.1. *Let $\xi \in [0,1]$ be a number with purely periodic continued fraction expansion*

$$\xi = [0,a_1,a_2,\ldots,a_k,a_1,a_2,\ldots,a_k,\ldots].$$

Then the (quadratic irrational) number ξ satisfies $H(\xi) \leq q_k$.

Lemma B.2. *If $x,y \in [0,1]$ have the same first k partial quotients $a_1a_2\cdots a_k$, then*

$$|x-y| \leq \frac{1}{q_k^2}.$$

Proof. : 1. By assumption, $\xi = [0,a_1,a_2,\ldots,a_k,\frac{1}{\xi}]$ that is $\xi = \dfrac{p_k+\xi p_{k-1}}{q_k+\xi q_{k-1}}$ and

$$q_{k-1}\xi^2+\xi(q_k-p_{k-1})-p_k=0.$$

It follows that

$$H(\xi) \leq \max(q_{k-1},|q_k-p_{k-1}|,p_k) \leq q_k,$$

since $p_n \leq q_n$ for all $n \geq 1$.

2. Since $p_k/q_k=[0,a_1,a_2,\ldots,a_k]$, we have simultaneously $|x-p_k/q_k| \leq 1/q_k^2$ and $|y-p_k/q_k| \leq 1/q_k^2$ by (B.11). Moreover, $x-p_k/q_k$ et $y-p_k/q_k$ having same sign depending on k only, we may write : $|x-y| = \Big| |x-p_k/q_k|-|y-p_k/q_k| \Big| \leq 1/q_k^2$. □

Now, we refer to section 5.3 for definitions and notations related to the substitutions. If $M:=M(\zeta)$ is the matrix of the primitive substitution ζ on A, and if θ is the Perron-Frobenius eigenvalue of M, recall that $d=(d_\alpha)_{\alpha\in A}$ denotes the positive associated eigenvector corresponding to θ and normalized by $\sum_{\alpha\in A} d_\alpha = 1$; d is thus the vector of frequencies of letters. Also, g denotes the corresponding left eigenvector uniquely defined by $\sum_{\alpha\in A} g_\alpha d_\alpha = 1$.

Definition B.4. If $W = w_1 \dots w_\ell \in \mathscr{L}(\zeta)$, the *weighted norm* of W is,

$$||W|| := \sum_{1 \le i \le \ell} g_{w_i} = \lim_{n \to \infty} \frac{|\zeta^n(W)|}{\theta^n}.$$

The equivalence between both writings results from corollary 5.1. Indeed, for every $\alpha, \beta \in A$, it is proved that $d_\alpha \cdot g_\beta = \lim_{n\to\infty} m^{(n)}_{\alpha\beta}/\theta^n$ where $m^{(n)}_{\alpha\beta} := (M^n)_{\alpha\beta}$ is nothing but the occurrence number of α in the word $\zeta^n(\beta)$. It follows that

$$g_\beta = \lim_{n \to \infty} \sum_{\alpha \in A} \frac{m^{(n)}_{\alpha\beta}}{\theta^n} = \lim_{n \to \infty} \frac{|\zeta^n(\beta)|}{\theta^n}$$

whence the second form of the weighted norm.

The following theorem states one of the first results in this direction. We identify x with its expansion.

Theorem B.5. *Let $x \in [0,1]$ be a non-quadratic number whose continued fraction expansion is the fixed point of some primitive substitution ζ. If there exists a prefix of x, $U := W_1W_2W_1$ where the words W_1 and W_2 satisfy $||W_1|| > ||W_2||$, then x is a transcendental number.*

Proof. : From our assumption and because $x = \zeta^n(x)$ for each $n \ge 1$, x begins for every n with the word $V'_n = \zeta^n(W_1W_2)$ whose length is denoted by ℓ_n. If ξ_n is the purely periodic sequence with period V'_n, ξ_n is a *reduced* quadratic irrational and, by lemma B.1, we have $H(\xi_n) \le q_{\ell_n}$.

Now, since $W_1W_2W_1$ is a prefix of x, x begins with $V_n = \zeta^n(W_1W_2W_1)$ for every n, so that x and ξ_n have the same $L_n = |V_n|$ first partial quotients. It follows from lemma B.2, that

$$|x - \xi_n| \le \frac{1}{q^2_{L_n}}.$$

If there exists some number $\theta > 3$ such that $q^\theta_{\ell_n} < q^2_{L_n}$ for infinitely many n, then

$$|x - \xi_n| \le \frac{1}{q^2_{L_n}} < q^{-\theta}_{\ell_n} \le H(\xi_n)^{-\theta},$$

and the theorem is a direct consequence of W.M. Schmidt's theorem.

We are left with estimates of both q_{ℓ_n} and q_{L_n}.

First of all, observe that, by primitivity, $\beta := \lim_{n\to\infty} \frac{1}{n} \log q_n$ exists. More generally we have the following.

Lemma B.3. *Let (Ω, S, ν) be the uniquely ergodic substitution system, associated with ζ, and, for every $\omega \in \Omega$, denote by $x_\omega \in [0,1]$ the number whose continued fraction expansion is $[0; \omega_1, \omega_2, \dots]$; then,*

$$\lim_{n \to \infty} \frac{1}{n} \log q_n(\omega) \quad \textit{exists for every} \quad \omega \in \Omega.$$

Moreover, all the numbers of the closed orbit are Lévy numbers with the same Lévy constant.

Put now $\rho = \lim_{n\to\infty} |\zeta^n(W_2)|/|\zeta^n(W_1)|$; under our assumption on the words W_i, we have $\rho < 1$ and

$$\lim_{n\to\infty} \frac{L_n}{\ell_n} = \lim_{n\to\infty} \frac{2+\dfrac{|\zeta^n(W_2)|}{|\zeta^n(W_1)|}}{1+\dfrac{|\zeta^n(W_2)|}{|\zeta^n(W_1)|}} = \frac{2+\rho}{1+\rho} > \frac{3}{2}.$$

Combining these remarks, we obtain that

$$\lim_{n\to\infty} \frac{\log q_{\ell_n}}{\log q_{L_n}} = \lim_{n\to\infty} \frac{\ell_n}{L_n} < \frac{2}{3}$$

which ends the proof.

Proof of the lemma. Actually, we can see more precisely that, for every $\omega \in \Omega$, $\beta(x_\omega)$ exists and is equal to

$$\beta = -\lim_{n\to\infty} \sum_{\alpha_1\ldots\alpha_n\in\Omega_n} \log[0;\alpha_1,\ldots,\alpha_n]\ \nu([\alpha_1\ldots\alpha_n]), \tag{B.12}$$

where Ω_n is the set of n-factors of elements in Ω. If φ is the natural map $\omega \in \Omega \mapsto x_\omega = [0;\omega_1,\omega_2,\ldots] \in [0,1]$, φ is continuous by lemma B.2 and $\varphi(\Omega)$ is a compact subset of $[0,1]$ avoiding 0. Hence the function $\log\varphi$ is continuous on Ω and obviously, $\varphi\circ S = T\circ\varphi$. As a consequence of Oxtoby's ergodic theorem,

$$\lim_{N\to\infty} \frac{1}{N} \sum_{n<N} \log\varphi(S^n\omega) = \lim_{N\to\infty} \frac{1}{N} \sum_{n<N} \log T^n\varphi(\omega)$$

exists for every $\omega \in \Omega$ and is equal to $\int_\Omega \log\varphi(\omega)\, d\nu(\omega)$. It follows that every x_ω admits a Lévy constant

$$\beta(x_\omega) = \lim_{n\to\infty} \frac{1}{n} \log q_n(\omega) = -\int_\Omega \log\varphi(\omega)\, d\nu(\omega).$$

By using the following remark : $|\log y - \log(p_n(y)/q_n(y))| \le \frac{1}{2^{n-2}}$ for $n \ge 1$ and $y \in \mathscr{X}$ [29], and invoking the uniform convergence, we get

$$\begin{aligned}\int_\Omega \log\varphi(\omega)\, d\nu(\omega) &= \lim_{n\to\infty} \int_\Omega \log\frac{p_n(\omega)}{q_n(\omega)}\, d\nu(w)\\ &= \lim_{n\to\infty} \int_\Omega \log[0;\omega_1,\ldots,\omega_n]\, d\nu(\omega).\end{aligned}$$

The lemma is proved. □

The proof of the theorem is complete. □

The statement can be simplified when the continued fraction expansion of x is an automatic sequence.

Corollary B.1. *: Let* $x \in]0,1[$ *be a non-quadratic number whose continued fraction expansion is the fixed point of some primitive substitution of constant length. If there exists a prefix of* x, $U = W_1W_2W_1$ *where the words* W_i *satisfy* $|W_1| > |W_2|$, *then* x *is a transcendental number.*

Proof. In case of a constant-length substitution, $g_\alpha = 1$ for every α, so that the weight norm of the word $W = w_1 \ldots w_l$ reduces to $|W|$. □

Example B.1. Most of the classical examples of substitutive sequences satisfy this criterion and provide transcendental numbers. However, once again, the Rudin-Shapiro automaton seems to escape this procedure and the algebraic nature of the Rudin-Shapiro continued fractions will be deduced from a next result.

Remark B.1. The method developed above involves properties of prefixes of the fixed point $\zeta^\infty(a)$, and, consequently, cannot be used to establish the transcendence of any limit point in the closed orbit x in $[0,1]$, although all these numbers are Lévy numbers.

As a generalization of this method, the next theorem establishes the transcendence for a broader class of sequences enjoying suitable combinatorial properties ([11] and see also [14]).

Theorem B.6. *Let* $x \in [0,1]$ *be an irrational number with continued fraction expansion :* $x = [0; a_1, a_2, \ldots]$ *and denominators* $(q_n)_{n\geq 0}$. *We suppose that the sequence* $(a_n)_{n\geq 1}$ *contains infinitely many prefixes of the form* U_kV_k *such that*
(i) $\lim_{k\to\infty} |U_k| = +\infty$,
(ii) V_k *itself is a prefix of* U_k.
We put $\gamma = \liminf_{k\to\infty}(|U_k| + |V_k|)/|U_k| \geq 1$, $M = \limsup_{k\to\infty} q_{|U_k|}^{1/|U_k|}$ *and* $m = \liminf_{k\to\infty} q_{|U_kV_k|}^{1/|U_kV_k|}$. *Then, if* $\gamma > 3\log M / 2\log m$, *the number* x *is transcendental.*

Example B.2. One can check that sturmian sequences, in turn, provide transcendental numbers since they meet the conditions of theorem B.6.

B.2.2 The Thue-Morse Continued Fraction

We consider the specific number whose continued fraction expansion is the Thue-Morse sequence on the alphabet $\{a,b\}$, where a,b are integers ≥ 2. We already observed that this number must be transcendental and a quick proof of this fact will be given in the next section. But, because of the rigidity and of the symmetry of this sequence, we are able to get rather precise growth estimates of the associated denominators [204].

We denote by A and B the matrices $\begin{pmatrix} a & 1 \\ 1 & 0 \end{pmatrix}$ and $\begin{pmatrix} b & 1 \\ 1 & 0 \end{pmatrix}$ respectively. If $||.||$ is the operator norm on matrices, recall that, for symmetric matrices, $||X|| = \rho(X)$, where $\rho(X)$ is the spectral radius of X, that is $\rho(X) = \sup\{|\lambda|, \lambda$ eigenvalue of $X\}$ for the symmetric matrix X; in particular $\rho(A) = \frac{a+\sqrt{a^2+4}}{2}$.

The first lemma below already appears in [67]. We always identify the number x and the sequence of its partial quotients.

Lemma B.4. *If the letters a and b occur in x with frequency α and β respectively, then*

$$\limsup q_n^{\frac{1}{n}} \leq ||A||^{\alpha} ||B||^{\beta}.$$

Proof. For $n \geq 1$, let u_n be the vector $\begin{pmatrix} q_n \\ q_{n-1} \end{pmatrix}$, with $u_0 = \begin{pmatrix} 1 \\ 0 \end{pmatrix}$, so that

$$u_n = \begin{pmatrix} a_n & 1 \\ 1 & 0 \end{pmatrix} \begin{pmatrix} a_{n-1} & 1 \\ 1 & 0 \end{pmatrix} \cdots \begin{pmatrix} a_1 & 1 \\ 1 & 0 \end{pmatrix} u_0 =: W_n u_0,$$

where $W_n := W_n(A,B)$ is the product of matrices. It follows that

$$||u_n|| \leq ||W_n|| \leq ||A||^m ||B||^{n-m}$$

if m is the occurrence number of A in W_n. Thus $||u_n||^{\frac{1}{n}} \leq ||A||^{\frac{m}{n}} ||B||^{1-\frac{m}{n}}$ and

$$\limsup ||u_n||^{\frac{1}{n}} \leq ||A||^{\alpha} ||B||^{\beta}.$$

Since $q_n < ||u_n|| < \sqrt{2}\, q_n$, we get

$$\frac{1}{\sqrt{2}} ||u_n|| < q_n < ||u_n||, \tag{B.13}$$

whence the lemma. □

We deduce the first estimate for the Thue-Morse number .

Proposition B.3. *If $(q_n)_n$ is the sequence of denominators of the Thue-Morse sequence on $\{a,b\}$, then*

$$\limsup_{n \to \infty} q_n^{1/n} \leq \sqrt{||AB||}.$$

Proof. Since the Thue-Morse sequence on $\{a,b\}$ is a fixed point of ζ on the alphabet $\{ab, ba\}$ as well, the proposition follows from the previous lemma applied with W_{2n} and W_{2n+1}. □

For the lower bound, we shall make use of the trace map equation (see appendix A). Let us denote by $\zeta^n(A) = A_n$ the product of the matrices corresponding to the word $\zeta^n(a)$ and by $\zeta^n(B) = B_n$ the analogue for $\zeta^n(b)$. Then,

$A_0 = A$, $B_0 = B$, and we obviously have $A_{n+1} = B_nA_n$ and $B_{n+1} = A_nB_n$ for $n \geq 0$. It is easily checked by induction on n that A_n and B_n are symmetric if n is even, and that $B_n = A_n^*$ otherwise, X^* denoting the transpose of X.

Lemma B.5. *For every* $n \geq 1$, *we have* $\rho(A_n) \geq \left(\rho(AB)\right)^{2^{n-1}}$.

Proof. The lemma follows from some properties of the spectral radius of square matrices. For any such X and Y, then $\rho(XY) = \rho(YX)$ and $\rho(XX^*) = \|XX^*\| = \|X\|^2 \geq \rho(X)^2$.

Suppose first that $n = 2k+2$. Thus,

$$\rho(A_{2k+2}) = \rho(B_{2k+1}A_{2k+1}) = \rho(A_{2k+1}^*A_{2k+1}) \geq \rho(A_{2k+1})^2.$$

Suppose now $n = 2k+1$. Then

$$\begin{aligned}\rho(A_{2k+1}) &= \rho(B_{2k}A_{2k}) = \rho(B_{2k-1}A_{2k-1}A_{2k-1}B_{2k-1}) \\ &= \rho(B_{2k-1}^2A_{2k-1}^2) = \rho((A_{2k-1}^*)^2A_{2k-1}^2) \\ &= \rho((A_{2k-1}^2)^*A_{2k-1}^2) \geq \rho(A_{2k-1}^2)^2 = \rho(A_{2k-1})^4.\end{aligned}$$

A combination of those inequalities gives the result since : $\rho(A_{2k+1}) \geq \rho(A_{2k-1})^4 \geq \rho(A_1)^{2^{2k}} = \rho(A_1)^{2^{n-1}}$ if $n = 2k+1$, and $\rho(A_{2k+2}) \geq \rho(A_{2k+1})^2 \geq \rho(A_1)^{2^{2k+1}} = \rho(A_1)^{2^{n-1}}$ if $n = 2k+2$. □

We deduce the following lower estimate.

Proposition B.4. *If* $(q_n)_n$ *is the sequence of denominators of the Thue-Morse sequence on* $\{a,b\}$, *and* $c_k = 5.2^k$, *then* $\liminf_{k\to\infty} q_{c_k}^{1/c_k} \geq \sqrt{\rho(AB)}$.

Proof. We start with the obvious inequality : $q_n > Tr(W_n)/2$. Indeed, remember that $\begin{pmatrix} q_n & p_n \\ q_{n-1} & p_{n-1}\end{pmatrix} = W_n \begin{pmatrix} 1 & 0 \\ 0 & 1\end{pmatrix}$ since $p_0 = 0$ and $p_{-1} = 1$, so that $2q_n > q_n + q_{n-1} \geq q_n + p_{n-1} = Tr(W_n)$, whence the claim.

But the trace of a product of matrices in $SL(2,\mathbf{Z})$ is computable by iteration of the Cayley-Hamilton identity, as we already did in appendix A. Denote by α_k and β_k the trace of A_k and B_k respectively; then, $\alpha_0 = a, \beta_0 = b$, and $\alpha_k = \beta_k$ for $k \geq 1$. If $X \in SL(2,\mathbf{R}^+)$ is such that $\rho(X) > 1$, then $Tr(X) \geq \rho(X)$; this shows that $\lim_{k\to\infty} \alpha_k = +\infty$. Now, put $Z_k = B_kA_kB_kB_kA_k$, associated with the word $\zeta^k(abbab)$ of length c_k, so that, with our notations, $Z_k = W_{c_k}$. Then we readily obtain with help of (A.5)

$$\begin{aligned}Tr(Z_k) &= Tr(A_{k+1}B_kA_{k+1}) = Tr(A_{k+1}^2B_k) \\ &= \alpha_{k+1}Tr(A_{k+1}B_k) - Tr(B_k) \\ &= \alpha_{k+1}Tr(A_kB_k^2) - Tr(B_k) \\ &= \alpha_{k+1}\alpha_kTr(A_kB_k) - \alpha_k - \alpha_{k+1}\alpha_k,\end{aligned}$$

finally

$$Tr(Z_k) = \alpha_{k+1}^2 \alpha_k - \alpha_{k+1}\alpha_k - \alpha_k. \tag{B.14}$$

Since α_k tends to infinity with k, we deduce from (B.14) that $Tr(Z_k) \sim \alpha_{k+1}^2 \alpha_k$ as k goes to $+\infty$. Keeping in mind that $q_{c_k} \geq \frac{1}{2} Tr(Z_k)$, we may write

$$\begin{aligned}\liminf q_{c_k}^{\frac{1}{c_k}} &\geq \liminf Tr(Z_k)^{\frac{1}{c_k}} \geq \liminf \left(\rho(A_{k+1})^2 \rho(A_k)\right)^{\frac{1}{c_k}} \\ &\geq \liminf \rho(AB)^{\frac{5.2^{2k-1}}{c_k}} = \sqrt{\rho(AB)},\end{aligned}$$

since, by lemma B.5,

$$\rho(A_{k+1})^2 \rho(A_k) \geq \rho(A_1)^{2^{2k+1}+2^{2k-1}} = \rho(A_1)^{5.2^{2k-1}}.$$

□

Corollary B.2. *The Lévy constant β of the Thue-Morse number satisfies*

$$\log \sqrt{\rho(AB)} \leq \beta \leq \log \sqrt{\|AB\|}$$

When $a = 1$, and $b = 2$ for example, those estimates lead to $0.658 \leq \beta \leq 0.676$. By applying the formula (B.12) with $n = 5$, we get the approximation $\beta \simeq 0.669$.

B.3 Schmidt Subspace Theorem

B. Adamczewski and Y. Bugeaud have obtained in [4] significant improvements of theorem B.6 by using a more powerful theorem of W. Schmidt, usually called the Schmidt subspace theorem [222].

Theorem B.7 (W.M. Schmidt). *Let $m \geq 2$ be an integer and $\varepsilon > 0$. Let $L_j, 1 \leq j \leq m$, be m linear forms on $\mathbf{R}^m$, with algebraic coefficients and linearly independent on $\overline{\mathbf{Q}}$. Then, the solutions $x = (x_1, \ldots, x_m)$ in $\mathbf{Z}^m$ to the inequality*

$$|L_1(x) \cdots L_m(x)| \leq \max\{|x_1|, \ldots, |x_m|\}^{-\varepsilon}$$

lie in finitely many proper subspaces of $\mathbf{Q}^m$.

The benefit of the Schmidt subspace theorem, as explained by the authors of [4], rests in the arbitrary number of concerned linear forms. Actually, taking $m = 2$ in this statement leads to the famous Roth theorem on algebraic numbers. Indeed, let α be an irrational algebraic number and suppose that there exist infinitely many distinct solutions (p_n/q_n) to the inequality $0 < |\alpha - \frac{p}{q}| < \frac{1}{q^{2+\varepsilon}}$; this condition can be expressed in the form $q_n|q_n\alpha - p_n| < q_n^{-\varepsilon}$, and for n large enough, modifying ε if necessary,

$$q_n|q_n\alpha - p_n| < \max\{|p_n|, q_n\}^{-\varepsilon}$$

since $|p_n| \leq q_n(|\alpha|+1)$ as soon as $|\alpha - p_n/q_n| < 1$. We can write this inequality as

$$|L_1(p_n,q_n)||L_2(p_n,q_n)| < \max\{|p_n|, q_n\}^{-\varepsilon},$$

where $L_1(x,y) = x - \alpha y$, $L_2(x,y) = y$ are linear forms on $\mathbf{R}^m$, with algebraic coefficients and linearly independent on $\mathbf{Q}(\alpha)$. By applying the previous theorem, we deduce that infinitely many p_n/q_n lie in a same vector line of $\mathbf{Q}^2$: one can find two integers x, y, $(x,y) \neq (0,0)$ such that $xp_n + yq_n = 0$ for those n; but there is only a finite number of such rational numbers whence a contradiction. We have thus got the following famous result :

Theorem B.8 (Roth). *If α is such that, for every $\varepsilon > 0$, the inequality*

$$0 < |\alpha - \frac{p}{q}| < \frac{1}{q^{2+\varepsilon}}$$

holds for infinitely many rational numbers, then α is transcendental.

But Roth's theorem is of no use for the transcendence of irrational numbers in BAD.
We select below a few of the results in [3,4] related to our purpose.

B.3.1 Transcendence and Repetitions

If V is a word on the alphabet A and $\alpha \geq 1$, we denote by V^α a partial repetition of V : $V^\alpha = VV'$ where V' is a prefix of V with $|VV'|/|V| = \alpha$. A square corresponds to $\alpha = 2$. The first improvement is the next theorem that we admit [4].

Theorem B.9 (Adamczewski & Bugeaud). *Let x be the real number identified with its continued fraction expansion $x = [0; a_1, a_2, \ldots]$ and suppose that x is neither quadratic nor rational. Then x is transcendental in both cases :*

(i) x begins with arbitrary large squares.

(ii) The sequence $(q_n^{1/n})_{n\geq 0}$ is bounded and there exists a rational number $\omega > 1$ such that x begins with arbitrary large words V_n^ω.

Remark B.2. 1. The condition on the sequence of denominators in (ii) is satisfied for almost all x by Lévy's theorem for instance. In addition, it is trivially true for numbers in BAD since the sequence then is lacunary.
2. In this version, the repetitions, once more, must appear at the very beginning of the infinite word and this constraint has to be released if one is interested in points of the closed orbit for instance. The authors obtained a second version

where a controlled shift of the repetitions is allowed; in counterpart, the numbers are supposed to be Lévy's numbers and the constraint on the growth of the denominators must be increased.

We deduce from theorem B.9 (ii) the following result which solves the question of morphic numbers in the constant-length case.

Theorem B.10 (Adamczewski & Bugeaud). *The continued fraction expansion of an algebraic number of degree at least three cannot be a recurrent fixed point of some constant-length substitution.*

Proof. Let ζ be a constant-length substitution on the alphabet A and suppose that $a = (a_j)_{j\geq 1}$ is a fixed point of ζ; necessarily, a_1 is the first letter of $\zeta(a_1)$ and by definition of a recurrent sequence, a_1 occurs in a at least twice; in particular, there exists $W \in \mathscr{L}_a$ such that the factor $a_1 W a_1$ appears as a prefix of a. Clearly, the assumptions of theorem B.9 (ii) are satisfied by taking $V_n = \zeta^n(a_1 W)$. Note that this result admits an extension to automatic sequences, i.e. the image letter-by-letter of such fixed points. □

Example B.3. The image of the Rudin-Shapiro sequence by some map $\{\pm 1\} \mapsto \{a,b\}$, where a,b are distinct integers ≥ 2, cannot be the continued fraction expansion of an algebraic number.

B.3.2 Transcendence and Palindromes

In 1840, Liouville has shown that e is not a quadratic irrational number by exhibiting "good" simultaneous rational approximations of e and e^{-1}. In the prolongation of Liouville's result, one can prove the following by using the Schmidt subspace theorem.

Theorem B.11 (W.M. Schmidt). *Let ξ be an irrational number. Suppose that ξ and ξ^2 admit good simultaneous rational approximations in sense that there exist infinitely many distinct rational numbers (p_n/q_n), (p'_n/q_n) such that*

$$|q_n\xi - p_n| < \frac{1}{q_n^\sigma} \quad \text{and} \quad |q_n\xi^2 - p'_n| < \frac{1}{q_n^\tau},$$

where $\sigma + \tau > 1$; then ξ is either quadratic or transcendental.

Proof. Suppose ξ to be an algebraic number of degree ≥ 3. We manage to make use of the Schmidt subspace theorem, so we consider the three linear forms on $\mathbf{Q}$

$$\begin{cases} L_1(X_1,X_2,X_3) = X_1, \\ L_2(X_1,X_2,X_3) = \xi X_1 - X_2, \\ L_3(X_1,X_2,X_3) = \xi^2 X_1 - X_3. \end{cases}$$

From our assumption, those forms have algebraic coefficients and are linearly independent on $\mathbf{Q}$. If the sequences (p_n/q_n), (p'_n/q_n) satisfy

$$|q\xi - p| < \frac{1}{q^\sigma}, \quad |q\xi^2 - p'| < \frac{1}{q^\tau}$$

with $\sigma + \tau > 1$ as assumed in the statement, we get,

$$\prod_{1 \le j \le 3} |L_j(q_n, p_n, p'_n)| < \frac{1}{q_n^{\sigma+\tau-1}}.$$

Now, applying Schmidt's theorem, we conclude that infinitely many (q_n, p_n, p'_n) lie in a same vector plane of $\mathbf{Q}^3$: one can thus find an infinite sequence (n_k) and three integers x_1, x_2, x_3 with $(x_1, x_2, x_3) \neq (0,0,0)$ such that, for every $n \in (n_k)$, $x_1 q_n + x_2 p_n + x_3 p'_n = 0$, or, equivalently,

$$x_1 + x_2 \frac{p_n}{q_n} + x_3 \frac{p'_n}{q_n} = 0.$$

Taking the limit as $k \to \infty$ we get $x_1 + x_2\xi + x_3\xi^2 = 0$, whence a contradiction. □

As a corollary, following [5], we shall establish a property of palindromic continuous fractions.

Recall that a palindrome is a symmetric word :

$$W := a_1 a_2 \ldots a_{n-1} a_n = a_n a_{n-1} \ldots a_2 a_1 =: \overline{W}.$$

If W is a palindrome and $p_n/q_n = [0; a_1 a_2 \ldots a_{n-1} a_n]$, then we have $p_n/q_n = q_{n-1}/q_n$ from (B.2) so that $p_n = q_{n-1}$. If now x begins with W, this remark can be used to provide an excellent simultaneous approximation to the number x and its square. Indeed, since $|x - p_n/q_n| < 1/q_n^2$, then

$$\begin{aligned}
|x^2 - \frac{p_{n-1}}{q_n}| &= |x^2 - \frac{p_{n-1}}{q_{n-1}} \frac{p_n}{q_n}| \\
&\le |x - \frac{p_n}{q_n}||x + \frac{p_{n-1}}{q_{n-1}}| + \frac{1}{q_n q_{n-1}} \le \frac{a_n + 3}{q_n^2} = \frac{a_1 + 3}{q_n^2}.
\end{aligned}$$

It follows that the infinite sequences (p_n/q_n) and (p_{n-1}/q_n) satisfy theorem B.11 and x must be either quadratic or transcendental. This proves the next statement.

Theorem B.12 (Adamczewski & Bugeaud). *The continued fraction expansion of an algebraic number of degree at least three never begins with arbitrarily large palindromes.*

Example B.4. This theorem provides a very short proof of the transcendence of the Thue-Morse continued fractions.

References

1. J. AARONSON, *An introduction to infinite ergodic theory*, Mathematical Surveys and Monographs, 50. American Mathematical Society, Providence, RI, 1997.
2. B. ADAMCZEWSKI and J.-P. ALLOUCHE, Reversals and palindromes in continued fractions, *Theoret. Comput. Sci.* **380** (2007), no. 3, 220–237.
3. B. ADAMCZEWSKI and Y. BUGEAUD, On the complexity of algebraic numbers. I. Expansions in integer bases, *Ann. of Math. (2)* **165** (2007), no. 2, 547–565.
4. B. ADAMCZEWSKI and Y. BUGEAUD, On the complexity of algebraic numbers. II. Continued fractions, *Acta Math.* **195** (2005), 1–20.
5. B. ADAMCZEWSKI and Y. BUGEAUD, Palindromic continued fractions, *Ann. Inst. Fourier* **57** (2007), no. 5, 1557–1574.
6. B. ADAMCZEWSKI, Y. BUGEAUD and L. DAVISON, Continued fractions and transcendental numbers. Numération, pavages, substitutions, *Ann. Inst. Fourier* **56** (2006), no. 7, 2093–2113.
7. O. N. AGEEV, Dynamical systems with a Lebesgue component of even multiplicity in the spectrum, *Translation in Math. USSR-Sb.* **64** (1989), no. 2, 305–317
8. J.-P. ALLOUCHE, Somme des chiffres et transcendance, *Bull. Soc. Math. France* **110** (1982), 279–285.
9. J.-P. ALLOUCHE, Schrödinger operators with Rudin-Shapiro potentials are not palindromic, *J. Math. Phys.* **38** (1997), no. 4, 1843–1848.
10. J.-P. ALLOUCHE and M. COSNARD, Itérations de fonctions unimodales et suites engendrées par automates, *C.R. Acad. Sc. Paris* **296** (1983), 159–162.
11. J.-P. ALLOUCHE, J.L. DAVISON, M. QUEFFÉLEC and L. ZAMBONI, Transcendence of Sturmian or morphic continued fractions, *J. Number Theory* **91** (2001), 33–66.
12. J.-P. ALLOUCHE and P. LIARDET, Generalized Rudin-Shapiro sequences, *Acta Arith.* **60** (1991), no. 1, 1–27.
13. J.-P. ALLOUCHE and M. MENDÈS FRANCE, Suites de Rudin-Shapiro et modèle d'Ising, *Bull. Soc. Math. France* **113** (1985), 273–283.
14. J.-P. ALLOUCHE and J. SHALLIT, *Automatic sequences : Theory and Applications*, Cambridge University Press, 2003.
15. J.-P. ALLOUCHE and L.Q. ZAMBONI, Algebraic irrational binary numbers cannot be fixed points of non-constant length or primitive morphisms, *J. Number Theory* **69** (1998), 119–124.
16. H. ANZAI, Ergodic skew product transformations on the torus, *Osaka Math. J.* **3** (1951), 83–99.
17. P. ARNOUX and G. RAUZY, Représentation géométrique de suites de complexité $2n+1$, *Bull. Soc. Math. France* **119** (1991), 199–215.
18. I. ASSANI *Wiener Wintner ergodic theorems*, World Scientific Publishing Co., Inc., River Edge, NJ, 2003.
19. J. AVRON and B. SIMON, Almost periodic Schrödinger operators. II. The integrated density of states, *Duke Math. J.* **50** (1983), no. 1, 369–391.

20. G. BARAT, V. BERTHÉ, P. LIARDET and J. THUSWALDNER, Dynamical directions in numeration. Numération, pavages, substitutions, *Ann. Inst. Fourier* **56** (2006), no. 7, 1987–2092.
21. M. BARGE and B. DIAMOND, Coincidence for substitutions of Pisot type, *Bull. Soc. Math. France* **130** (2002), 619–626.
22. A. BELLOW and V. LOSERT, The weighted pointwise ergodic theorem and the individual ergodic theorem along subsequences, *Trans. Amer. Math. Soc.* **288** (1985), no. 1, 307–345.
23. V. BERGELSON and E. LESIGNE, Van der Corput sets in $\mathbf{Z}^d$, *Colloq. Math.* **110** (2008), no. 1, 1–49.
24. M.-J. BERTIN, A. DECOMPS-GUILLOUX, M. GRANDET-HUGOT, M. PATHIAUX-DELEFOSSE and J.-P. SCHREIBER, *Pisot and Salem numbers*, Birkhaüser Verlag, Basel, 1992.
25. A. BERTRAND-MATHIS, Ensembles intersectifs et récurrence de Poincaré, *Israel J. Math.* **55** (1986), 184–198.
26. A. BERTRAND-MATHIS, Nombres de Perron et problèmes de rationnalité, *Astérisque* **198-200** (1991), 67–76.
27. J.-P. BERTRANDIAS, Espaces de fonctions bornées et continues en moyenne asymptotique d'ordre *p*, *Bull. Soc. Math. France* mémoire **5** (1966), 3–106.
28. J. BESINEAU, Indépendance statistique d'ensembles liés à la fonction "somme des chiffres", *Acta Arith.* **20** (1972), 401–416.
29. P. BILLINGSLEY, *Ergodic theory and information*, John Wiley and Sons, New York, 1965.
30. BLUM and EISENBERG, Generalized summing sequences and the mean ergodic theorem, *Proc. Amer. Math. Soc.* **42** (1974), 423–429.
31. J. R. BLUM and J. I. REICH, Strongly ergodic sequences of integers and the individual ergodic theorem, *Proc. Amer. Math. Soc.* **86** (1982), no. 4, 591–595.
32. E. BOMBIERI and J. BOURGAIN, On Kahane's ultraflat polynomials, preprint.
33. M. BOSHERNITZAN, G. KOLESNIK, A. QUAS and M. WIERDL, Ergodic averaging sequences, *J. Anal. Math.* **95** (2005), 63–103.
34. P. BOUGEROL and J. LACROIX, *Products of random matrices with applications to Schrödinger operators*, Progress in Probability and Statistics, No 8, Birkhäuser, Boston, 1985.
35. J. BOURGAIN, Ruzsa's problem on sets of recurrence, *Israel J. Math.* **59** (1987), no. 2, 150–166.
36. J. BOURGAIN, On the maximal ergodic theorem for certain subsets of the integers, *Israel J. Math.* **61** (1988), no. 1, 39–72.
37. J. BOURGAIN, An approach to pointwise ergodic theorems, in *Geometric aspects of functional analysis (1986/87)*, 204–223, Lecture Notes in Math., 1317, Springer, Berlin, 1988.
38. J. BOURGAIN, Pointwise ergodic theorems for arithmetic sets, with an appendix by the author, Harry Furstenberg, Yitzhak Katznelson and Donald S. Ornstein, *Inst. Hautes études Sci. Publ. Math.* **69** (1989), 5–45.
39. A. BOVIER and J.-M. GHEZ, Spectral properties of one-dimensional Schrödinger operators with potentials generated by substitutions, *Comm. Math. Phys.* **158** (1993), no. 1, 45–66. [Erratum in *Comm. Math. Phys.* **166** (1994), no. 2, 431–432.]
40. D.W. BOYD, J. COOK, P. MORTON, On sequences of 1's defined by binary patterns, *Dissertationes Math.* **283** (1989).
41. M. BOYLE and D. HANDELMAN, The spectra of nonnegative matrices via symbolic dynamics, *Ann. of Math.* **133** (1991), no. 2, 249–316.
42. X. BRESSAUD, F. DURAND and A. MAASS, Necessary and sufficient conditions to be an eigenvalue for linearly recurrent dynamical Cantor systems, *J. London Math. Soc.* **72** (2005), no. 3, 799–816.
43. J. BRILLHART, On the Rudin-Shapiro polynomials, *Duke Math. J.* **40** (1973), 335–353.
44. J. BRILLHART and L. CARLITZ, Note on the Shapiro polynomials, *Proc. Amer. Math. Soc.* **25** (1970), 114–118.
45. J. BRILLHART, P. ERDÒS and P. MORTON, On sums of Rudin-Shapiro coefficients II, *Pac. J. Math.* **107** (1983), 39–69.

46. A. BROISE, Transformations dilatantes de l'intervalle et théorèmes limites. Etudes spectrales d'opérateurs de transfert et applications, *Astérisque* **238** (1996), 1–109.
47. G. BROWN and W. MORAN, A dichotomy for infinite convolutions of discrete measures, *Proc. Camb. Phil. Soc.* **73** (1973), 307–316.
48. G. BROWN and W. MORAN, On orthogonality of Riesz products. *Proc. Cambridge Philos. Soc.* **76** (1974), 173–181.
49. G. BROWN and W. MORAN, Coin tossing and powers of singular measures, *Math. Proc. Cambridge Philos. Soc.* **77** (1975), 349–364.
50. G. BROWN, W. MORAN and C.E.M. PIERCE, Riesz products and normal numbers, *J. London Math. Soc.* **32** (1985), 12–18.
51. Y. BUGEAUD, *Approximation by algebraic numbers*, Cambridge Tracts in Mathematics, 160. Cambridge University Press, Cambridge, 2004.
52. V. CANTERINI and A. SIEGEL, Geometric representation of substitutions of Pisot type, *Trans. Amer. Math. Soc.* **353** (2001), no. 12, 5121–5144 (electronic).
53. J. CASSAIGNE, Constructing infinite words of intermediate complexity, *Developments in language theory*, 173–184, Lecture Notes in Comput. Sci., 2450, Springer, Berlin, 2003.
54. R.V. CHACON, Weakly mixing transformations which are not strongly mixing, *Proc. Amer. Math. Soc.* **22** (1969), 559–562.
55. G. CHRISTOL, T. KAMAE, M. MENDÈS FRANCE and G. RAUZY, Suites algébriques, automates et substitutions, *Bull. Soc. Math. France* **108** (1980), 401–419.
56. A. COBHAM, On the base-dependence of sets of numbers recognizable by finite automata, *Math. Systems Theory* **3** (1969), 186–192.
57. A. COBHAM, Uniform tag sequences, *Math. Systems Theory* **6** (1972), 164–192.
58. J. COQUET, A summation formula related to the binary digit, *Inv. Math.* **73** (1983), 107–115.
59. J. COQUET, T. KAMAE and M. MENDÈS FRANCE, Sur la mesure spectrale de certaines suites arithmétiques, *Bull. Soc. Math. France* **105** (1977), 369–384.
60. J. COQUET and P. LIARDET, Répartitions uniformes de suites et indépendance statistique, *Comp. Math.* **51** (1984), 215–236.
61. J.P. CORNFELD, S.V. FOMIN and Y.G. SINAI, *Ergodic Theory*, Springer Verlag, Berlin, 1982.
62. E.M. COVEN and G.A. HEDLUND, Sequences with minimal block growth, *Math. Syst. Theory* **7** (1973), 138–153.
63. E.M. COVEN and M. KEANE, The structure of substitution minimal sets, *Trans. Amer. Math. Soc.* **162** (1971), 89–102.
64. E.M. COVEN, M. KEANE and M. LEMASURIER, A characterization of the Morse minimal set up to topological conjugacy, *Erg. Theory Dyn. Systems* **28** (2008), no. 5, 1443–1451.
65. M.J. CRABB, J. DUNCAN and C.M. McGREGOR, Finiteness and recognazibility problems for substitution maps on two symbols, to appear.
66. D. CRISP, W. MORAN, A. POLLINGTON and P. SHIUE, Substitution invariant cutting sequences, *J. Théor. Nombres Bordeaux* **5** (1993), no. 1, 123–137.
67. J.L. DAVISON, A class of transcendental numbers with bounded partial quotients, in *Number theory and applications* (Banff, AB, 1988), 365–371, NATO Adv. Sci. Inst. Ser. C Math. Phys. Sci., 265, Kluwer Acad. Publ., Dordrecht, 1989.
68. F.M. DEKKING, The spectrum of dynamical systems arising from substitutions of constant length, *Zeit. Wahr. Verw. Gebiete* **41** (1978), 221–239.
69. F.M. DEKKING, Regularity and irregularity of sequences generated by automata, *Seminar on Number Theory* (1979–1980), Exp. No. 9, Univ. Bordeaux I, Talence, 1980.
70. F.M. DEKKING, On the distribution of digits in arithmetic sequences, *Seminar on Number Theory* (1982–1983), Exp. No. 32, Univ. Bordeaux I, Talence, 1983.
71. F.M. DEKKING and M. KEANE, Mixing properties of substitutions, *Zeit. Wahr.* **42** (1978), 23–33.
72. F.M. DEKKING and M. MENDÈS FRANCE, Uniform distribution modulo one, a geometrical viewpoint, *J. für dir Reine Angew. Math.* **329** (1981), 143–153.
73. J.-M. DUMONT, Discrépance des progressions arithmétiques dans la suite de Morse, *C.R. Acad. Sc. Paris* **297** (1983), 145–148.

74. J.-M. DUMONT and A. THOMAS, Systèmes de numération et fonctions fractales relatifs aux substitutions, *Theoret. Comput. Sc.* **65** (1989), 153–169.
75. D.F. DUNKL and D. RAMIREZ, Bounded projections on Fourier-Stieltjes transforms, *Proc. Amer. Math. Soc.* **31** (1972), 122–126.
76. F. DURAND, A characterization of substitutive sequences using return words, *Discrete Math.* **179** (1998), 89–101.
77. F. DURAND, A generalization of Cobham's theorem, *Theory Comput. Syst.* **31** (1998), no. 2, 169–185.
78. F. DURAND, Sur les ensembles d'entiers reconnaissables, *J. Théor. Nombres Bordeaux* **10** (1998), no. 1, 65–84.
79. W.F. EBERLEIN, Abstract ergodic theorems and weak almost periodic functions, *Trans. Amer. Math. Soc.* **67** (1949), 217–240.
80. S. EILENBERG, *Automata, languages and machines, Vol. A*, Academic Press, New York, 1974.
81. W.J. ELLISON *Les nombres premiers*. Publications de l'Institut de Mathématique de l'Université de Nancago, No. IX. Actualités Scientifiques et Industrielles, No. 1366. Hermann, Paris, 1975.
82. C. FAIVRE, Distribution of Lévy constants for quadratics numbers, *Acta Arith.* **41** (1992), 13–34.
83. S. FERENCZI, Bounded Remainder sets, *Acta Arith.* **61** (1992), 319–326.
84. S. FERENCZI, Les transformations de Chacon : combinatoire, structure géométrique, lien avec les systèmes de complexité $2n+1$, *Bull. Soc. Math. France* **123** (1995), 271–292.
85. S. FERENCZI, C. HOLTON and L. ZAMBONI, Structure of three-interval exchange transformations III: ergodic and spectral properties, *J. Anal. Math.* **93** (2004), 103–138.
86. S. FERENCZI and C. MAUDUIT, Transcendence of numbers with a low complexity expansion, *J. Number Theory* **67** (1998), 146–161.
87. S. FERENCZI, C. MAUDUIT and A. NOGUEIRA, Substitution dynamical systems : algebraic characterization of eigenvalues, *Ann. Sc. École Norm. Sup.* **29** (1996), 519–533.
88. N.P. FOGG, *Substitutions in dynamics, arithmetics and combinatorics.* Edited by V. Berthé, S. Ferenczi, C. Mauduit and A. Siegel. Lecture Notes in Mathematics, 1794. Springer-Verlag, Berlin, 2002.
89. C. FOIAS and S. STRATILA, Ensembles de Kronecker dans la théorie ergodique, *C.R. Acad. Sc. Paris* **267** (1968), 166–168.
90. H. FURSTENBERG, Strict ergodicity and transformation of the torus, *Amer. J. Math.* **83** (1961), 573–601.
91. H. FURSTENBERG, Disjointness in ergodic theory, minimal sets, and a problem in Diophantine approximation, *Math. Systems Theory* **1** (1967), 1–49.
92. H. FURSTENBERG, Ergodic behaviour of diagonal measures and a theorem of Szemeredi on arithmetic progressions and transformation of the torus, *J. Anal. Math.* **31** (1977), 204–256.
93. H. FURSTENBERG, Poincaré recurrence and number theory, *Bull. Amer. Math. Soc.* **5** (1981), 211–234.
94. H. FURSTENBERG, *Recurrence in ergodic theory and combinatorial number theory*, Princeton University Press, 1981.
95. H. FURSTENBERG and H. KESTEN, Products of random matrices, *Ann. Math. Statist* **31** (1960), 457–469.
96. H. FURSTENBERG, H. KEYNES and L. SHAPIRO, Prime flows in topological dynamics, *Israel J. Math.* **14** (1973), 26–38.
97. H. FURSTENBERG and B. WEISS, The finite multipliers of infinite ergodic transformations, Attractors in ergodic theory, *Lecture Notes Math.* **668** (1977), 127–132.
98. H. FURSTENBERG and B. WEISS, Topological dynamics and combinatorial number theory, *J. Anal. Math.* **34** (1978), 61–85.
99. A.O. GEL'FOND, Sur les nombres qui ont des propriétés additives et multiplicatives données, *Acta Arith.* **13** (1967/1968), 259–265.
100. G. R. GOODSON, Approximations and spectral theory of finite skew products, *J. London Math. Soc* **14** (1976), 249–259.

101. G. R. GOODSON, On the spectral multiplicity of a class of finite rank transformations, *Proc. Amer. Math. Soc.* **93** (1985), no. 2, 303–306.
102. G. R. GOODSON and M. LEMAŇCZYK, On the rank of a class of bijective substitutions, *Studia Math.* **96** (1990), no. 3, 219–230.
103. G. R. GOODSON and P.N. WHITMAN, On the spectral properties of a class of special flows, *J. London Math. Soc* **21** (1980), 567–576.
104. W.H. GOTTSCHALK, Substitution minimal sets, *Trans. Amer. Math. Soc.* **109** (1963), 467–491.
105. W.H. GOTTSCHALK and G. HEDLUND, A characterization of the Morse minimal set, *Proc. Amer. Math. Soc.* **15** (1964), 70–74.
106. M. GUENAIS, Singularité des produits de Anzai associés aux fonctions caractéristiques d'un intervalle, *Bull. Soc. Math. France* **127** (1999), no. 1, 71–93.
107. M. GUENAIS, Morse cocycles and simple Lebesgue spectrum, *Ergodic Theory Dynam. Systems* **19** (1999), no. 2, 437–446.
108. C. GUILLE-BIEL, Sparse Schrödinger operators, *Rev. Math. Phys.* **9** (1997), no. 3, 315–341.
109. M. GUENAIS and F. PARREAU, Eigenvalues of transformations arising from irrational rotations and step functions, *ArXiv* (2003).
110. J.V. GUIRSANOV, On spectrum of dynamical systems generated by Gauss stationery process, *Dokl. Acad.Sc. USSR* **119** (1958), 851–853.
111. P.R. HALMOS, *Lectures on ergodic theory*, Chelsea Publishing Co, New York, 1960.
112. G.H. HARDY and E.M. WRIGHT, *An introduction to the theory of numbers*, Clarendon Press, Oxford Univ. Press, 1979.
113. H. HELSON, Cocycles on the circle, *J. Operator Theory* **16** (1986), no. 1, 189–199.
114. M.R. HERMAN, L^2 regularity of measurable solutions of a finite-difference equation of the circle, *Ergodic Theory Dynam. Systems* **24** (2004), no. 5, 1277–1281.
115. A. HOF, Some remarks on discrete aperiodic Schrödinger operators, *J. Statist. Phys.* 72 (1993), no. 5-6, 1353–1374.
116. A. HOF, O. KNILL and B.SIMON, Singular continuous spectrum for palindromic Schrödinger operators, *Comm. Math. Phys.* 174 (1995), no. 1, 149–159.
117. M. HOLLANDER and B. SOLOMYAK, Two-symbol Pisot subtitutions have pure discrete spectrum, *Erg. Theory Dyn. Systems* **23** (2003), 533–540.
118. C. HOLTON and L. ZAMBONI, Geometric realization of substitutions, *Bull. Soc. Math. France* **126** (1998), 149–179.
119. B. HOST, Valeurs propres des systèmes dynamiques définis par des substitutions de longueur variable, *Erg. Theory Dyn. Systems* **6** (1986), 529–540.
120. B. HOST, Mixing of all orders and pairwise independent joinings of systems with singular spectrum, *Israel J. Math.* **76** (1991), no. 3, 289–298.
121. B. HOST, Représentation géométrique des substitutions sur deux lettres. Unpublished manuscript, 1992.
122. B. HOST, Substitution subshifts and Bratteli diagrams, in *Topics in symbolic dynamics and applications* (Temuco, 1997), London Math. Soc. Lecture Note Ser. **279**, Cambridge Univ. Press, 2000.
123. B. HOST, J.-F. MELA and F. PARREAU, *Analyse harmonique des mesures*. Astérisque **135–136** (1986).
124. B. HOST and F. PARREAU, Orthogonalité et propriétés spectrales dans les algèbres de convolution de mesures, *preprint* (1987).
125. B. HOST and F. PARREAU, The generalized purity law for ergodic measures: a simple proof, *Colloq. Math.* **60/61** (1990), no. 1, 205–212.
126. P. HUBERT, Complexité de suites définies par des billards rationnels, *Bull. Soc. Math. France* **123** (1995), no. 2, 257–270.
127. B. JESSEN and A. WINTNER, Distribution functions and the Riemann zeta function, *Trans. Amer. Math. Soc.* **38** (1935), no. 1, 48–88.
128. J.-P. KAHANE and R. SALEM, Distribution modulo 1 and sets of uniqueness, *Bull. Amer. Math. Soc.* **70** (1964), 259–261.

129. J. P. KAHANE and R. SALEM, *Ensembles parfaits et séries trigonométriques*, Hermann, Paris, second edition 1994.
130. S. KAKUTANI, Examples of ergodic measure-preserving transformations which are weakly mixing but not strongly mixing, *Lecture Notes Math.* **318** (1973), 143–149.
131. S. A. KALIKOW, Twofold mixing implies threefold mixing for rank one transformations, *Erg. Theory Dyn. Systems* **4** (1984), no. 2, 237–259.
132. T. KAMAE, Spectrum of a substitution minimal set, *J. Math. Soc. Japan* **22** (1970), 567–578.
133. T. KAMAE, Sum of digits to different bases and mutual singularity of their spectral measures, *Osaka J. Math.* **15** (1978), 569–574.
134. T. KAMAE, Mutual singularity of spectra of dynamical systems given by sums of digits to different bases, Dynamical Systems I, Warsaw, *Astérisque* **49** (1978), 109–116.
135. T. KAMAE, Spectral properties of automaton-generating sequences, unpublished.
136. T. KAMAE, Cyclic extensions of odometer transformations and spectral disjointness, *Israel J. Math.* **59** (1987), 41–63.
137. T. KAMAE, Number-theoretic problems involving two independent bases. Number theory and cryptography (Sydney, 1989), 196–203, *London Math. Soc. Lecture Note Ser.* **154**, Cambridge Univ. Press, Cambridge, 1990.
138. T. KAMAE and M. MENDÈS FRANCE, Van der Corput's difference theorem, *Israel J. Math.* **31** (1978), 335–342.
139. A.B. KATOK and B. HASSELBLATT, *Introduction to the modern theory of dynamical systems*, Cambridge University Press, 1995.
140. A.B. KATOK and A.M. STEPIN, Approximations in ergodic theory, *Russian Math. Survey* **22** (1967), 77–102.
141. Y. KATZNELSON, *An introduction to harmonic analysis*, Cambridge University Press, third edition, 2004.
142. Y. KATZNELSON and B. WEISS, The construction of quasi-invariant measures, *Israel J. Math.* **12** (1972), 1–4.
143. M. KEANE, Generalized Morse sequences, *Zeit. Wahr. Verw. Gebiete* **10** (1968), 335–353.
144. M. KEANE, Strongly mixing g-measures, *Inv. Math.* **16** (1972), 309–324.
145. M. KEANE, Ergodic theory and subshifts of finite type, in *Ergodic theory, symbolic dynamics, and hyperbolic spaces* (1989), 35–70, Oxford Sci. Publ., Oxford Univ. Press, New York, 1991.
146. H. KESTEN, On a conjecture of Erdös and Szüsz related to uniform distribution mod1, *Acta Arith.* **12** (1966/1967), 193–212.
147. A.Y. KHINTCHINE, *Continued Fractions*, P. Noordhoff, Ltd., Groningen, 1963.
148. J. F. C. KINGMAN, The ergodic theory of subadditive stochastic processes, *J. Roy. Statist. Soc. Ser. B* **30** (1968), 499–510.
149. S. KOTANI, Jacobi matrices with random potentials taking finite many values, *Rev. Math. Phys.* **1** (1988), 129–133.
150. S. KOTANI, One-dimensional Schrödinger operators with stationary deterministic potentials, in *Probability theory (Singapore, 1989)*, 105–108, de Gruyter, Berlin, 1992.
151. U. KRENGEL, *Ergodic theorems*, with a supplement by Antoine Brunel, de Gruyter Studies in Mathematics, 6. Walter de Gruyter & Co., Berlin, 1985.
152. W. KRIEGER, On entropy and generators of measure-preserving transformations, *Trans. Amer. Math. Soc.* **149** 1970 453–464.
153. L. KUIPERS and H. NIEDERREITER, *Uniform distribution of sequences*, Dover Publications, Inc., New York, second edition, 2002.
154. J. KWIATKOWSKI, Spectral isomorphism of Morse dynamical systems, *Bull. Acad. Pol. Sc.* **29** (1981), 105–114.
155. J. KWIATKOWSKI, Isomorphism of regular Morse dynamical systems, *Studia Math.* **62** (1982), 59–89.
156. J. KWIATKOWSKI and A. SIKORSKI, Spectral properties of G-symbolic Morse shifts, *Bull. Soc. Math. France* **115** (1987), 19–33.
157. Y. LAST and B. SIMON, Eigenfunctions, transfer matrices, and absolutely continuous spectrum of one-dimensional Schrödinger operators, *Invent. Math.* 135 (1999), no. 2, 329–367.

158. F. LEDRAPPIER, Des produits de Riesz comme mesures spectrales, *Ann. Inst. H. Poincaré. Probab. Stat.* **4** (1970), 335–344.
159. M. LEMANCZYK, *Ergodic properties of Morse sequences*. Dissertation, Nicholas Copernicus University, Torún, 1985.
160. M. LEMANCZYK, Toeplitz $\mathbf{Z}_2$-extensions, *Ann. Inst. H. Poincaré. Probab. Stat.* **24** (1987), 1–43.
161. M. LEMANCZYK, E. LESIGNE, F. PARREAU, D. VOLNỲ and M. WIERDL, Random ergodic theorems and real cocycles, *Israel J. Math.* **130** (2002), 285–321.
162. M. LEMANCZYK and M. K. MENTZEN, Generalized Morse sequences on n symbols and m symbols are not isomorphic, *Bull. Acad. Pol. Sc.* **29** (1981), 105–114.
163. E. LESIGNE, Un théorème de disjonction de systèmes dynamiques et une généralisation du théorème ergodique de Wiener-Wintner, *Erg. Theory Dyn. Systems* **10** (1990), no. 3, 513–521.
164. E. LESIGNE, Ergodic theorem along a return time sequence in *Ergodic theory and related topics*, III (Gstrow, 1990), Lecture Notes in Math., 1514, Springer, Berlin, 1992.
165. E. LESIGNE, C. MAUDUIT and B. MOSSÉ, Le théorème ergodique le long d'une suite q-multiplicative, *Compositio Math.* **93** (1994), no. 1, 49–79.
166. P. LIARDET, Propriétés dynamiques des suites arithmétiques, *Journées mathématiques S.M.F.–C.N.R.S.*, Caen, 1980.
167. P. LIARDET, Regularities of distribution, *Comp. Math.* **61** (1988), 267–293.
168. D.A. LIND, The entropies of topological Markov shifts and a related class of algebraic integers, *Erg. Theory Dyn. Systems* **4** (1984), no. 2, 283–300.
169. A.N. LIVSHITS, Sufficient conditions for weak mixing of substitutions and of stationary adic transformations, *Mat. Zametki* **44** (1988), 785–793, 862. English translation : *Math. Notes* **44** (1988), 920–925.
170. R. LYONS, *A characterization of measures whose Fourier-Stieltjes transforms vanish at infinity*. Dissertation, University of Michigan, 1983.
171. R. LYONS, Mixing and asymptotic distribution modulo one, *Erg. Theory Dyn. Systems* **8** (1988), no. 4, 597–619.
172. R. LYONS, Seventy years of Rajchman measures, Proceedings of the Conference in Honor of Jean-Pierre Kahane (Orsay, 1993), *J. Fourier Anal. Appl.* (1995), Special Issue, 363–377.
173. J.C. MARTIN, Substitution minimal flows, *Amer. J. Math.* **93** (1971), 503–526.
174. J.C. MARTIN, Minimal flows arising from substitutions of non-constant length, *Math. Syst. Theory* **7** (1973), 73–82.
175. J.C. MARTIN, Generalized Morse sequences on n symbols, *Proc. Amer. Math. Soc.* **54** (1976), 379–383.
176. J.C. MARTIN, The structure of generalized Morse minimal sets on n symbols, *Trans. Amer. Math. Soc.* **232** (1977), 343–355.
177. J. MATHEW and M.G. NADKARNI, A measure-preserving transformation whose spectrum has Lebesgue component of multiplicity two, *Bull. London Math. Soc* **16** (1984), 402–406.
178. J. MATHEW and M.G. NADKARNI, A measure-preserving transformation whose spectrum has finite Lebesgue component, *preprint* (1984).
179. C. MAUDUIT, Automates finis et ensembles normaux, *Ann. Inst. Fourier* **36** (1986), 1–25.
180. C. MAUDUIT, Caractérisation des ensembles normaux substitutifs, *Invent. Math.* **95** (1989), 133–147.
181. C. MAUDUIT and J. RIVAT, Sur un problème de Gelfond : la somme des chiffres des nombres premiers, to appear in *Annals of Math.*
182. J.-F. MÉLA, Groupes de valeurs propres des systèmes dynamiques et sous-groupes saturés du cercle, *C.R. Acad. Sc. Paris* **296** (1985), 419–422.
183. M. MENDÈS FRANCE, Nombres normaux. Application aux fonctions pseudo-aléatoires, *J. Anal. Math.* **20** (1967), 1–56.
184. M. MENDÈS FRANCE and G. TENENBAUM, Dimension des courbes planes, papiers pliés et suites de Rudin-Shapiro, *Bull. Soc. Math. France* **109** (1981), 207–215.
185. K.D. MERRILL, Cohomology of step functions under irrational rotations, *Israel J. Math.* **52** (1985), no. 4, 320–340.

186. Y. MEYER, Les produits de Riesz sont des Bernoulli-shifts, *preprint* (1974).
187. Y. MEYER, *Algebraic Numbers and Harmonic Analysis*, North-Holland mathematical library, **2**, North-Holland Publishing Company, Amsterdam-London, 1972.
188. P. MICHEL, Stricte ergodicité d'ensembles minimaux de substitutions, *C.R. Acad. Sc. Paris* **278** (1974), 811–813.
189. P. MICHEL, Coincidence values and spectra of substitutions, *Zeit. Wahr. Verw. Gebiete* **42** (1978), 205–227.
190. H.M. MORSE, Recurrent geodesics on a surface of negative curvature, *Trans. Amer. Math. Soc.* **22** (1921), 84–100.
191. B. MOSSÉ, Puissances de mots et reconnaissabilité des points fixes d'une substitution, *Theoret. Comput. Sc.* **99** (1992), 327–334.
192. B. MOSSÉ, Reconnaissabilité des substitutions et complexité des suites automatiques, *Bull. Soc. Math. France* **124** (1996), 329–346.
193. M.G. NADKARNI, *Spectral theory of dynamical systems*, Birkhäuser Verlag, Basel, 1998.
194. D.G. NEWMAN, On the number of binary digits in a multiple of three, *Proc. Amer. Math. Soc.* **21** (1969), 719–721.
195. D. ORNSTEIN, D. RUDOLPH and B. WEISS, *Equivalence of measure-preserving transformations*, Memoirs Amer. Math. Soc. **262**, Providence, RI, 1982.
196. F. PARREAU, Ergodicité et pureté des produits de Riesz, *Ann. Inst. Fourier* **40** (1990), no. 2, 391–405.
197. W. PARRY, *Topics in ergodic theory*, Cambridge University Press, 1981.
198. C.E.M. PEARCE and M. KEANE, On normal numbers, *J. Austr. Math. Soc.* **32** (1982), 79–87.
199. K. PETERSEN, On a series of cosecants related to a problem in ergodic theory, *Compositio Math.* **26** (1973), 313–317.
200. K. PETERSEN, *Ergodic theory*, Cambridge University Press, second edition 1989.
201. J. PEYRIÈRE, Etude de quelques propriétés des produits de Riesz, *Ann. Inst. Fourier* **25** (1975), no. 2, 127–169.
202. M. QUEFFÉLEC, Mesures spectrales associées à certaines suites arithmétiques, *Bull. Soc. Math. France* **107** (1979), 385–421.
203. M. QUEFFÉLEC, Une nouvelle propriété des suites de Rudin-Shapiro, *Ann. Inst. Fourier* **37** (1987), 115–138.
204. M. QUEFFÉLEC, Transcendance des fractions continues de Thue-Morse, *J. Number Theory* **73** (1998), 201–211.
205. H. QUEFFÉLEC and B. SAFFARI, On Bernstein's inequality and Kahane's ultraflat polynomials, *J. Fourier Anal. Appl.* **2** (1996), no. 6, 519–582.
206. G. RAUZY, *Propriétés statistiques de suites arithmétiques*, Presses Universitaires de France, Collection SUP, Le Mathématicien No 15, 1976.
207. G. RAUZY, Répartition modulo un, *Astérisque, Soc. Math. France* **41–42** (1977), 81–101.
208. G. RAUZY, Nombres algébriques et substitutions, *Bull. Soc. Math. France* **110** (1982), 147–178.
209. G. RAUZY, Suites à termes dans un alphabet fini, *Séminaire de théorie des nombres 1982–1983* Univ. Bordeaux I, (1983), Exp. No 25.
210. G. RAUZY, Rotations sur les groupes, nombres algébriques et substitutions, *Séminaire de théorie des nombres 1987–1988* Univ. Bordeaux I, (1988), Exp. No 21.
211. D. RIDER, Transformations of Fourier coefficients, *Pac. J. Math.* **19** (1966), 347–355.
212. G.W. RILEY, On spectral properties of skew products over irrational rotations, *J. London Math. Soc* **17** (1978), 152–160.
213. G.W. RILEY, Approximations and the spectral properties of measure-preserving group actions, *Israel J. Math.* **33** (1979), no. 1, 9–31.
214. E.A. ROBINSON Jr, Ergodic measure-preserving transformations with arbitrary finite spectral multiplicities, *Inv. Math.* **72** (1983), 299-314.
215. E.A. ROBINSON Jr, On uniform convergence in the Wiener-Wintner theorem, *J. London Math. Soc.* **49** (1994), no. 3, 493–501.
216. V. A. ROKHLIN, On endomorphisms of compact commutative groups, *Izv. Akad. Nauk SSSR, Ser. Mat.*, 13, 329340 (1949)

217. W. RUDIN, Some theorems on Fourier coefficients, *Proc. Amer. Math. Soc.* **10** (1959), 855–859.
218. W. RUDIN, *Fourier analysis on groups*, Interscience Tracts in Math. No 12, John Wiley and Sons, New York, 1962.
219. I. RUSZA, On difference sets, *Studia Sc. Math. Hung.* **13** (1978), 319–326.
220. B. SAFFARI, Une fonction extrèmale liée à la suite de Rudin-Shapiro, *C.R. Acad. Sc. Paris* **303** (1986), 97–100.
221. R. SALEM, *Algebraic numbers and Fourier analysis*, D. C. Heath and Co., Boston, Mass. 1963.
222. W. SCHMIDT, On simultaneous approximations of two algebraic numbers by rationals, *Acta Math.* **119** (1967), 27–50.
223. K. SCHMIDT, *Cocycles on ergodic transformation groups*, Macmillan Lectures in Mathematics, Vol. 1. Macmillan Company of India, Ltd., Delhi, 1977.
224. E. SENETA, *Nonnegative matrices and Markov chains*, Springer Verlag, New York, second edition, 1981.
225. J.-P. SERRE, *Linear representations of finite groups*, Graduate Texts in Mathematics, Vol. 42. Springer-Verlag, New York-Heidelberg, 1977.
226. H.S. SHAPIRO, *Extremal problems for polynomials and power series*, PhD thesis, Massachusetts Institute of Technology, 1951.
227. A. SIEGEL, *Représentation géométrique, combinatoire et arithmétique de substitutions de type Pisot*, PhD thesis, Université de la Méditerranée, 2000.
228. A. SIEGEL, Représentation des systèmes dynamiques substitutifs non unimodulaires, *Erg. Theory Dyn. Systems* **23** (2003), 1247–1273.
229. B. SIMON, Operators with singular continuous spectrum. I. General operators, *Ann. of Math. (2)* **141** (1995), no. 1, 131–145.
230. B. SOLOMYAK, Substitutions, adic transformations and beta-expansions, *Contemp. Math.* **135** (1992), 361–372.
231. A. SÜTÖ, Singular continuous spectrum on a Cantor set of zero Lebesgue measure for the Fibonacci Hamiltonian, *J. Statist. Phys.* **56** (1989), no. 3-4, 525–531.
232. J.L. TAYLOR, *Measure algebras*, Regional conference series in math., Amer. Math. Soc., 1972.
233. G. TENENBAUM, Sur la non-dérivabilité de fonctions périodiques associées à certaines formules sommatoires, in *The mathematics of Paul Erdös, I*, 117–128, Algorithms Combin., 13, Springer, Berlin, 1997.
234. A. THUE, Über unendliche Zeichenreihen, *Selected mathematical papers*, Universitetsforlaget, Oslo, (1977), 139–158.
235. A. J. Van der POORTEN, An introduction to continued fractions in *Diophantine analysis* (Kensington, 1985), 99–138, London Math. Soc. Lecture Note Ser., 109, Cambridge Univ. Press, Cambridge, 1986.
236. W.A. VEECH, Strict ergodicity in zero dimensional dynamical systems and the Kronecker-Weyl theorem modulo two, *Trans. Amer. Math. Soc.* **140** (1969), 1–33.
237. A.M. VERSHIK, On the theory of normal dynamical systems, *Sov. Math. Dokl.* **3** (1962), 625–628.
238. A.M. VERSHIK, Spectral and metric isomorphism of some normal dynamical systems, *Sov. Math. Dokl.* **3** (1962), 693–696.
239. A.M. VERSHIK and A. LIVSHITS, Adic models of ergodic transformations, spectral theory, and related topics, *Adv. Soviet Math.* **9** (1992), 185–204.
240. J. Von NEUMANN, Zur operatorenmethode in der klassichen mechanik, *Ann. Math.* **33** (1932), 587–642.
241. P. WALTERS, *An introduction to ergodic theory*, Springer Verlag, New York, 1982.
242. Z.-X. WEN and Z.-Y. WEN, Remarques sur la suite engendrée par des substitutions composées, *Ann. Fac. Sci. Toulouse Math.* **9** (1988), no. 1, 55–63.
243. Z.-X. WEN and Z.-Y. WEN, Some properties of the singular words of the Fibonacci word, *European J. Combin.* **15** (1994), 587–598.

244. H. WEYL, Über die gleichverteilung von zahlen modulo eins, *Math. Ann.* **77** (1916), 313–352.
245. N. WIENER, Generalized harmonic analysis, *Acta. Math.* **55** (1930), 117–258.
246. N. WIENER and A. WINTNER, Harmonic analysis and ergodic theory, *Amer. J. Math.* **63** (1941), 415–426.
247. L. ZAMBONI, Une généralisation du théorème de Lagrange sur le développement en fraction continue, *C. R. Acad. Sci. Paris Sér. I Math.* **327** (1998), no. 6, 527–530.
248. A. ZYGMUND, *Trigonometric series*, Cambridge university press, third edition, 2003.

Glossary

$L_C(B)$	occurrence number of C in B, 130
$A^* = \cup_{k\geq 0} A^k$	all words on the alphabet A, 97
$C(\zeta) = (C_{\alpha\beta}^{\gamma\delta})$	coincidence matrix, 244
$H(\mu)$	translation group of μ, 58
$H(\xi)$	height of the algebraic number ξ, 324
$M(\mathbf{T})$	algebra of the regular Borel complex measures on $\mathbf{T}$, 1
$M(\zeta)$	composition matrix of ζ, 131
$M_0(\mathbf{T})$	Rajchman measures, 3
$M_c(\mathbf{T})$	convolution-ideal of continuous measures on $\mathbf{T}$, 2
$M_d(\mathbf{T})$	sub-algebra of discrete measures in $M(\mathbf{T})$, 2
$O(u)$	orbit of u, 98
R_θ	irrational rotation, 29
$SL(2,\mathbf{R})$	group of 2×2-real matrices with ± 1 determinant, 295
S_q	q-shift, 52
U_T	operator $f \to f \circ T$ on L^2, 21
$Z = (\sigma_{\alpha\beta}^{\gamma\delta})$	bi-correlation matrix, 256
$[U,f]$	cyclic subspace spanned by $f \in H$, 22
$[\alpha_0\alpha_1\cdots\alpha_k]$	cylinder set, 85
$[\mu]$	type of the measure μ, 31
$[\sigma_{\max}]$	maximal spectral type, 34
Δ	Gelfand spectrum of the Banach algebra $M(\mathbf{T})$, 6
$\Sigma = (\sigma_{\alpha\beta})_{\alpha,\beta\in A}$	correlation matrix, 197
$\chi = (\chi_\mu)_{\mu\in M(\mathbf{T})}$	generalized character, 7
δ_t	Dirac measure at t, 2
$\ell^2(\mathbf{Z})$	space of square summable bi-infinite sequences, 293
$\ell^\infty(\mathbf{Z})$	space of bounded bi-infinite sequences, 300
$\hat{\mu}(n)$	n^{th} Fourier coefficient of the measure μ, 2

$\mathbf{T}$	$\mathbf{R}\backslash 2\pi\mathbf{Z}$, 1
$\mu * \nu$	convolution of μ and ν, 1
$\mu \ll \nu$	μ is absolutely continuous with respect to ν, 2
$\mu \perp \nu$	μ and ν are mutually singular, 3
$\mu \sim \nu$	μ and ν are equivalent, 2
$\overline{\Gamma}$	closure of Γ in Δ, 9
$\rho(\mu,\nu)$	affinity between μ and ν, 3
σ_0	reduced maximal spectral type, 49
σ_f	spectral measure of $f \in H$, 21
$\sigma_{f,g}$	spectral measure of $f,g \in H$, 22
$h(\zeta)$	height of ζ, 162
h_d	discrete idempotent in Δ, 15
$m(U)$	spectral multiplicity of U, 37
$sp(A)$	spectrum of the operator A, 21
$u_{[m,n]}$	factor $u_m u_{m+1} \cdots u_n, m \leq n$, 98
$\mathbf{1}^{\perp}$	orthogonal of the constants, 49
$\mathscr{L}(X)$	langage of X, 98
$\mathscr{L}(u)$	langage of u, 98
$\mathscr{L}_{\zeta}$	language of the substitution ζ, 125
$\mathscr{S}$	Wiener space, 104
$\mathbf{Z}_2$	2-adic integers, 85
$\mathbf{Z}_q$	q-adic integers, 226
BAD	badly approximable numbers, 323

Index

Lecture Notes in Mathematics

For information about earlier volumes please contact your bookseller or Springer
LNM Online archive: springerlink.com

Vol. 1807: V. D. Milman, G. Schechtman (Eds.), Geometric Aspects of Functional Analysis. Israel Seminar 2000-2002 (2003)

Vol. 1808: W. Schindler, Measures with Symmetry Properties (2003)

Vol. 1809: O. Steinbach, Stability Estimates for Hybrid Coupled Domain Decomposition Methods (2003)

Vol. 1810: J. Wengenroth, Derived Functors in Functional Analysis (2003)

Vol. 1811: J. Stevens, Deformations of Singularities (2003)

Vol. 1812: L. Ambrosio, K. Deckelnick, G. Dziuk, M. Mimura, V. A. Solonnikov, H. M. Soner, Mathematical Aspects of Evolving Interfaces. Madeira, Funchal, Portugal 2000. Editors: P. Colli, J. F. Rodrigues (2003)

Vol. 1813: L. Ambrosio, L. A. Caffarelli, Y. Brenier, G. Buttazzo, C. Villani, Optimal Transportation and its Applications. Martina Franca, Italy 2001. Editors: L. A. Caffarelli, S. Salsa (2003)

Vol. 1814: P. Bank, F. Baudoin, H. Föllmer, L.C.G. Rogers, M. Soner, N. Touzi, Paris-Princeton Lectures on Mathematical Finance 2002 (2003)

Vol. 1815: A. M. Vershik (Ed.), Asymptotic Combinatorics with Applications to Mathematical Physics. St. Petersburg, Russia 2001 (2003)

Vol. 1816: S. Albeverio, W. Schachermayer, M. Talagrand, Lectures on Probability Theory and Statistics. Ecole d'Eté de Probabilités de Saint-Flour XXX-2000. Editor: P. Bernard (2003)

Vol. 1817: E. Koelink, W. Van Assche (Eds.), Orthogonal Polynomials and Special Functions. Leuven 2002 (2003)

Vol. 1818: M. Bildhauer, Convex Variational Problems with Linear, nearly Linear and/or Anisotropic Growth Conditions (2003)

Vol. 1819: D. Masser, Yu. V. Nesterenko, H. P. Schlickewei, W. M. Schmidt, M. Waldschmidt, Diophantine Approximation. Cetraro, Italy 2000. Editors: F. Amoroso, U. Zannier (2003)

Vol. 1820: F. Hiai, H. Kosaki, Means of Hilbert Space Operators (2003)

Vol. 1821: S. Teufel, Adiabatic Perturbation Theory in Quantum Dynamics (2003)

Vol. 1822: S.-N. Chow, R. Conti, R. Johnson, J. Mallet-Paret, R. Nussbaum, Dynamical Systems. Cetraro, Italy 2000. Editors: J. W. Macki, P. Zecca (2003)

Vol. 1823: A. M. Anile, W. Allegretto, C. Ringhofer, Mathematical Problems in Semiconductor Physics. Cetraro, Italy 1998. Editor: A. M. Anile (2003)

Vol. 1824: J. A. Navarro González, J. B. Sancho de Salas, $\mathscr{C}^\infty$ – Differentiable Spaces (2003)

Vol. 1825: J. H. Bramble, A. Cohen, W. Dahmen, Multiscale Problems and Methods in Numerical Simulations, Martina Franca, Italy 2001. Editor: C. Canuto (2003)

Vol. 1826: K. Dohmen, Improved Bonferroni Inequalities via Abstract Tubes. Inequalities and Identities of Inclusion-Exclusion Type. VIII, 113 p, 2003.

Vol. 1827: K. M. Pilgrim, Combinations of Complex Dynamical Systems. IX, 118 p, 2003.

Vol. 1828: D. J. Green, Gröbner Bases and the Computation of Group Cohomology. XII, 138 p, 2003.

Vol. 1829: E. Altman, B. Gaujal, A. Hordijk, Discrete-Event Control of Stochastic Networks: Multimodularity and Regularity. XIV, 313 p, 2003.

Vol. 1830: M. I. Gil', Operator Functions and Localization of Spectra. XIV, 256 p, 2003.

Vol. 1831: A. Connes, J. Cuntz, E. Guentner, N. Higson, J. E. Kaminker, Noncommutative Geometry, Martina Franca, Italy 2002. Editors: S. Doplicher, L. Longo (2004)

Vol. 1832: J. Azéma, M. Émery, M. Ledoux, M. Yor (Eds.), Séminaire de Probabilités XXXVII (2003)

Vol. 1833: D.-Q. Jiang, M. Qian, M.-P. Qian, Mathematical Theory of Nonequilibrium Steady States. On the Frontier of Probability and Dynamical Systems. IX, 280 p, 2004.

Vol. 1834: Yo. Yomdin, G. Comte, Tame Geometry with Application in Smooth Analysis. VIII, 186 p, 2004.

Vol. 1835: O.T. Izhboldin, B. Kahn, N.A. Karpenko, A. Vishik, Geometric Methods in the Algebraic Theory of Quadratic Forms. Summer School, Lens, 2000. Editor: J.-P. Tignol (2004)

Vol. 1836: C. Năstăsescu, F. Van Oystaeyen, Methods of Graded Rings. XIII, 304 p, 2004.

Vol. 1837: S. Tavaré, O. Zeitouni, Lectures on Probability Theory and Statistics. Ecole d'Eté de Probabilités de Saint-Flour XXXI-2001. Editor: J. Picard (2004)

Vol. 1838: A.J. Ganesh, N.W. O'Connell, D.J. Wischik, Big Queues. XII, 254 p, 2004.

Vol. 1839: R. Gohm, Noncommutative Stationary Processes. VIII, 170 p, 2004.

Vol. 1840: B. Tsirelson, W. Werner, Lectures on Probability Theory and Statistics. Ecole d'Eté de Probabilités de Saint-Flour XXXII-2002. Editor: J. Picard (2004)

Vol. 1841: W. Reichel, Uniqueness Theorems for Variational Problems by the Method of Transformation Groups (2004)

Vol. 1842: T. Johnsen, A. L. Knutsen, K_3 Projective Models in Scrolls (2004)

Vol. 1843: B. Jefferies, Spectral Properties of Noncommuting Operators (2004)

Vol. 1844: K.F. Siburg, The Principle of Least Action in Geometry and Dynamics (2004)

Vol. 1845: Min Ho Lee, Mixed Automorphic Forms, Torus Bundles, and Jacobi Forms (2004)

Vol. 1846: H. Ammari, H. Kang, Reconstruction of Small Inhomogeneities from Boundary Measurements (2004)

Vol. 1847: T.R. Bielecki, T. Björk, M. Jeanblanc, M. Rutkowski, J.A. Scheinkman, W. Xiong, Paris-Princeton Lectures on Mathematical Finance 2003 (2004)
Vol. 1848: M. Abate, J. E. Fornaess, X. Huang, J. P. Rosay, A. Tumanov, Real Methods in Complex and CR Geometry, Martina Franca, Italy 2002. Editors: D. Zaitsev, G. Zampieri (2004)
Vol. 1849: Martin L. Brown, Heegner Modules and Elliptic Curves (2004)
Vol. 1850: V. D. Milman, G. Schechtman (Eds.), Geometric Aspects of Functional Analysis. Israel Seminar 2002-2003 (2004)
Vol. 1851: O. Catoni, Statistical Learning Theory and Stochastic Optimization (2004)
Vol. 1852: A.S. Kechris, B.D. Miller, Topics in Orbit Equivalence (2004)
Vol. 1853: Ch. Favre, M. Jonsson, The Valuative Tree (2004)
Vol. 1854: O. Saeki, Topology of Singular Fibers of Differential Maps (2004)
Vol. 1855: G. Da Prato, P.C. Kunstmann, I. Lasiecka, A. Lunardi, R. Schnaubelt, L. Weis, Functional Analytic Methods for Evolution Equations. Editors: M. Iannelli, R. Nagel, S. Piazzera (2004)
Vol. 1856: K. Back, T.R. Bielecki, C. Hipp, S. Peng, W. Schachermayer, Stochastic Methods in Finance, Bressanone/Brixen, Italy, 2003. Editors: M. Fritelli, W. Runggaldier (2004)
Vol. 1857: M. Émery, M. Ledoux, M. Yor (Eds.), Séminaire de Probabilités XXXVIII (2005)
Vol. 1858: A.S. Cherny, H.-J. Engelbert, Singular Stochastic Differential Equations (2005)
Vol. 1859: E. Letellier, Fourier Transforms of Invariant Functions on Finite Reductive Lie Algebras (2005)
Vol. 1860: A. Borisyuk, G.B. Ermentrout, A. Friedman, D. Terman, Tutorials in Mathematical Biosciences I. Mathematical Neurosciences (2005)
Vol. 1861: G. Benettin, J. Henrard, S. Kuksin, Hamiltonian Dynamics – Theory and Applications, Cetraro, Italy, 1999. Editor: A. Giorgilli (2005)
Vol. 1862: B. Helffer, F. Nier, Hypoelliptic Estimates and Spectral Theory for Fokker-Planck Operators and Witten Laplacians (2005)
Vol. 1863: H. Führ, Abstract Harmonic Analysis of Continuous Wavelet Transforms (2005)
Vol. 1864: K. Efstathiou, Metamorphoses of Hamiltonian Systems with Symmetries (2005)
Vol. 1865: D. Applebaum, B.V. R. Bhat, J. Kustermans, J. M. Lindsay, Quantum Independent Increment Processes I. From Classical Probability to Quantum Stochastic Calculus. Editors: M. Schürmann, U. Franz (2005)
Vol. 1866: O.E. Barndorff-Nielsen, U. Franz, R. Gohm, B. Kümmerer, S. Thorbjønsen, Quantum Independent Increment Processes II. Structure of Quantum Lévy Processes, Classical Probability, and Physics. Editors: M. Schürmann, U. Franz, (2005)
Vol. 1867: J. Sneyd (Ed.), Tutorials in Mathematical Biosciences II. Mathematical Modeling of Calcium Dynamics and Signal Transduction. (2005)
Vol. 1868: J. Jorgenson, S. Lang, $Pos_n(R)$ and Eisenstein Series. (2005)
Vol. 1869: A. Dembo, T. Funaki, Lectures on Probability Theory and Statistics. Ecole d'Eté de Probabilités de Saint-Flour XXXIII-2003. Editor: J. Picard (2005)
Vol. 1870: V.I. Gurariy, W. Lusky, Geometry of Müntz Spaces and Related Questions. (2005)
Vol. 1871: P. Constantin, G. Gallavotti, A.V. Kazhikhov, Y. Meyer, S. Ukai, Mathematical Foundation of Turbulent Viscous Flows, Martina Franca, Italy, 2003. Editors: M. Cannone, T. Miyakawa (2006)
Vol. 1872: A. Friedman (Ed.), Tutorials in Mathematical Biosciences III. Cell Cycle, Proliferation, and Cancer (2006)
Vol. 1873: R. Mansuy, M. Yor, Random Times and Enlargements of Filtrations in a Brownian Setting (2006)
Vol. 1874: M. Yor, M. Émery (Eds.), In Memoriam Paul-André Meyer - Séminaire de Probabilités XXXIX (2006)
Vol. 1875: J. Pitman, Combinatorial Stochastic Processes. Ecole d'Eté de Probabilités de Saint-Flour XXXII-2002. Editor: J. Picard (2006)
Vol. 1876: H. Herrlich, Axiom of Choice (2006)
Vol. 1877: J. Steuding, Value Distributions of *L*-Functions (2007)
Vol. 1878: R. Cerf, The Wulff Crystal in Ising and Percolation Models, Ecole d'Eté de Probabilités de Saint-Flour XXXIV-2004. Editor: Jean Picard (2006)
Vol. 1879: G. Slade, The Lace Expansion and its Applications, Ecole d'Eté de Probabilités de Saint-Flour XXXIV-2004. Editor: Jean Picard (2006)
Vol. 1880: S. Attal, A. Joye, C.-A. Pillet, Open Quantum Systems I, The Hamiltonian Approach (2006)
Vol. 1881: S. Attal, A. Joye, C.-A. Pillet, Open Quantum Systems II, The Markovian Approach (2006)
Vol. 1882: S. Attal, A. Joye, C.-A. Pillet, Open Quantum Systems III, Recent Developments (2006)
Vol. 1883: W. Van Assche, F. Marcellàn (Eds.), Orthogonal Polynomials and Special Functions, Computation and Application (2006)
Vol. 1884: N. Hayashi, E.I. Kaikina, P.I. Naumkin, I.A. Shishmarev, Asymptotics for Dissipative Nonlinear Equations (2006)
Vol. 1885: A. Telcs, The Art of Random Walks (2006)
Vol. 1886: S. Takamura, Splitting Deformations of Degenerations of Complex Curves (2006)
Vol. 1887: K. Habermann, L. Habermann, Introduction to Symplectic Dirac Operators (2006)
Vol. 1888: J. van der Hoeven, Transseries and Real Differential Algebra (2006)
Vol. 1889: G. Osipenko, Dynamical Systems, Graphs, and Algorithms (2006)
Vol. 1890: M. Bunge, J. Funk, Singular Coverings of Toposes (2006)
Vol. 1891: J.B. Friedlander, D.R. Heath-Brown, H. Iwaniec, J. Kaczorowski, Analytic Number Theory, Cetraro, Italy, 2002. Editors: A. Perelli, C. Viola (2006)
Vol. 1892: A. Baddeley, I. Bárány, R. Schneider, W. Weil, Stochastic Geometry, Martina Franca, Italy, 2004. Editor: W. Weil (2007)
Vol. 1893: H. Hanßmann, Local and Semi-Local Bifurcations in Hamiltonian Dynamical Systems, Results and Examples (2007)
Vol. 1894: C.W. Groetsch, Stable Approximate Evaluation of Unbounded Operators (2007)
Vol. 1895: L. Molnár, Selected Preserver Problems on Algebraic Structures of Linear Operators and on Function Spaces (2007)
Vol. 1896: P. Massart, Concentration Inequalities and Model Selection, Ecole d'Été de Probabilités de Saint-Flour XXXIII-2003. Editor: J. Picard (2007)
Vol. 1897: R. Doney, Fluctuation Theory for Lévy Processes, Ecole d'Été de Probabilités de Saint-Flour XXXV-2005. Editor: J. Picard (2007)

Vol. 1898: H.R. Beyer, Beyond Partial Differential Equations, On linear and Quasi-Linear Abstract Hyperbolic Evolution Equations (2007)
Vol. 1899: Séminaire de Probabilités XL. Editors: C. Donati-Martin, M. Émery, A. Rouault, C. Stricker (2007)
Vol. 1900: E. Bolthausen, A. Bovier (Eds.), Spin Glasses (2007)
Vol. 1901: O. Wittenberg, Intersections de deux quadriques et pinceaux de courbes de genre 1, Intersections of Two Quadrics and Pencils of Curves of Genus 1 (2007)
Vol. 1902: A. Isaev, Lectures on the Automorphism Groups of Kobayashi-Hyperbolic Manifolds (2007)
Vol. 1903: G. Kresin, V. Maz'ya, Sharp Real-Part Theorems (2007)
Vol. 1904: P. Giesl, Construction of Global Lyapunov Functions Using Radial Basis Functions (2007)
Vol. 1905: C. Prévôt, M. Röckner, A Concise Course on Stochastic Partial Differential Equations (2007)
Vol. 1906: T. Schuster, The Method of Approximate Inverse: Theory and Applications (2007)
Vol. 1907: M. Rasmussen, Attractivity and Bifurcation for Nonautonomous Dynamical Systems (2007)
Vol. 1908: T.J. Lyons, M. Caruana, T. Lévy, Differential Equations Driven by Rough Paths, Ecole d'Été de Probabilités de Saint-Flour XXXIV-2004 (2007)
Vol. 1909: H. Akiyoshi, M. Sakuma, M. Wada, Y. Yamashita, Punctured Torus Groups and 2-Bridge Knot Groups (I) (2007)
Vol. 1910: V.D. Milman, G. Schechtman (Eds.), Geometric Aspects of Functional Analysis. Israel Seminar 2004-2005 (2007)
Vol. 1911: A. Bressan, D. Serre, M. Williams, K. Zumbrun, Hyperbolic Systems of Balance Laws. Cetraro, Italy 2003. Editor: P. Marcati (2007)
Vol. 1912: V. Berinde, Iterative Approximation of Fixed Points (2007)
Vol. 1913: J.E. Marsden, G. Misiołek, J.-P. Ortega, M. Perlmutter, T.S. Ratiu, Hamiltonian Reduction by Stages (2007)
Vol. 1914: G. Kutyniok, Affine Density in Wavelet Analysis (2007)
Vol. 1915: T. Bıyıkoğlu, J. Leydold, P.F. Stadler, Laplacian Eigenvectors of Graphs. Perron-Frobenius and Faber-Krahn Type Theorems (2007)
Vol. 1916: C. Villani, F. Rezakhanlou, Entropy Methods for the Boltzmann Equation. Editors: F. Golse, S. Olla (2008)
Vol. 1917: I. Veselić, Existence and Regularity Properties of the Integrated Density of States of Random Schrödinger (2008)
Vol. 1918: B. Roberts, R. Schmidt, Local Newforms for GSp(4) (2007)
Vol. 1919: R.A. Carmona, I. Ekeland, A. Kohatsu-Higa, J.-M. Lasry, P.-L. Lions, H. Pham, E. Taflin, Paris-Princeton Lectures on Mathematical Finance 2004. Editors: R.A. Carmona, E. Çinlar, I. Ekeland, E. Jouini, J.A. Scheinkman, N. Touzi (2007)
Vol. 1920: S.N. Evans, Probability and Real Trees. Ecole d'Été de Probabilités de Saint-Flour XXXV-2005 (2008)
Vol. 1921: J.P. Tian, Evolution Algebras and their Applications (2008)
Vol. 1922: A. Friedman (Ed.), Tutorials in Mathematical BioSciences IV. Evolution and Ecology (2008)
Vol. 1923: J.P.N. Bishwal, Parameter Estimation in Stochastic Differential Equations (2008)
Vol. 1924: M. Wilson, Littlewood-Paley Theory and Exponential-Square Integrability (2008)
Vol. 1925: M. du Sautoy, L. Woodward, Zeta Functions of Groups and Rings (2008)
Vol. 1926: L. Barreira, V. Claudia, Stability of Nonautonomous Differential Equations (2008)
Vol. 1927: L. Ambrosio, L. Caffarelli, M.G. Crandall, L.C. Evans, N. Fusco, Calculus of Variations and Non-Linear Partial Differential Equations. Cetraro, Italy 2005. Editors: B. Dacorogna, P. Marcellini (2008)
Vol. 1928: J. Jonsson, Simplicial Complexes of Graphs (2008)
Vol. 1929: Y. Mishura, Stochastic Calculus for Fractional Brownian Motion and Related Processes (2008)
Vol. 1930: J.M. Urbano, The Method of Intrinsic Scaling. A Systematic Approach to Regularity for Degenerate and Singular PDEs (2008)
Vol. 1931: M. Cowling, E. Frenkel, M. Kashiwara, A. Valette, D.A. Vogan, Jr., N.R. Wallach, Representation Theory and Complex Analysis. Venice, Italy 2004. Editors: E.C. Tarabusi, A. D'Agnolo, M. Picardello (2008)
Vol. 1932: A.A. Agrachev, A.S. Morse, E.D. Sontag, H.J. Sussmann, V.I. Utkin, Nonlinear and Optimal Control Theory. Cetraro, Italy 2004. Editors: P. Nistri, G. Stefani (2008)
Vol. 1933: M. Petkovic, Point Estimation of Root Finding Methods (2008)
Vol. 1934: C. Donati-Martin, M. Émery, A. Rouault, C. Stricker (Eds.), Séminaire de Probabilités XLI (2008)
Vol. 1935: A. Unterberger, Alternative Pseudodifferential Analysis (2008)
Vol. 1936: P. Magal, S. Ruan (Eds.), Structured Population Models in Biology and Epidemiology (2008)
Vol. 1937: G. Capriz, P. Giovine, P.M. Mariano (Eds.), Mathematical Models of Granular Matter (2008)
Vol. 1938: D. Auroux, F. Catanese, M. Manetti, P. Seidel, B. Siebert, I. Smith, G. Tian, Symplectic 4-Manifolds and Algebraic Surfaces. Cetraro, Italy 2003. Editors: F. Catanese, G. Tian (2008)
Vol. 1939: D. Boffi, F. Brezzi, L. Demkowicz, R.G. Durán, R.S. Falk, M. Fortin, Mixed Finite Elements, Compatibility Conditions, and Applications. Cetraro, Italy 2006. Editors: D. Boffi, L. Gastaldi (2008)
Vol. 1940: J. Banasiak, V. Capasso, M.A.J. Chaplain, M. Lachowicz, J. Miękisz, Multiscale Problems in the Life Sciences. From Microscopic to Macroscopic. Będlewo, Poland 2006. Editors: V. Capasso, M. Lachowicz (2008)
Vol. 1941: S.M.J. Haran, Arithmetical Investigations. Representation Theory, Orthogonal Polynomials, and Quantum Interpolations (2008)
Vol. 1942: S. Albeverio, F. Flandoli, Y.G. Sinai, SPDE in Hydrodynamic. Recent Progress and Prospects. Cetraro, Italy 2005. Editors: G. Da Prato, M. Röckner (2008)
Vol. 1943: L.L. Bonilla (Ed.), Inverse Problems and Imaging. Martina Franca, Italy 2002 (2008)
Vol. 1944: A. Di Bartolo, G. Falcone, P. Plaumann, K. Strambach, Algebraic Groups and Lie Groups with Few Factors (2008)
Vol. 1945: F. Brauer, P. van den Driessche, J. Wu (Eds.), Mathematical Epidemiology (2008)
Vol. 1946: G. Allaire, A. Arnold, P. Degond, T.Y. Hou, Quantum Transport. Modelling, Analysis and Asymptotics. Cetraro, Italy 2006. Editors: N.B. Abdallah, G. Frosali (2008)

Vol. 1947: D. Abramovich, M. Mariño, M. Thaddeus, R. Vakil, Enumerative Invariants in Algebraic Geometry and String Theory. Cetraro, Italy 2005. Editors: K. Behrend, M. Manetti (2008)
Vol. 1948: F. Cao, J-L. Lisani, J-M. Morel, P. Musé, F. Sur, A Theory of Shape Identification (2008)
Vol. 1949: H.G. Feichtinger, B. Helffer, M.P. Lamoureux, N. Lerner, J. Toft, Pseudo-Differential Operators. Quantization and Signals. Cetraro, Italy 2006. Editors: L. Rodino, M.W. Wong (2008)
Vol. 1950: M. Bramson, Stability of Queueing Networks, Ecole d'Eté de Probabilités de Saint-Flour XXXVI-2006 (2008)
Vol. 1951: A. Moltó, J. Orihuela, S. Troyanski, M. Valdivia, A Non Linear Transfer Technique for Renorming (2009)
Vol. 1952: R. Mikhailov, I.B.S. Passi, Lower Central and Dimension Series of Groups (2009)
Vol. 1953: K. Arwini, C.T.J. Dodson, Information Geometry (2008)
Vol. 1954: P. Biane, L. Bouten, F. Cipriani, N. Konno, N. Privault, Q. Xu, Quantum Potential Theory. Editors: U. Franz, M. Schuermann (2008)
Vol. 1955: M. Bernot, V. Caselles, J.-M. Morel, Optimal Transportation Networks (2008)
Vol. 1956: C.H. Chu, Matrix Convolution Operators on Groups (2008)
Vol. 1957: A. Guionnet, On Random Matrices: Macroscopic Asymptotics, Ecole d'Eté de Probabilités de Saint-Flour XXXVI-2006 (2009)
Vol. 1958: M.C. Olsson, Compactifying Moduli Spaces for Abelian Varieties (2008)
Vol. 1959: Y. Nakkajima, A. Shiho, Weight Filtrations on Log Crystalline Cohomologies of Families of Open Smooth Varieties (2008)
Vol. 1960: J. Lipman, M. Hashimoto, Foundations of Grothendieck Duality for Diagrams of Schemes (2009)
Vol. 1961: G. Buttazzo, A. Pratelli, S. Solimini, E. Stepanov, Optimal Urban Networks via Mass Transportation (2009)
Vol. 1962: R. Dalang, D. Khoshnevisan, C. Mueller, D. Nualart, Y. Xiao, A Minicourse on Stochastic Partial Differential Equations (2009)
Vol. 1963: W. Siegert, Local Lyapunov Exponents (2009)
Vol. 1964: W. Roth, Operator-valued Measures and Integrals for Cone-valued Functions and Integrals for Cone-valued Functions (2009)
Vol. 1965: C. Chidume, Geometric Properties of Banach Spaces and Nonlinear Iterations (2009)
Vol. 1966: D. Deng, Y. Han, Harmonic Analysis on Spaces of Homogeneous Type (2009)
Vol. 1967: B. Fresse, Modules over Operads and Functors (2009)
Vol. 1968: R. Weissauer, Endoscopy for GSP(4) and the Cohomology of Siegel Modular Threefolds (2009)
Vol. 1969: B. Roynette, M. Yor, Penalising Brownian Paths (2009)
Vol. 1970: M. Biskup, A. Bovier, F. den Hollander, D. Ioffe, F. Martinelli, K. Netočný, F. Toninelli, Methods of Contemporary Mathematical Statistical Physics. Editor: R. Kotecký (2009)
Vol. 1971: L. Saint-Raymond, Hydrodynamic Limits of the Boltzmann Equation (2009)
Vol. 1972: T. Mochizuki, Donaldson Type Invariants for Algebraic Surfaces (2009)
Vol. 1973: M.A. Berger, L.H. Kauffmann, B. Khesin, H.K. Moffatt, R.L. Ricca, De W. Sumners, Lectures on Topological Fluid Mechanics. Cetraro, Italy 2001. Editor: R.L. Ricca (2009)
Vol. 1974: F. den Hollander, Random Polymers: École d'Été de Probabilités de Saint-Flour XXXVII – 2007 (2009)
Vol. 1975: J.C. Rohde, Cyclic Coverings, Calabi-Yau Manifolds and Complex Multiplication (2009)
Vol. 1976: N. Ginoux, The Dirac Spectrum (2009)
Vol. 1977: M.J. Gursky, E. Lanconelli, A. Malchiodi, G. Tarantello, X.-J. Wang, P.C. Yang, Geometric Analysis and PDEs. Cetraro, Italy 2001. Editors: A. Ambrosetti, S.-Y.A. Chang, A. Malchiodi (2009)
Vol. 1978: M. Qian, J.-S. Xie, S. Zhu, Smooth Ergodic Theory for Endomorphisms (2009)
Vol. 1979: C. Donati-Martin, M. Émery, A. Rouault, C. Stricker (Eds.), Śeminaire de Probablitiés XLII (2009)
Vol. 1980: P. Graczyk, A. Stos (Eds.), Potential Analysis of Stable Processes and its Extensions (2009)
Vol. 1981: M. Chlouveraki, Blocks and Families for Cyclotomic Hecke Algebras (2009)
Vol. 1982: N. Privault, Stochastic Analysis in Discrete and Continuous Settings. With Normal Martingales (2009)
Vol. 1983: H. Ammari (Ed.), Mathematical Modeling in Biomedical Imaging I. Electrical and Ultrasound Tomographies, Anomaly Detection, and Brain Imaging (2009)
Vol. 1984: V. Caselles, P. Monasse, Geometric Description of Images as Topographic Maps (2010)
Vol. 1985: T. Linß, Layer-Adapted Meshes for Reaction-Convection-Diffusion Problems (2010)
Vol. 1986: J.-P. Antoine, C. Trapani, Partial Inner Product Spaces. Theory and Applications (2009)
Vol. 1987: J.-P. Brasselet, J. Seade, T. Suwa, Vector Fields on Singular Varieties (2010)
Vol. 1988: M. Broué, Introduction to Complex Reflection Groups and Their Braid Groups (2010)

Recent Reprints and New Editions

Vol. 1702: J. Ma, J. Yong, Forward-Backward Stochastic Differential Equations and their Applications. 1999 – Corr. 3rd printing (2007)
Vol. 830: J.A. Green, Polynomial Representations of GL_n, with an Appendix on Schensted Correspondence and Littelmann Paths by K. Erdmann, J.A. Green and M. Schoker 1980 – 2nd corr. and augmented edition (2007)
Vol. 1693: S. Simons, From Hahn-Banach to Monotonicity (Minimax and Monotonicity 1998) – 2nd exp. edition (2008)
Vol. 470: R.E. Bowen, Equilibrium States and the Ergodic Theory of Anosov Diffeomorphisms. With a preface by D. Ruelle. Edited by J.-R. Chazottes. 1975 – 2nd rev. edition (2008)
Vol. 523: S.A. Albeverio, R.J. Høegh-Krohn, S. Mazzucchi, Mathematical Theory of Feynman Path Integral. 1976 – 2nd corr. and enlarged edition (2008)
Vol. 1764: A. Cannas da Silva, Lectures on Symplectic Geometry 2001 – Corr. 2nd printing (2008)
Vol. 1670: J.W. Neuberger, Sobolev Gradients and Differential Equations 1997 – 2nd edition (2010)
Vol. 1294: M. Queffélec, Substitution Dynamical Systems – Spectral Analysis – Second Edition (2010)

LECTURE NOTES IN MATHEMATICS

Springer

Edited by J.-M. Morel, F. Takens, B. Teissier, P.K. Maini

Editorial Policy (for the publication of monographs)

1. Lecture Notes aim to report new developments in all areas of mathematics and their applications - quickly, informally and at a high level. Mathematical texts analysing new developments in modelling and numerical simulation are welcome.

   Monograph manuscripts should be reasonably self-contained and rounded off. Thus they may, and often will, present not only results of the author but also related work by other people. They may be based on specialised lecture courses. Furthermore, the manuscripts should provide sufficient motivation, examples and applications. This clearly distinguishes Lecture Notes from journal articles or technical reports which normally are very concise. Articles intended for a journal but too long to be accepted by most journals, usually do not have this "lecture notes" character. For similar reasons it is unusual for doctoral theses to be accepted for the Lecture Notes series, though habilitation theses may be appropriate.

2. Manuscripts should be submitted either online at www.editorialmanager.com/lnm to Springer's mathematics editorial in Heidelberg, or to one of the series editors. In general, manuscripts will be sent out to 2 external referees for evaluation. If a decision cannot yet be reached on the basis of the first 2 reports, further referees may be contacted: The author will be informed of this. A final decision to publish can be made only on the basis of the complete manuscript, however a refereeing process leading to a preliminary decision can be based on a pre-final or incomplete manuscript. The strict minimum amount of material that will be considered should include a detailed outline describing the planned contents of each chapter, a bibliography and several sample chapters.

   Authors should be aware that incomplete or insufficiently close to final manuscripts almost always result in longer refereeing times and nevertheless unclear referees' recommendations, making further refereeing of a final draft necessary.

   Authors should also be aware that parallel submission of their manuscript to another publisher while under consideration for LNM will in general lead to immediate rejection.

3. Manuscripts should in general be submitted in English. Final manuscripts should contain at least 100 pages of mathematical text and should always include

   - a table of contents;
   - an informative introduction, with adequate motivation and perhaps some historical remarks: it should be accessible to a reader not intimately familiar with the topic treated;
   - a subject index: as a rule this is genuinely helpful for the reader.

   For evaluation purposes, manuscripts may be submitted in print or electronic form (print form is still preferred by most referees), in the latter case preferably as pdf- or zipped ps-files. Lecture Notes volumes are, as a rule, printed digitally from the authors' files. To ensure best results, authors are asked to use the LaTeX2e style files available from Springer's web-server at:

   ftp://ftp.springer.de/pub/tex/latex/svmonot1/ (for monographs) and
   ftp://ftp.springer.de/pub/tex/latex/svmultt1/ (for summer schools/tutorials).

Additional technical instructions, if necessary, are available on request from: lnm@springer.com.

4. Careful preparation of the manuscripts will help keep production time short besides ensuring satisfactory appearance of the finished book in print and online. After acceptance of the manuscript authors will be asked to prepare the final LaTeX source files and also the corresponding dvi-, pdf- or zipped ps-file. The LaTeX source files are essential for producing the full-text online version of the book (see http://www.springerlink.com/openurl.asp?genre=journal&issn=0075-8434 for the existing online volumes of LNM).
   The actual production of a Lecture Notes volume takes approximately 12 weeks.
5. Authors receive a total of 50 free copies of their volume, but no royalties. They are entitled to a discount of 33.3% on the price of Springer books purchased for their personal use, if ordering directly from Springer.
6. Commitment to publish is made by letter of intent rather than by signing a formal contract. Springer-Verlag secures the copyright for each volume. Authors are free to reuse material contained in their LNM volumes in later publications: a brief written (or e-mail) request for formal permission is sufficient.

Addresses:

Professor J.-M. Morel, CMLA,
École Normale Supérieure de Cachan,
61 Avenue du Président Wilson, 94235 Cachan Cedex, France
E-mail: Jean-Michel.Morel@cmla.ens-cachan.fr

Professor F. Takens, Mathematisch Instituut,
Rijksuniversiteit Groningen, Postbus 800,
9700 AV Groningen, The Netherlands
E-mail: F.Takens@rug.nl

Professor B. Teissier, Institut Mathématique de Jussieu,
UMR 7586 du CNRS, Équipe "Géométrie et Dynamique",
175 rue du Chevaleret,
75013 Paris, France
E-mail: teissier@math.jussieu.fr

For the "Mathematical Biosciences Subseries" of LNM:

Professor P.K. Maini, Center for Mathematical Biology,
Mathematical Institute, 24-29 St Giles,
Oxford OX1 3LP, UK
E-mail: maini@maths.ox.ac.uk

Springer, Mathematics Editorial, Tiergartenstr. 17,
69121 Heidelberg, Germany,
Tel.: +49 (6221) 487-259
Fax: +49 (6221) 4876-8259
E-mail: lnm@springer.com